COMPUTATION
for HUMANITY
Information Technology
to Advance Society

Computational Analysis, Synthesis, and Design of Dynamic Systems Series

Series Editor
Pieter J. Mosterman

MathWorks, Inc.
Natick, Massachusetts

McGill University
Montréal, Québec

Computation for Humanity: Information Technology to Advance Society,
edited by Justyna Zander and Pieter J. Mosterman

Discrete-Event Modeling and Simulation: A Practitioner's Approach,
Gabriel A. Wainer

Discrete-Event Modeling and Simulation: Theory and Applications,
edited by Gabriel A. Wainer and Pieter J. Mosterman

Model-Based Design for Embedded Systems,
edited by Gabriela Nicolescu and Pieter J. Mosterman

Model-Based Testing for Embedded Systems,
edited by Justyna Zander, Ina Schieferdecker, and Pieter J. Mosterman

Multi-Agent Systems: Simulation & Applications,
edited by Adelinde M. Uhrmacher and Danny Weyns

Real-Time Simulation Technologies: Principles, Methodologies, and Applications,
edited by Katalin Popovici and Pieter J. Mosterman

COMPUTATION for HUMANITY
Information Technology to Advance Society

Edited by
JUSTYNA ZANDER
PIETER J. MOSTERMAN

CRC Press
Taylor & Francis Group
Boca Raton London New York

CRC Press is an imprint of the
Taylor & Francis Group, an **informa** business

CRC Press
Taylor & Francis Group
6000 Broken Sound Parkway NW, Suite 300
Boca Raton, FL 33487-2742

First issued in paperback 2017

© 2014 by Taylor & Francis Group, LLC
CRC Press is an imprint of Taylor & Francis Group, an Informa business

No claim to original U.S. Government works

ISBN-13: 978-1-4398-8327-3 (hbk)
ISBN-13: 978-1-138-07341-8 (pbk)

Library of Congress Cataloging-in-Publication Data

Computation for humanity : information technology to advance society / editors, Justyna Zander, Pieter J. Mosterman.
      pages cm. -- (Computational analysis, synthesis, and design of dynamic systems)
      Summary: "This book discusses various aspects computational science and engineering in combination with applied science. It also highlights sustainability to demonstrate how computation can improve different aspects of life. The editors provide a collection of numerous computation-related projects that form a foundation to cross-pollinate between different disciplines and further extensive collaboration. They also present a clear and profound understanding of computing in today's world. The detailed application examples provide the reader \with contributions that behold approaches to provide fundamental solutions to some of the most pertinent humanity-related problems"-- Provided by publisher.
      Includes bibliographical references and index.
      ISBN 978-1-4398-8327-3 (hardback)
      1. Social sciences--Data processing. I. Zander, Justyna. II. Mosterman, Pieter J.

H61.3.C6446 2013
303.48'33--dc23   2013017594

Visit the Taylor & Francis Web site at
http://www.taylorandfrancis.com

and the CRC Press Web site at
http://www.crcpress.com

About the Cover

On the cover picture: Jerzy Kukuczka, Nanga Parbat, 1985.

Jerzy Kukuczka (March 24, 1948 – October 24, 1989) was born in Katowice, Poland. He was a Polish alpine and high-altitude climber. On September 18, 1987, he became the second man, after Reinhold Messner, to climb all fourteen eight thousanders in the world. He was also the first man to climb three of the fourteen in winter: Dhaulagiri with Andrzej Czok in 1985, Kangchenjunga with Krzysztof Wielicki in 1986 and Annapurna I with Artur Hajzer in 1987.

Contents

Part IV   Citizen and Artisan Computing

Part V Visionary Pondering and Outlook on Computation for Humanity

Preface

Through the ages, indeed millennia, humans have been exceptionally ingenious and adept at exploiting nature to develop tools for their benefit. Understanding nature better is thus nothing short of a guarantee for the development of novel tools. Never before had this been put on display as prominently as with the formulation of a fundamental scientific approach to the study of nature (Chapter 13). The systematic development of knowledge that derives from such study provided a rapidly expanding body of intelligence to the avail of technological exploitation. It should come as no surprise that humanity indeed capitalized on the technological opportunities like never before in history (e.g., steam engine, reciprocating engine, and pendulum clock).

A critical element of each scientific approach comprises a formal theory with hypotheses to be falsified. The development of mathematical theories such as calculus was indispensable to the advances in science and, consequently, in the engineering of technical advances. Because of the foundational character, any progress in automating mathematical problem-solving would be destined to surpass even the avalanche of technological advances that resulted from adopting the scientific method. And so it happened that among the spectrum of technologies sprouting on the fertile soil of abundant scientific discoveries, one took a special place: computing. What had been incidental exercises in developing machines to aid with arithmetic became focused endeavors to automate computing, even if not a formally defined concept at first.

Whereas computing was a menial task all through the first half of the previous century (e.g., the human computers for the Manhattan project), during the Second World War much effort was dedicated to automate computing, for example, in the field of encryption. The momentum built during the years of war carried over into the consecutive decades and combined with emerging semiconductor technology that enabled extreme miniaturization, and humanity witnessed an exponential explosion of automatic computing power.

And society never looked back, with computation changing our lives in many ways on time scales that have never been seen before. The changes are on display in our everyday interactions as the increasingly "smart" systems, that is, systems that incorporate some logic, are made operational by computation. Much like machination rendered power a commodity to make tools "active" and ultimately supplanted human and animal labor for all practical purposes, computation has commoditized reasoning and enabled tools to take "initiative." An important consequence of the ability to automatically solve logic problems was the foundation that it provided for network communication because of computation-enabled protocols that facilitate an otherwise unattainable quality of service. The now ubiquitous communication capabilities allow for a redefinition of computer-based solutions to provide a new and better world as, for example, evidenced by the penetration of wireless communication services in African countries compared to their wired counterparts (Aker and Mbiti, 2010).

So let us pause for a moment of reflection to put in context the development of computation and to be in awe of the network and information technology that has come about because of it. Where we are today would have been unfathomable only a generation before.

And yet, for all the benefit derived from network and information technology in our everyday lives, the truly stupendous impact of computation has been in its autocatalytic nature. Where computing was the class standout of an abundance of technologies that emerged because of systematic studies, in addition to experiment and theory, computing came to advance this very scientific discovery (e.g., the instrumental use of computation in work by Nobel laureate Kenneth G. Wilson). Not only does computation serve to formulate hypotheses that may invalidate theories, it even helps formulate theories themselves in a formal computational sense as well as conduct experiments to develop theories and to test hypotheses in a pure computational sense.

And even that is not where the seemingly unbounded value of computing is exposed to its limit. Computing not only accelerated our understanding of nature, it also helped automate the development and exploitation of new technologies. The engineering of technical systems that extensively rely on computing has come to critically depend on computational approaches in effective and efficient design space exploration. The difficulty of overstating the impact of computing is finally demonstrated by noticing how systems to automate the design are, in turn, technical systems themselves. So, advances in computation help build better systems to study, invent, and develop new technologies and systems, including systems that help build more such better systems. This autocatalytic effect has caused some to speculate about a point of singularity where systems have become sufficiently "intelligent" so as to be more adept at engineering systems than humans (*IEEE Spectrum*, 2008). Indeed, there is historical precedence with labor having come to be supplanted by machination, and there is little doubt (if not evidence) that human reasoning is facing a similar fate in a steadily increasing number of tasks.

To substantiate that the value of computing is tackling a broad array of tasks, observe that these tasks have not remained limited to hard disciplines such as electrical engineering, computer engineering, and mechanical engineering. While it is only natural for a core discipline to first adopt novel approaches and methodologies, quickly followed by adjacent disciplines, it is only when distinct utility derives from an approach, though it may be longer, that it will ultimately be adopted by more distant disciplines. In the case of computation, this "osmosis" can be observed by the emerging use of computation in the arts, for example, history (Chapter 17), where otherwise unknown correlations are identified and made explicit by utilizing computation in a specific domain.

As a result, computing now plays an essential role in advancing research across almost every scientific discipline. The fact that everyday computing is becoming exponentially cheaper holds the promise of vastly increasing manifold data flows and revolutionizing the practice of science and engineering. This book is a guide for the creation of services, products, and tools based on computation that facilitates, supports, and enhances the progress of humanity toward a more sustainable life.

Looking forward, the ubiquitous network that has become a reality has caused an inundation of data. Combined with the exponential growth of computation power, a critical need for understanding computation fundamentals and their application has emerged. Also, the next wave of discovery and innovation will comprise powerful computation, data mining, and knowledge and wisdom management. These trends culminate in a still increasing momentum of computational modeling, simulation, and prediction-based reasoning across an expanding spectrum of disciplines and applications.

The *computation for humanity* idea was conceived in 2009 as we became familiar with then current trends presented by thought leaders in exponential technologies at Singularity University (singularity.org), Moffett Field, California. As the technological part matured over the years that followed, the book's content came to include the human aspect and the perspective of looking at crowds of people from different cultural backgrounds. In a sense, the book comprises contributions that reflect the international contacts that we have maintained. These contributors made a profound impression over the course of our lives in various countries in Europe (Germany, Poland, and the Netherlands) and the United States as well as during our extensive travels worldwide. By inviting the brightest and most advanced authors with the charge to create a discussion on a topic that mattered to them and humanity together, we were able to compile this exciting collection of chapters with a very personal attachment.

Because the format of the contributions was left almost exclusively to the authors, we noticed a distinct difference in the manner in which certain cultures write about progress and project the future. For example, U.S. authors frequently refer to spectacular achievements in light-hearted essays. From there, spreading the take-home message is achieved similar to Technology, Entertainment, Design (TED) talks. As an example of a distinctly different style, German authors mostly attain excellence by introducing well-structured organization and thorough analyses. By having the various types of presentations gel at the content level, the overall collection merges the individual parts into a whole that is nothing short of a spectacular piece of work.

To summarize, the content of this book discusses various aspects of computational science and engineering in combination with arts and applied science. This makes it easier for scientists, entrepreneurs, engineers, and researchers to capitalize on a readily available foundation of state-of-the-art information. The *key benefit* is not only to provide a collection of numerous computation-related projects that form a foundation to cross-pollinate between different disciplines and further extensive collaboration, but also to present a clear and profound understanding of computing in today's world while identifying and revealing related advances.

Computation for Humanity: Information Technology to Advance Society provides a collection of the work of internationally renowned experts on current technological achievements to show how computation can be applied in engineering, science, and the arts to drive positive changes in society. Each of the chapters contributes significantly to an understanding of this nexus. Not only do the contributors excel in their expertise, they also add interdisciplinary aspects and draw from synthesis studies based on the merging of disciplines. Overall, this creates an opportunity to analyze, discover, and build novel technical opinions leading to extraordinary solutions, including human grand challenges solutions.

Justyna Zander

Pieter J. Mosterman

Descriptive Characteristics of the book

Following are some of the *intended features* of this book:

- Provides a deep understanding of the practical applications of computation to solve human-related problems, involving either machines or humans, or both
- Delivers insight into theoretical approaches in an accessible manner
- Provides a comprehensive overview of computational science and engineering applications in selected disciplines
- Crosses the boundaries between different domains and shows how they interrelate and complement one another
- Focuses on grand challenges and issues that matter for the future of humanity
- Presents case studies of complex systems
- Examines a variety of technologies and applications
- Shows different perspectives of computational thinking, understanding, and reasoning
- Provides a basis for scientific discoveries and enables adopting scientific theories and engineering practices from other disciplines

- Takes a step back to provide a human-related abstraction level that is not ultimately seen in pure technological elaborations/collections.
- Includes a website with media, presentations, and audio material that aid in understanding the contents of the book (see https://sites. google.com/site/computationofthings/book).

References

Aker J. C. and Mbiti I. M. (2010), Mobile phones and economic development in Africa, *Journal of Economic Perspectives*, 24(3), Summer 2010, 207–232, http://sites.tufts. edu/jennyaker/files/2010/09/aker_mobileafrica.pdf
IEEE Spectrum (2008), June, 2008, Issue, Special report: The singularity.

Technical Review Committee

- Brad Allenby, PhD, Professor, Arizona State University, USA
- Christina Cogdell, PhD, Professor, University of California, Davis, USA
- David Eichmann, PhD, Associate Professor, University of Iowa, USA
- Artur Ekert, PhD, Professor, University of Oxford, UK
- Gregg Jaeger, PhD, Associate Professor, Boston University, USA
- Lisa C. Kaczmarczyk, PhD, bestselling author of *Computers and Society—Computing for Good*, USA
- Dina Shona Laila, PhD, University of Southampton, UK
- Manja Lohse, PhD, University Bielefeld, Germany
- Kirill Alexandrovich Mechitov, PhD, University of Illinois at Urbana-Champaign, USA
- Pieter J. Mosterman, PhD, The MathWorks, Natick, Massachusetts, USA
- Taro Narahara, PhD, New Jersey Institute of Technology, USA
- Eva Navarro, PhD, The University of Manchester, UK
- Cezary Orłowski, PhD, Professor, Gdańsk University of Technology, Poland
- Lee Anna Rasar, PhD, Professor, University of Wisconsin, USA
- Simon Rogers, *The Guardian*, UK
- Diana Serbanescu, PhD, TestingTech GmbH, Berlin, Germany
- Paweł Skruch, PhD, Delphi Poland SA, Poland

- Robert Villa, PhD, University of Sheffield, UK
- Justyna Zander, PhD, The MathWorks, Natick, Massachusetts, USA; Gdańsk University of Technology, Poland
- Mary Lou Zeeman, PhD, Professor, Bowdoin College, Brunswick, Maine, USA

MATLAB® is a registered trademark of The MathWorks, Inc. For product information, please contact:

The MathWorks, Inc.
3 Apple Hill Drive
Natick, MA 01760-2098 USA
Tel: 508-647-7000
Fax: 508-647-7001
E-mail: info@mathworks.com
Web: www.mathworks.com

Acknowledgments

We are thrilled and honored by the participation of all the contributing experts. It has been a pleasure to collect, review, and edit the material, and we sincerely hope that the reader will find this endeavor of intellectual excellence as enjoyable, gratifying, and valuable as we have.

We express our genuine appreciation and gratitude for all the time and effort that each of the authors has put in. We gladly recognize that the high quality of this book is solely thanks to all the collective effort, collaboration, and communication. In addition, we acknowledge the volunteer services of those who joined the technical review committee and extend our sincere appreciation for their involvement. Many thanks to each and every one of you! We hope the end result will do you proud.

Editors

 **Justyna Zander** has acted as a senior technical education evangelist at The MathWorks in Natick, Massachusetts, since October 2012. She has been an assistant professor at Gdańsk University of Technology in Poland since 2011. Prior to this, she founded SIMULATEDWAY that she recently sold. She served as a postdoctoral research scientist at Harvard University in Cambridge, Massachusetts, from 2009 to 2012. She also served as a project manager at the Fraunhofer Institute for Open Communication Systems in Berlin, Germany, from 2004 to 2010.

Dr. Zander earned her doctorate of engineering science and master of science degrees in the fields of computer science and electrical engineering from Technical University of Berlin, Germany, in 2008 and 2005, respectively; a bachelor of science degree in computer science in 2004 and a bachelor of science degree in environmental protection from Gdańsk University of Technology, Poland, in 2003.

In 2009, she graduated from Singularity University (SU), Moffett Field, California, where she participated in a program on exponentially growing technologies as one of the 40 participants selected from 1200 applications. In 2010 and 2011, Dr. Zander was a teaching fellow at SU. Prior to this, she was a visiting scholar at the University of California in San Diego, California, in 2007 and a visiting researcher at The MathWorks in Natick, Massachusetts, in 2008.

Dr. Zander has received grants, awards, and scholarships from such institutions as the Polish Prime Ministry (1999–2000); Polish Ministry of Education and Sport, awarded to 0.04% of students in Poland (2001–2004); German Academic Exchange Service (2002); European Union (2003–2004); Hertie Foundation (2004–2005); IFIP TC6 (2005); IEEE (2006); Siemens (2007); Metodos y Tecnologia (2008); Singularity University (2009); Fraunhofer Gesellschaft (2009–2010); Alexander von Humboldt Foundation (2011); German Academic International Network (2011); German Federal Ministry of Economics and Technology (2011); Hertie Foundation (2012); and the U.S. National Institute of Standards and Technology (2012). Her doctoral thesis on model-based testing was supported by the German National Academic Foundation with a grant awarded to 0.31% of students in Germany (2005–2008). Dr. Zander is also certified by the International Software Quality Institute.

Her web page is https://sites.google.com/site/justynazander/. Her e-mail address is justyna.zander@gmail.com.

Pieter J. Mosterman is a senior research scientist at The MathWorks in Natick, Massachusetts, where he works on computational modeling, simulation, and code generation technologies. He also holds an adjunct professor position at the School of Computer Science at McGill University. Prior to this, he was a research associate at the German Aerospace Center (DLR) in Oberpfaffenhofen. He earned his PhD in electrical and computer engineering from Vanderbilt University in Nashville, Tennessee, and his MSc in electrical engineering from the University of Twente, the Netherlands. His primary research interests are in computer-automated multiparadigm modeling (CAMPaM) with principal applications in design automation, training systems, and fault detection, isolation, and reconfiguration.

Dr. Mosterman designed the electronics laboratory simulator that was nominated for The Computerworld Smithsonian Award by Microsoft Corporation in 1994. In 2003, he was awarded the IMechE Donald Julius Groen Prize for his paper on the hybrid bond graph modeling and simulation environment HyBrSim. In 2009, he received the Distinguished Service Award of The Society for Modeling and Simulation International (SCS) for his services as editor in chief of *SIMULATION: Transactions of SCS*. Dr. Mosterman was guest editor for special issues on CAMPaM of *SIMULATION*, *IEEE Transactions on Control Systems Technology*, and *ACM Transactions on Modeling and Computer Simulation*. He has chaired over 30 scientific events, served on more than 100 international program committees, published over a 100 peer-reviewed papers, and is the inventor on over 40 awarded patents.

His web page is http://msdl.cs.mcgill.ca/people/mosterman/. His e-mail address is pmosterman@yahoo.com.

Contributors

Irena Bach-Dąbrowska
Faculty of Management and
    Economics
Gdańsk University of Technology
Gdańsk, Poland

Dines Bjørner
Professor Emeritus
Holte, Denmark

Thomas Boettcher
HumanLabs, Inc.
Wilmington, Delaware

Joseph F. Fitzsimons
Centre for Quantum Technologies
National University of Singapore
Singapore

Alessio Gabriele
CRM-Centro Ricerche Musicali
Rome, Italy

Alon Halevy
Google Research
Mountain View, California

Frank Hopfgartner
CLARITY: Centre for Sensor Web
    Technologies
Dublin City University
Dublin, Ireland

Kasper Hornbæk
University of Copenhagen
Copenhagen, Denmark

Paola Inverardi
Department of Information
    Engineering, Computer Science
    and Mathematics
University of L'Aquila
L'Aquila, Italy

Daniel Janies
Department of Bioinformatics and
    Genomics
University of North Carolina at
    Charlotte
Charlotte, North California

Steve Johnson
Wave Semiconductor
San Francisco, California

Azam Khan
University of Copenhagen
Copenhagen, Denmark
and
Autodesk Research
Toronto, Ontario, Canada

Alexandra Kirsch
Department of Computer Science
University of Tübingen
Tübingen, Germany

Kincho H. Law
Stanford University
Stanford, California

Michelangelo Lupone
CRM-Centro Ricerche Musicali
Rome, Italy

Walter Mattauch
Fraunhofer FOKUS
Berlin, Germany

Susan McGregor
Columbia School of Journalism
New York, New York

Björn H. Menze
Institut für Bildverarbeitung
Eidgenössische Technische
 Hochschule
Zürich, Switzerland

Pieter Mosterman
The MathWorks
Natick, Massachusetts

and

School of Computer Science
McGill University
Montreal, Quebec, Canada

Ross Mounce
Department of Biology and
 Biochemistry
University of Bath
Bath, United Kingdom

Taro Narahara
College of Architecture and Design
New Jersey Institute of Technology
Newark, New Jersey

Eva M. Navarro-López
School of Computer Science
The University of Manchester
Manchester, United Kingdom

Scott Nesbit
University of Richmond
Richmond, Virginia

Patrizio Pelliccione
Department of Information
 Engineering, Computer Science
 and Mathematics
University of L'Aquila
L'Aquila, Italy

Jim Pinto
Technology Futurist
Carlsbad, California

Eleanor G. Rieffel
NASA Ames Research
Moffett Field, California

Valerio Scarani
Centre for Quantum Technologies
and
Department of Physics
National University of Singapore
Singapore

Ina Schieferdecker
Fraunhofer FOKUS
and
Freie Universität
Berlin, Germany

Kay Smarsly
Department of Civil Engineering
Berlin Institute of Technology
Berlin, Germany

Justyna Zander
The MathWorks
Natick, Massachusetts

and

Gdansk University of Technology
Gdansk, Poland

Alexander Ziegler
Department of Organismic and
 Evolutionary Biology
Museum of Comparative Zoology
Harvard University
Cambridge, Massachusetts

Part I

Context and Intent

1

Introduction

Justyna Zander and Pieter J. Mosterman

CONTENTS

This book compiles achievements and progress of computation in applications that span an unusually broad spectrum of domains. The immediate goal is to document how computation in its omnipresence is affecting people in most all facets of their everyday life. From here, the intent is to motivate and inspire innovative applications of computation to successfully take on the grand challenges for humanity. The combined content should set us off on a road that leads beyond today's problems and solution approaches. As such, the material is targeted to capture and answer issues regarding the computed and computable world that is currently being developed but also to inspire new ideas and visions so as to help create a better future.

1.1 Content

Computation finds its application everywhere in everyday life, not only in commercial, government, and academic organizations but also in every household, in every appliance, device, and machine around us, including cars, airplanes, and mobile devices, to name a few. Moreover, by means of computational simulation, computation has become indispensable in the design of such products, rendering it essential to their very existence.

Beyond everyday life, the effects are no less profound. In science, computation has come to contribute on par with experimentation to further our understanding of nature. In engineering, computation not only has been key in developing new technologies, it also has become the very foundation and core of new technologies in many domains such as communications (e.g., mobile terminals, traffic light control), transportation (e.g., automotive, railroad, and avionics devices), energy and environmental systems (e.g., disaster prediction systems, climate control systems), and health care (e.g., medical devices, electronic patient dossiers).

The material in this book is scoped to pertain to and highlight how computational progress is helpful for society in developing advanced human skills, applying computational results, and using them for the improvement of the quality of life. The main goal of the collected contributions is to provide a background in computation and a clear understanding of its multitude of applications and methodologies to develop those applications, all of it from a wide range of perspectives. This book then provides a thorough understanding of problems, solutions, and opportunities in terms of computational technologies as they affect humanity, including promising developments in foundations of computation. All this is aimed at in the context of humanity and specifically those characteristics that contribute to making a positive impact.

To truly valorize computational advances, it is necessary to interface both with the physical world as well as with humans. The former is discussed by creating an improved and sustainable environment and by developing machines with the objective to serve humans. The latter is treated from the perspective of interfacing with humans by machines, virtual environments, and domain-specific information technology. Tailoring information technology to specific domains is a particularly important enabler because not only does it enhance everyday activity and craftsmanship, but it also provides the opportunity for a boundless suite of versatile tools to support discovery in arts and science in all its many guises.

As computation is destined to be the defining notion of our foreseeable future, this book serves as a means to introduce computation in arts, science, and engineering in a manner that is easily accessible to society and that makes an individual the key recipient of the knowledge on technological progress. It merges the inevitably increasing presence of computation in our life with personal development along all its axes (physical, spiritual, mental, etc.) to ultimately affect, impact, and shape humanity as such.

Technology is supposed to provide a happier life; hence, it should serve human beings rather than relegating them to becoming serfs to progress. The obligation of educators and the mission of this book is to guide the reader in how Science and Technology can contribute to finding personal positive fulfillment. The collected material helps understand the technology trends directed toward humanity's goodness and make the right and balanced decisions for ourselves.

After an extensive survey and overview of the benefits of computation, in fields related to advancing humanity, the selected contributions present successful approaches where different algorithms, methodologies, tools, and techniques result in important advances of human well-being.

1.2 Organization

This book is to provide a broad overview of the current state of computational advances related to humanity, including the anticipated directions, potential breakthroughs, the challenges, and the achievements observed from numerous perspectives. To attain this objective, the book offers a compilation of 20 high-quality contributions from internationally renowned industrial and academic authors. These chapters are grouped into five parts as illustrated in Figure 1.1 that includes (I) Context and Intent, (II) Computing for a Sustainable Society, (III) Computational Discovery and Ingenuity, (IV) Citizen and Artisan Computing, and (V) Visionary Pondering and Outlook on Computation for Humanity:

- Part I comprises the intent, introduction, and the context. Key remarks about the process of content creation are provided. Computational approaches are then reviewed, including those presented in the

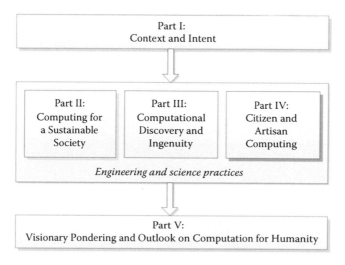

FIGURE 1.1
Book organization.

collection but extending to others as well. This forms a computation taxonomy investigated under a structured framework. Theoretical, technical, and societal perspectives are explored.

- In Parts II–IV, different engineering and science practices are discussed for various applications. Ingenuity is a strong element for each of them. And so, in Part II, computation usage examples for the sake of planet Earth, urban design, and societal sustainability are provided.

- Part III contains hard-science contributions on various topics from robotics through image processing for zoology to quantum engineering. Attention is directed toward domain-based reasoning, modeling, simulation, and educational implications.

- Part IV is composed of contributions that illustrate how art (e.g., music) and citizen science (e.g., data journalism) are advanced by computational progress as well as how computation helps solve challenges in humanities studies.

- Finally, Part V highlights the visionary statements and outlook on the future of humanity based on the discovery of today's technologies.

Following an overview of the overarching idea that defines the spirit of this book, a brief introduction of the contents of each of the individual chapters for each of the parts is presented.

Today's computationally skilled society has power and influence on our future like never before, and this effect is only going to grow. A single person, a corporation, nonprofit organization, or an entrepreneur involved in computing-centric activities has great power to benefit or harm the society or the environment along an abundance of avenues, including directions such as cutting-edge medicine and health care, educational innovation, and cultural adaptation in a developing country.

In *Computers and Society: Computing for Good*, Lisa Kaczmarczyk (2011) focuses on the positive computation applications and tackles topics such as poverty alleviation in the Peruvian Andes, earthquake modeling, improving patient care, and internet voting [cf., Figure 1.2 (Bhalla, 2012)]. And so is the intent of this book. In its first part, the trends that are going to be explored in the following parts are presented, and the context of this exploration is provided.

1.2.1 Part I: Context and Intent

- Chapter 1, Introduction by Justyna Zander and Pieter J. Mosterman. In this chapter the main intent of the collection is defined and overviews and context for the subsequent contributions are provided.

- Chapter 2, Taxonomy of Computation: Capabilities and Challenges by Justyna Zander and Pieter J. Mosterman.

FIGURE 1.2
Transforming the African Internet voting. (From ThirdFactor, Biometrics, emerging tech: Transforming the African vote, http://www.thirdfactor.com/2012/11/20/biometrics-emerg ing-tech-transforming-the-african-vote (November 20, 2011.))

This is a survey that spans different types of computation. It provides a synthesis for understanding computation from a technical user, a citizen scientist, and a researcher perspective. Current trends are identified and a novel framework for classifying them is provided. This chapter summarizes the capabilities and challenges of currently known computational approaches and concludes with a set of guidelines for researchers and practitioners. Technical and social demeanor is included in the analysis. The notion of cyber-physical systems (PCAST, 2007) is evolving as a consequence of computation convergence with other engineering domains. Technical methodologies, a consistent product family, and a set of models based on a case study are outlined.

1.2.2 Part II: Computing for a Sustainable Society

- Chapter 3, Sustainability through Computation by Azam Khan and Kasper Hornbæk

  This chapter describes how computation can improve individuals, groups, and societies by guiding them to live sustainably. Sustainability as a term refers to the planet capacity to support humans while accounting for Earth's existing climate and biodiversity. Measures for assessing system sustainability are proposed. The status quo is characterized and the effects of an intervention are quantified. Further, strategies for computing sustainability while applying multiple models, simulations, and parameters are described and illustrated with compelling case studies.

- Chapter 4, Computer as a Tool for Creative Adaptation: Biologically Inspired Simulation for Architecture and Urban Design by Taro Narahara

  In this work, a 4D design strategy and thinking are proposed as a response to ongoing radical population growth and environmental changes. The purpose is to develop more flexible and adaptable architecture in an urban environment and broaden the territory of design to include systems issues. In particular, computational approaches are introduced to relate natural processes in design using conceptual models. In addition, attention is dedicated to self-organizing systems in nature and exemplified with case studies.

- Chapter 5, Advanced Structural Health Monitoring Based on Multi-Agent Technology by Kay Smarsly and Kincho H. Law

  As safety and durability of civil structures impact society's economic and industrial prosperity, this chapter proposes how to reliably assess the conditions of spatial structures to comply with the safety standards throughout the operational live of a structure. A structural health monitoring system is explored as a solution for this problem. Such a system comprises sensors (e.g., accelerometers, temperature sensors, displacement transducers) that are installed in the structure to collect response measurements caused by external and internal stimuli. For structural health monitoring to be most successful, a multi-agent technology is deployed in the implementation. This enhances the modularity, robustness, accuracy, and reliability of the system activities. Here, self-contained cooperating software entities are adopted to perform the monitoring tasks in an autonomous manner. Mobile software agents in a wireless sensor network are proposed to monitor a wind turbine and a structural specimen in a laboratory setting.

- Chapter 6, ICT for Smart Cities: Innovative Solutions in the Public Space by Ina Schieferdecker and Walter Mattauch

  Urbanization, globalization, and demographic and climate changes are the motivators to study a city as a system and explore opportunities that arise from technological advances. As a basis for this exploration, the integration of information and communication systems into the various technical systems and infrastructures of a city is proposed. This enables a more flexible control of supply and disposal networks, especially for electricity, water, and gas. Furthermore, novel solutions for mobility, administration, and public safety in the city can be easily anticipated. Citizens, businesses, institutions, and government are to be in a constant collaboration to jointly increase the quality of life. The city then becomes a sophisticated service provider for its citizens and businesses. This chapter emphasizes the relevance of data, information, and communication between various stakeholders, systems, and processes in a smooth and integrated manner.

1.2.3 Part III: Computational Discovery and Ingenuity

- Chapter 7, Domain Science & Engineering: A Foundation for Computation for Humanity by Dines Bjørner

  Domain science and domain-based engineering are the topics discussed in this contribution. It is proposed to qualify the emerging trend of domain specificity in the information space as a new type of science. The definition of a domain and its role in software engineering are provided. Further, the discussion convincingly illustrates the necessity to introduce computational thinking in the current education system. In addition to mathematics and natural and life sciences, the logics and logical reasoning should be studied. Thus, it is encouraged to include studies of man-made domains in the early-age curricula.

- Chapter 8, Human-Robot Interaction by Alexandra Kirsch

  When humans interact with one another, there is always a social component to the interaction. In this chapter, it is argued that for a successful development of interacting robots, a similar reasoning must be applied. Hence, a classification of interactive robot applications is provided to understand the social and technical challenges. In addition, case studies borrowed from health care and biomechanics are discussed. Robotic simulations are also briefly touched upon.

- Chapter 9, Multimedia Retrieval and Adaptive Systems by Frank Hopfgartner

  A paradigm change from a passive information consumption habit to a more active information search is recently observed, especially with the rise of YouTube and TED talks. As a result, a new landscape for business and innovation opportunities in multimedia content and technologies has naturally emerged. In this chapter, video retrieval technologies are presented to argue how personalized media content can be offered to a selected user based on her/his interests. Ranking algorithms are being invented to find the best manner of interaction with a receiving user. Recommendation and user feedback methods are reviewed to find the best methods for selected applications.

- Chapter 10, Accelerated Acquisition, Visualization, and Analysis of Zoo-Anatomical Data by Alexander Ziegler and Björn H. Menze

  Computed tomography and magnetic resonance imaging allow gathering digital anatomical data about species in a noninvasive manner, at high speed, with the promise that accompanying software enables the analysis of the resulting images in real time and in three dimensions. This chapter introduces this notion into the study of zoological specimens. The interdisciplinarity challenges are discussed demonstrating technical problems and high investments in educational efforts and equipment. It is also illustrated how computational algorithms performing shape recognition can be adapted to large-scale anatomical studies in zoology.

- Chapter 11, Quantum Frontier by Joseph F. Fitzsimons, Eleanor G. Rieffel, and Valerio Scarani

As computation is too a physical process, in this chapter, a picture of quantum computating, a new form of computing, is presented. Quantum computating gives rise to a variety of faster algorithms, novel cryptographic mechanisms, and alternative methods of communication. Some aspects of quantum mechanics that are at the heart of a quantum information processing are discussed. Multiple examples and applications illustrate the concepts and provide a thorough overview on this fascinating topic.

- Chapter 12, DYVERSE: From Formal Verification to Biologically-Inspired Real-Time Self-Organizing Systems by Eva M. Navarro-López

Today's safety-critical and resilient systems require reliable validation and verification methods to ensure their stability. In this chapter, a computational-dynamical framework for modeling, analysis, and control of complex control systems is applied to verify a given type of systems. The developed flow of thoughts is inspired by nature formalized in disciplines such as biology and chemistry. Then, the reasoning is reversed and the use of engineering, computational, and systems thinking is explored to influence and better understand natural processes. Finally, a hybrid system framework that combines continuous and discrete elements is presented to model complex systems.

- Chapter 13, Computing for Models of the World by Steve Johnson

This chapter that presents the scientific method, the notions of abstraction, modeling, and prediction based not only on historical data while exploring the potential of social networking. Unconscious use of the computationally injected language and metaphor is made more conscious to better focus on the inner and outer reality surrounding the human. In this light, provocative questions arise challenging the human race and its directions.

1.2.4 Part IV: Citizen and Artisan Computing

- Chapter 14, Ad-Opera: Music-Inspired Self-Adaptive Systems by Paola Inverardi and Patrizio Pelliccione, Michelangelo Lupone, and Alessio Gabriele

This is a vision to investigate adaptive music that is synthesized through a dynamic process that constantly renovates itself. It is argued that music as a reference model with its synchronous and diachronic sounds can lead to new self-organizing approaches in software engineering. An empathic system engineering method is presented to create new dimensions for business and technology but also to enhance the human-related concerns, such as satisfaction, emotions, and creativity of the software users. As a first application domain candidate public and urban context is explored.

- Chapter 15, Fuzzy Methods and Models for a Team-Building Process by Irena Bach-Dąbrowska

Agile philosophy for team formation and project management is a common practice in selected industries (e.g., IT outsourcing companies in East Europe). Projects are often lead by self-organizing, small teams characterized by transfer of decision powers and responsibilities from a single leader to all team members. In such conditions task execution effectiveness depends mostly on interpersonal relationships inside the team. Thus, to obtain improved performance, it is beneficial to develop better team member selection methods, including assessing the behavioral, cultural, and psychological makeup of the entire team. To capitalize maximally on collaborating human brains, a fuzzy logic method is proposed in this chapter. The method provides the capability to define and measure imprecise information about behavioral and psychological features. Though imprecise, it aids human resources in finding the right matches for particular types of jobs using computationally aided automated selection methods.

- Chapter 16, Opportunities and Challenges in Data Journalism by Susan McGregor and Alon Halevy

As more than one-quarter of the world's population is either consuming or creating content on the web, personal journalism has changed its definition and scale. Today any organization can easily reach an audience of millions. In this chapter, the challenges related to this disruptive change are discussed. The notion of big data is introduced in a new manner. It is argued that the most relevant measure of big data today is not the size of the file but the extent of the network. For example, a single YouTube video becomes big data by generating 100 million page views independently of other criteria. Technical and societal directions are discussed to introduce a new type of journalism quality.

- Chapter 17, Visualizing Emancipation: Mapping the End of Slavery in the American Civil War by Scott Nesbit

Historians have been studying the timing of emancipation for years. This chapter introduces an approach for organizing knowledge about emancipation-related events in a spatial manner. The work uses algorithmically generated data to create maps of the interaction between armies and enslaved refugees during the U.S. civil war. The power of open-access projects demonstrates the public value of humanities research while finding ways to integrate the knowledge of citizen scientists into scholarly conversations. The resulting dataset for the emancipation example discussed in the chapter includes a list of approximately 40,000 data points describing the movements of 3,500 regiments over the course of the war.

- Chapter 18, Synergistic Sharing of Data and Tools to Enable Team Science by Ross Mounce and Daniel Janies

In this chapter, examples of projects in biology, medicine, and paleontology are reviewed to explain different models of interaction among scientific teams. Fostering development of computationally oriented team science that makes use of synergistic sharing is emphasized as a vehicle for driving innovation. Examples of open data projects are provided such as a Smithsonian field research team in Guyana using social media for rapid fish identification or increased usage of microarray datasets in genetics. It is argued that data publication could enhance knowledge synthesis and reuse if proper standards are followed.

1.2.5 Part V: Visionary Pondering and Outlook on Computation for Humanity

- Chapter 19, Evolution of the Techno-Human by Jim Pinto

This chapter presents a vision how technology acceleration is going to influence humans. Focused on cheap and abundant computing power, the evolutionary consequences are forecast based on the significant changes in the human landscape that have occurred in recent decades.

- Chapter 20, Computation in Human Context: Personal Dataspace as Informational Construct

In this contribution, a vision on a human as a computational construct is documented. A person is defined by the computational capabilities and the behavioral models derived from their use. Thus, the expression of humanity as such occurs through the surrounding information and the structures that allow for the data exchange. A notion of an *Infoindividual* is introduced through application of two components: the ultimate private act and a public assertion of self.

1.3 Target Audience

This book intends to be accessible as much as valuable to those interested in interdisciplinary and computational ways to solve everyday issues and humanity's grand challenges. The book is directed toward industry-related professionals, academic experts, students, entrepreneurs, and futurists. The book is, furthermore, intended for engineers, analysts, and scientists involved in the analysis and development of solutions that are dedicated to help humanity achieve sustainable growth.

Also, those interested in computer-based sciences will find fascinating applications of computational fundamentals in various disciplines and recognize the breakthroughs at the confluence of their adjacent work. The book constitutes a solid and inspiring effort to promote popular science to society and to provide a broad understanding of the world of technological progress in an easily accessible manner without sacrificing rigor where necessary.

References

Bhalla, J. (2012), CNN Opinion, Can tech revolutionize African elections? http://www.cnn.com/2012/11/17/opinion/sierra-leone-election-biometric/ (November 17, 2012).

Kaczmarczyk, L. C. (2011), *Computers and Society: Computing for Good*, 305pp., ISBN-10: 1439810885, CRC Press, Boca Raton, FL.

President's Council of Advisors on Science and Technology (PCAST). (2007), Leadership under challenge: Information technology R&D in a competitive world, An Assessment of the Federal Networking and Information Technology R&D Program, http://www.nitrd.gov/pcast/reports/PCAST-NIT-FINAL.pdf.

ThirdFactor (2011), Biometrics, emerging tech: Transforming the African vote, http://www.thirdfactor.com/2012/11/20/biometrics-emerging-tech-transforming-the-african-vote (November 20, 2011).

2

Computation Taxonomy: Capabilities and Challenges

Justyna Zander and Pieter J. Mosterman

CONTENTS

2.1 Introduction

A new kind of science called *computational science* has been established over the past decades (Wolfram 2002). The definition of this *science* according to the collective report by Lazowska et al. (2005) recognizes the elements that combined span the range of algorithms, software, architectures, applications, and infrastructure. *Computational science* is a rapidly growing multidisciplinary field that relies on advanced computing capabilities to understand and help solve complex problems in a broad range of disciplines (e.g., biological, physical, and social). As such, computational science comprises a multitude of elements: (1) *algorithms, modeling, and simulation software* for solving engineering and humanities problems; (2) *computer science* (CS) *advances* to develop and optimize system hardware, software, networking, and data management components necessary to handle computationally demanding problems; and (3) *the computing infrastructure* supporting the aforementioned two elements. Key to bringing to bear these computational advances for the better of humanity is understanding how to successfully leverage the respective elements.

Although applying computation for human needs is an art in and by itself as reported by Zander and Mosterman (2012), computing plays a critical role in advancing research across almost every scientific discipline. For example, exponentially cheaper everyday computing hardware holds the promise to unlock a seemingly boundless potential of simulation (Kurzweil 2005) if only because applying simulation increases manifold data flows and revolutionizes the practice of science and engineering (Pollak 2006). The value of computation is highlighted when observing the trends including big data, cloud computing, open and interoperable interfaces, social computing (Lazer et al. 2009), pervasive computation, and cyber-physical intersections (PCAST 2007). Some of these trends will briefly be reviewed in the following to motivate the content of this and, indeed, every chapter of this collection.

The *performance* of computers has shown remarkable and steady growth, doubling every year and a half since the 1970s (Kurzweil 2005). Ever smaller and less power-intensive computing devices proliferate, paving the way for new mobile computing and communications applications that create the opportunity to collect and use data in real time (Computing Technology Review 2011). The consequence is an exponentially growing deluge data (MIT Spectrum 2010) that has the collection, management, and analysis of data, a fast-growing concern of many research groups (e.g., MIT Human Dynamics Lab). Strategies to handle *big data* are, therefore, required in many disciplines (Pollak 2006) exploiting foundational enablers such as automated analysis techniques (e.g., data mining and machine learning) to facilitate the transformation of data into knowledge, and of knowledge into action (Hey et al. 2009).

Cloud computing, in which users have Internet access to shared computing resources, software, and data, makes massive computational and storage capability available to everyone at a steadily decreasing cost, as documented by Hardesty (2010). Of specific importance are *interoperable interfaces* because of their infrastructure nature and corresponding multiplier. Enabling intelligent components to communicate provides an important stimulus for innovation and adoption. For example, electronic health records treated as common data stored in the cloud can be shared, added, and processed by many parties. Such interfaces may often be *open, public, and transparent,* with a crucial role for standardization to enable and foster an entirely new breed of businesses that at present cannot be nearly conceived.

Further, the emergence of *social computing* as communication and interaction through social networks makes mechanisms such as coordinating crowd-sourcing available (Wightman 2010). By transforming the manner in which people interact with each other, encyclopedias, translators, and smart cities can now be collaboratively invented, designed, created, and maintained.

Social computing adds to *the open source software* trend. *Open* denotes software of which the source code is available to others, often without financial charge. This software is created by volunteers or by companies who wish to make certain functionality widely available and enable others to read it, vet it, or change it.

The *ubiquitous network* that has become a reality has caused an inundation of data. Combined with the exponential growth of computing power, a critical need for understanding computation fundamentals and their application has emerged. The next wave of discovery and innovation comprises powerful computation, data mining, and knowledge and wisdom management (Surowiecki 2004). Abstract thinking, modeling, simulation, and prediction are receiving broad acceptance in multiple disciplines. Likewise, leveraging aspects such as behavioral model analysis, big data extraction, and human computation to tackle the gamut of challenges that humanity is facing is rapidly gaining momentum (Lazowska et al. 2005).

The nexus of the previously mentioned trends facilitates users in obtaining awareness about their surroundings and about themselves. For example, investigating the application side, Apple's virtual assistant *Siri* shows how software enables computers to merge into human collaborators (Simonite 2012). Similarly, Google's mobile assistant, *Google Now*, built into Android smartphones and tablets, works like a search engine in reverse. It predicts what information may be required by a user and offers up weather forecasts, traffic reports, or transit times. Likewise, Microsoft Research is building a *Lifebrowser* to tag and organize personal data from the past, showing that a model is not only learning about how a person thinks, but it attempts to understand what it means to capture human values (Horvitz 2007).

When building a *taxonomy for computation*, select branches naturally emerge. In first order, these branches constitute traditional approaches such as computation used to *advance engineering*, computation understood as *simulation* (and, thus, execution), and *high-performance computing*. In second order, these concepts are further extended to cover the emerging ideas such as *human-based computation, social computing, affective computing, knowledge-based computation*, and *nature-inspired computation* but also *participatory sensing* and *computational thinking*.

This chapter then provides a brief overview of the *state of the art* in the field of computation. It is a consolidation of selected disciplines that make use of computational advances to stimulate thought and extract future trends at the intersections between them. The collection of contributions in this book in general, and this chapter in particular, takes a step back to evaluate and assess computation and humanity from a broader perspective seeking recurring patterns between different disciplines. It aims to be a comprehensive survey that provides insight into *practice-based understanding* of computational theories, opportunities, and restrictions.

The survey in this chapter is not to be taken as exhaustive. The purpose is to shed light on selected *theoretical, technical*, and *societal* aspects of computation, from theory through applied research to everyday practice. First, a state-of-the-art *analysis* is performed and then a framework is created as a *synthesis.* This framework serves as a means to understand the current trends in multiple domains from the *CS*, interdisciplinary *engineering*, and *computational science* viewpoints. As a result of the investigations, a structure for the existing approaches is proposed to include the modern computational technologies from many fields in one common perspective. The knowledge organization succeeds at some levels of abstractions (e.g., in aspects such as scientific application, collaboration, and industrial application), whereas it shows potential for even more investigation at some other levels (e.g., in areas such as unified engineering integration and unified simulation semantics definition). The *capabilities* and *challenges* of today's computation are identified as a result of that given earlier.

In the next section, a multi-faceted *computation taxonomy* is presented and for readability purposes illustrated using three dimensions. The *capabilities* of computation are emphasized throughout the entire chapter, whereas the *challenges* are outlined mostly for the theoretical and technical dimensions. Then, in Section 2.3, a link to education is made while Section 2.4 completes this chapter with a summary.

2.2 Taxonomy of Computation

The classification of *computation* presented in Figure 2.1 intends to organize the knowledge about science and applications that primarily serve humanity and that are necessary for a proper synthesis. The classification is divided based on the *exploration perspective* and *computation abstraction*. A first-cut categorization comprises three dimensions: *theoretical, technical,* and *social and societal.* These dimensions illustrate how to drive innovation and benefit for humanity from very different perspectives and for different reasons. However, they all ultimately converge to create a positive impact multiplier.

The classification for computation is further detailed in Figures 2.2 and 2.3. Figure 2.2 includes three perspectives (presented vertically). The first one is called *theoretical and technical* and includes a number of disciplines (shown vertically) listed along with example concepts for each of them.

The *theoretical and technical* perspective is typically related to *specific domains, technologies,* and *tools.* Hence, a *tool-oriented* perspective serves as an illustration of the theory and technology in practice. Here, for each row, a select set of example tools is provided. For example, while expressing *computer science (CS)* that introduces concepts such as *abstractions, software engineering principles, model-based design* (Nicolescu and Mosterman 2009), and *model-based testing* (Zander et al. 2011), MATLAB® and Simulink® (MathWorks 2012) are products that provide a popular design platform, especially for embedded software systems. This platform then may include, for example, Stateflow® for supervisory control algorithms, SimEvents®

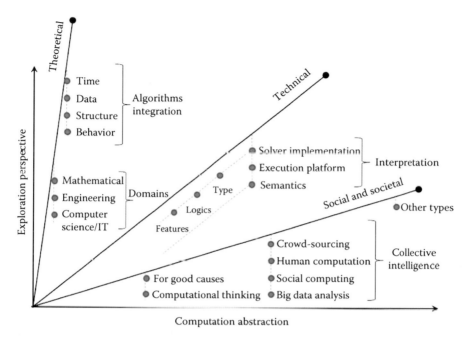

FIGURE 2.1
Exploration dimensions for creating computation taxonomy.

for network information flows, and Bioinformatics® Toolbox for analyzing genomic and proteomic data.

Next in the classification of computation, a perspective called *pillars of scientific inquiry* (Lazowska et al. 2005) is identified as consisting of theory, experimentation, and simulation. These three elements complement the elements in the other perspectives such that, in a horizontal sense across the diagram, each scientific inquiry pillar element is associated with a certain discipline and accompanying tool.

2.2.1 Theoretical and Technical Dimension

From the *theoretical viewpoint*, computation is a converging discipline requiring expertise from domains such as CS and information technology, electrical engineering (EE), physics, mechanical engineering, and mathematics. To illustrate, a complex system considered from the algorithmic viewpoint often relies on separation of concerns and aspects such as *time, data, structure (i.e., architecture)*, and *behavior* become critically important; however, they are treated differently within each of the previously mentioned domains (Lee and Sangiovanni-Vincentelli 1998). For example, time may have different meaning to various experts. The challenge is to integrate the algorithms dealing with all the various concerns within one domain but also between

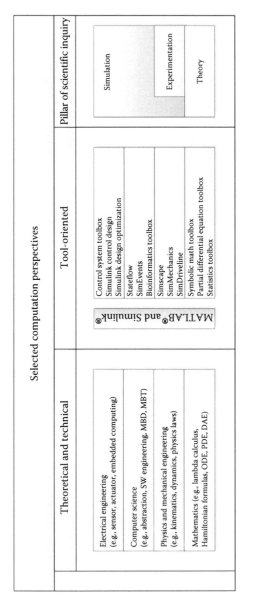

FIGURE 2.2
Selected perspectives on computation.

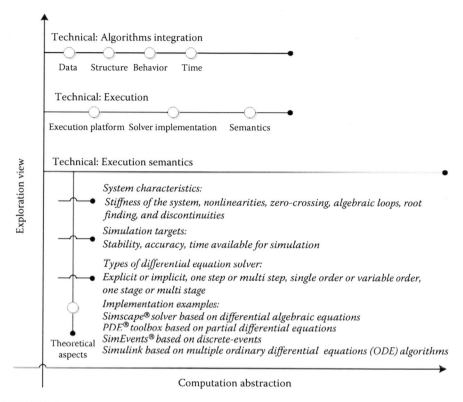

FIGURE 2.3
Technical exploration of computation abstractions.

disciplines as a crosscutting concept (Lee 2006). In terms of a merge between CS and EE, at times an abstract notion is required to obtain consensus in both fields (Zander et al. 2011).

In Figure 2.3, the *technical dimension* for computation is addressed. Here, computation is considered in terms of its execution. Aspects such as computationally-based implementation (e.g., *solver implementation*), *execution platform*, and the underlying *semantics* become relevant and important. A prominent example to illustrate the technical dimension of computation is a cyber-physical system (CPS). A CPS (Lee 2006, 2010) is frequently characterized as a smart system that includes digital cyber technologies, software, and physical components and is intelligently interacting with other systems across information and physical interfaces. As such, a CPS senses the external world and immediately reacts on the state of its surrounding, which includes effecting change in the physical world as well as sharing information with other CPS in an attempt to collaboratively achieve a given goal. The result expresses an emerging behavior and so a CPS creates live functionalities during deployment (PCAST 2007, Zander 2012).

Designing such an overall constellation of collaborating CPS operating in and on a physical world with near-instantaneous response times is of a complexity far beyond the current state of science, engineering, and technology (NIST 2013). In order to successfully and competitively design such complex systems, early and continuous consideration of design semantics and system quality is of vital importance. A successful technical merge of multiple disciplines, as are inherent in CPS, requires the understanding of semantics underlying the computational representation behind CPS components. To solve this type of convergence and integration complexity, *abstraction* is a natural fit in designing embedded computational features of a CPS. In the system design process, abstraction facilitates developing critical characteristics by providing high-level formalisms (e.g., UML®, Simulink blocks (MathWorks 2012), SCADE® blocks, and the UML® Testing Profile). These formalisms relate mostly to the problem space (business processes, control algorithms, signal processing algorithms, test suite creation, etc.), while the solution space is typically understood as code, which nowadays is frequently automatically synthesized (Nicolescu and Mosterman 2009). The abstraction results in *models* of the underlying logical, software, and hardware components, and so the corresponding approach to design is referred to as *model-based design*. It enables the study of challenges that arise from applying models at multiple time scales, with multiple levels of detail, and using a combination of multiple formalisms (Mosterman and Vangheluwe 2004). In this context, it is essential to concentrate on the formalization of the meaning of models that represent the physical world so as to correspond to formalizations typical for embedded computation.

The *technical dimension* of computation typically relates to *technical simulation*. The simulation logic is hidden in the equations *solver* that is implementing the execution of a computational model as represented by the equations (Cellier and Kofman 2006). From a CS viewpoint, a solver is a software component that determines the time of the next simulation step and applies a numerical method to solve a set of ordinary differential equations (ODEs) that represent the model. In the process of solving an initial value problem, the solver is designed to operate with a given accuracy (MathWorks 2012). There exists a multitude of solvers, and a selection of the proper one for the desired model execution is a trade-off between *simulation characteristics* such as stability, accuracy, and time available for simulation. Typically, it is recommended to use a computationally expensive variable-step solver to generate a reference solution. From this reference behavior, solvers that are more convenient for design studies because more efficient, though less accurate, may be employed while solutions are being compared to the reference. This allows, for example, investigating the model sensitivity (i.e., improve the model robustness) and ultimately compares different solver behaviors and their *characteristics (i.e., features)*.

Solvers handle issues such as *stiffness of the system, nonlinearities, zero-crossings, algebraic loops, root finding* (cf. numerical challenges discussed by Moler 1997), *and discontinuities.* Furthermore, if only considering the Simulink product family (MathWorks 2012), different blocksets use different algorithms that are implemented in separate code bases for solving certain types of equations. For example, the Simscape® solver is based on differential algebraic equations, the PDE® Toolbox on partial differential equations, SimEvents on a discrete event calendar, and Simulink itself on multiple ODE algorithms. Integrating all the various execution modalities remains a challenging endeavor that is subject to continued study (e.g., Mosterman 2007). In the following, for illustrative purposes, ODE solvers are explored in more detail.

Simulation expertise is required in order to match model characteristics with a proper solver to obtain desired analysis conditions. It is frequently a challenge for industry experts to match a model with an appropriate solver. For example, if, on the one hand, the system model must have a predictable execution time, however, the execution duration is not restricted, and a fixed-step differential equation solver may well serve as a means to experiment with integration order, step size, and accuracy of the simulation (i.e., prediction). On the other hand, if the system model must be simulated in a time duration that is as short as possible, a variable-step solver often is a better approach because the step size adjusts to the dynamics of the model, and the simulation typically executes faster with higher accuracy, though execution may take longer during certain periods within the overall duration. In the face of discontinuities in a model, good practice is to also employ a zero crossing solver to obtain quicker and more accurate results.

Another significant consideration in industry is *code generation for models* (Halbwachs et al. 1991). If it is planned to generate code from a model and run it in real time on a computer system, a fixed-step solver must be applied because the variable-step size cannot be uniformly mapped to the real-time clock (MathWorks 2012).

To conclude the results provided in Figure 2.3, *computation abstraction* and *exploration view* of computation semantics are the axes that represent the dimensions for the theoretical and technical analyses. The *exploration view* is then divided into areas such as *algorithms integration, execution,* and *execution semantics. Algorithms integration* means that computation formalization includes *data, structure, behavior,* and *time. Execution platform, solver implementation,* and the *semantics definition* constitute the analysis means for attaining the proper computational *execution.* Finally, elements such as *system characteristics, simulation targets, types of solver,* and *implementation examples* complete the categorization from the practical viewpoint.

In the following paragraphs, *challenges* related to the *theoretical and technical dimension* of computation are outlined. From the mathematical and technical viewpoints, many breakthroughs in solving scientific problems have been enabled by advances in algorithms. As a victim of its own success, in response, computational methods are often taken for granted because of the past (isolated or shared) achievements (NSF 2011). However, there is still an urgent and acute necessity to work on the *scalability* of computational modeling and simulation (M&S) methods. Systems under design today (e.g., at the petascale level) require advances, in particular, in relation to modeling *heterogeneity*, *multi-physics couplings*, *multi-scale* and *multi-rate* behavior, *uncertainty*, dynamically evolving and *emergent* behavior, and so on. Corresponding systematic research is imperative for developing the necessary next-generation M&S methodologies and technologies.

The challenge of *multi-domain design* and its corollary, the definition of a *unified M&S semantics*, can be identified as critical enablers. Simulation semantics (i.e., the execution) must be defined in such a manner that it is understandable for the community but also for the end user, in particular because this end user may eventually become a part of the community. A further implication of a multi-domain approach is the *interaction of approximations* in the various numerical algorithms as discussed in previous work of Mosterman et al. (2009) and Mosterman and Zander (2011). Such algorithms must be combined to solve differential equations, difference equations, algebraic equations, algebraic loops, root-finding inequalities, etc. (Mosterman et al. 2012). A precise formal semantics for such interactions, a common understanding of the notion of time, and its proper indexing in the solvers (cf. semantics domain) constitute specific items under this (absolute priority) umbrella.

From a technical system perspective, *the verification and validation of models and of simulation in its own right* are still challenges that have not yet been solved, although distinct progress has been reported (Zander et al. 2011). M&S enable *high-risk research*, holding the promise of dramatic breakthroughs with high technological and societal impact. In this sense, the *challenge* lies in capitalizing on the abundance of simulated realities to impact a *social good cause*.

Further, *execution and targeting hardware platforms* become increasingly important from the technical designer viewpoint (Mosterman et al. 2004). Nowadays, rapid prototyping based on M&S that includes hardware components (e.g., the Arduino platform) is possible and easier than ever before. It is becoming increasingly less expensive and scales up in the educational classroom (Arduino).

The next element to consider is the *Internet of Things* (Van Kranenburg 2008), which uses radiofrequency identification technology to connect an increasingly broad range of artifacts around us. In ever more business areas, companies depend on the Internet as a powerful, cost-effective medium for data transport. The *challenge* is to *interface Internet of Things with M&S* and leverage this combination to build *Computation of Things* (Zander and Mosterman 2010, Zander et al. 2010, Zander and Mosterman 2012), where artifacts are not only connected but actually process the data and come up with novel solutions while able to effect changes in the physical surroundings they operate in. This will open up the space for services that organize and automate governmental, societal, and engineering systems and structures even more and to an as-of-yet unimaginable extent.

2.2.2 Social and Societal Dimension

Ubiquitous communication capabilities allow for a redefinition of computer-based solutions to provide a new and better world (Diamandis and Kotler 2012). The *social and societal* dimension of computation relates to its application areas independently of the *theoretical* and *technical* dimensions. This dimension constitutes a factor that influences the overall impact for human kind. In Figure 2.4, selected human-related types of computation are presented. Here, abstract concepts such as *computational thinking* and *computation for good causes* are relevant in terms of humanity. Next, more specific types of *collective intelligence* are significant. According to Quinn and Bederson (2011), *human computation, social computing, crowd-sourcing,* and *big-data analysis* are the key ingredients of such collective intelligence. Further, *societal applications* are defined as those driven by industry and government. In the proceedings, selected types of socially/societally driven computation are described in more detail.

Computational thinking involves activities such as designing systems, understanding human behavior, and solving problems by drawing on the concepts fundamental to CS. It includes a range of mental tools that reflect the breadth of the field of CS and computational science. As such, computational thinking represents a universally applicable aptitude and skill set everyone, not

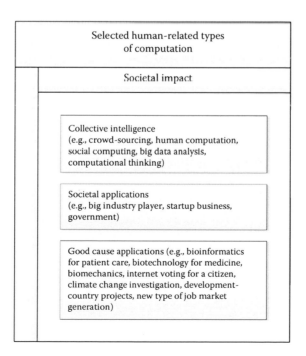

FIGURE 2.4
Computation as a factor creating social impact.

just computer scientists, would benefit from and be eager to learn and use. In its implementation, massive amounts of data are used to accelerate computation while making trade-offs between time and space and between processing power and storage capacity (Wing 2006).

Computation for good causes is a topic treated well by Kaczmarczyk (2011). She rightfully claims that until recently people who worked with computers were labeled as brainy but at times difficult to collaborate with. Computing was seen as an impersonal activity not related to your average citizen. However, this trend has changed. Computer use, or, implicit use of computers, has become an integral part of almost all facets of our everyday lives. As a result, when scientists, educators, and engineers use computing to support societal values and solve human problems, potential for amazing innovations appears (Gajos et al. 2012). As Duben (2012) admits, one of the most challenging topics to teach in the undergraduate computing curricula is the social and ethical ("do no evil") implications of computing. Thus, there is terrific opportunity today to look for computationally driven professions that are dedicated solely to good.

Human-based computation is a CS methodology in which certain computational activities are outsourced to humans. The aim is to attain a symbiotic human–computer interaction by applying differences in abilities and cost alternatives between humans and computer agents. Humans perform a major part of the computation and by that are an integral part of the overall computational system (von Ahn 2008). The initiative involving *reCAPTCHA* is a prominent example for human-based computation (von Ahn 2008). It unlocked the potential to digitize, on average, hundred million words a day that then translates to about two million books a year. Similarly, the newest game, called *Duolingo*, is intended to help translate the Internet contents with the help of the crowd (von Ahn 2011).

Social computing is a general term for an area of CS that intersects social behavior and computational systems. Similar to human-based computation, it pertains to supporting computing-related activities that are carried out by groups of people (Surowiecki 2004). Examples of social computing include collaborative filtering of data, prediction markets, reputation systems, computational social choice, and verification games (cf. Christakis Lab at Harvard 2012).

Participatory sensing relates to *crowd-sourcing* and colloquially means to observe, monitor, digitize, study, reflect on, and share the surrounding world (Burke et al. 2006). For example, urban sensing is related to a citizen equipped with currently available mobile technology who is participating in the process of collecting data about the city. As an example, the Personal Environmental Impact Report (Mun et al. 2009) is built using a participatory sensing systems platform implemented as an online tool that is based on current geo-location and human activity patterns. Continuous use of a mobile device allows exploring and sharing how an individual impacts the environment and vice versa. With participatory sensing, complex global issues such as climate change (Coyle and Meier 2009), obesity (Christakis and Fowler 2009), health (CureTogether service), and wellness can be studied.

Affective computing is the study and development of systems and devices that can recognize, interpret, process, and simulate human affections (Picard 1997). CS, psychology, and cognitive science are the constituents of this field. Emotions are studied to explore the ability to simulate empathy by a machine. In particular, the machine is intended to interpret the emotional state of humans by adapting its behavior and responding appropriately to the state of feelings (cf. Affective Computing Lab at MIT 2012).

Recently, leveraging aspects such as behavioral model simulation and *big data extraction and analysis* is gaining momentum. At the confluence of the afore-mentioned trends, mass-scale users receive awareness about their surroundings and themselves. To realize this purpose, Zander and Mosterman (2012) propose an online platform for M&S on demand. It allows citizen technologists to capitalize on broadly shared information and its analysis using scientifically founded predictions and extrapolations. This process occurs by leveraging crowd-sourcing along with clearly defined technical methodologies and social network-based processes. The platform aims at connecting users, developers, researchers, passionate citizens, and scientists in a professional network and opens the door to collaborative and multidisciplinary innovations.

2.3 Computing and Education

Education is a challenge in its *multi-domain cooperation* dimension. Collaboration on the level of scientific disciplines but also from an organizational viewpoint is a must. Hence, creating sustainable programs that promote cooperation across disciplines should be stimulated for academics to pursue their career. Institutional transformative change is likely to have to come from the level of governments, educational organizations, and industry (Ferguson 2004). Training to exploit careers in computational science and engineering is required for computation and, in particular, for computation to maintain its pace of growth and keep up with rapidly increasing demands. For example, European research on technology-enhanced learning investigates how information and communication technologies can be used to support learning and teaching, as well as more generally competence development throughout life (Bullinger et al. 2009). This is an opportunity for such instances as M&S to become part of such a learning initiative (e.g., through massive open online courses), in particular because M&S is the foundation for *interactive interfaces* with value that is difficult to overstate. Raising the interest in this context includes not only openness to industry consumers. It also calls for openness to the young people who should be offered *scientific and technical education* at the same time. *Education* constitutes one of the keys for the success of computation. In particular, broad application of M&S, including *technology democratization*, is significant if not critical as the driving force for human progress over decades to come.

The first (i.e., theoretical), second (i.e., empirical), and third (i.e., simulation) paradigms of science have successfully caused existing paradigms and technologies to progress incrementally (NSF report 2011). For revolutionary breakthroughs, novel transformational paradigms are required, though. In this light, embracing big data analysis can be considered the next, *fourth paradigm of science* according to Gray (Hey et al. 2009). His vision of this paradigm calls for a new scientific methodology focused on the power of *data-intensive science*. Computing technology, with its pervasive connectivity, already underpins almost all scientific study. Previously unimaginable amounts of data are captured in digital form, and so, data may help bring about a profound transformation of scientific research and insight (PCAST 2007).

Taking a step back and reflecting on the *third pillar of scientific inquiry*, it was Kenneth G. Wilson who first recognized computation to be of such value and importance (Lazowska et al. 2005). Analyzing a complex system is a difficult task as a large number of degrees of freedom in many possible states occur in it. Because of the improvement of computational engines, new possibilities emerged. High-precision numerical methods were developed to examine the relevant models that are prohibitively complex for analytic approaches. It then started to become clear that the base of the new discipline of numerical physics (i.e., the third pillar of the scientific discovery) extends beyond experiments and theory. For example, by using numerical calculations, the statistical description of multi-body systems became available. This then led to a renaissance of statistical physics as well as the evaluation of Monte Carlo methods, which provided special computational techniques for statistical physics simulations (Karsai 2009).

2.4 Summary

An overview on the state of the art of the broadly understood field of computation has been provided in this chapter. Opportunities and challenges have been analyzed. A synthesis of the results led to a formulation of taxonomies for computation from the theoretical, technical, and social points of view. In future work, more attention will be directed to the benefits and shortcomings of the proposed taxonomy framework by studying a comprehensive CPS case.

As converging technologies for behavioral model analysis, simulation, big data extrapolation, and human computation have been gaining momentum in the past decade (Levytskyy 2008, Levytskyy et al. 2009), further studies will be performed in this direction. Proper integration promises new advances in

CPS development that ultimately allow for raising human awareness of many unexplored and unexploited phenomena. Modeling computational approximations and sharing expertise among colleagues, potentially from other domains, are certainly one of the manners to further improve the growth of the discussed theories and technologies with a mutual and simultaneous acceptance of the human factors that are involved. In particular, the *theoretical and technical dimension* of computation will be explored next. Related work on building solvers in a declarative manner will be analyzed. In this respect, time monotonicity (Zander et al. 2011) and the concept of nested time will be addressed in further studies.

References

Affective Computing Lab at MIT: http://www.media.mit.edu/research/groups/affective-computing, 2013.

Broy, M. et al.: Economical impact of model-based development of embedded software systems in cars, *Elektronik Online Edition*, April 2011. http://www.scribd.com/doc/54417302/ATZ-Elektronik-Online-Apr-2011.

Bullinger, H.-J. et al.: Where the new growth comes from. Innovation-policy impetus for a strong Germany on the global stage, Media services of Fraunhofer Informationszentrum IRB, Germany 2009. http://www.forschungsunion.de/pdf/where_the_new_growth_comes_from.pdf

Burke, J., Estrin, D., Hansen, M., Parker, A., Ramanathan, N., Reddy, S., Srivastava, M.B. et al.: Participatory sensing. *World Sensor Web Workshop*, ACM Sensys 2006, Boulder, CO, 2006.

Cellier, F.E. and Kofman, E.: *Continuous System Simulation*, ISBN-10: 0387261028, Springer, New York, March 15, 2006.

Christakis Lab at Harvard University: http://christakis.med.harvard.edu/, 2013.

Christakis, N.A. and Fowler, J.H.: *Connected:* The Surprising Power of *Our Social Networks and How They Shape Our Lives, Little, Brown and Company*, New York, ISBN-10: 0316036145, 352p., 2009.

Computing Technology Review: 2011. http://www.technologyreview.com/business/40016/.

Coyle, D. and Meier, P.: *New Technologies in Emergencies and Conflicts: The Role of Information and Social Networks*, UN Foundation-Vodafone Partnership, Washington, DC and London, U.K., 2009.

Craglia, M. et al.: Next-generation digital earth, *International Journal of Spatial Data Infrastructures Research*, 3, 146–167, 2008.

CureTogether Service. http://www.curetogether.com/.

Diamandis, P.H. and Kotler, S.: *Abundance: The Future is Better Than You Think*, ISBN-10: 1451614217, ISBN-13: 978-1451614213, 400p., Free Press, London, U.K., February 21, 2013.

Ferguson, E.: Impact of offshore outsourcing on CS/IS curricula. *Journal of Computing Sciences in Colleges*, 19(4):68–77, April 2004. https://www.scss.tcd.ie/~smcginns/enrolments/p68-ferguson.pdf

Gajos, K.Z. et al.: Personalized dynamic accessibility. *Interactions*, 19(2):69–73, March 2012.

Halbwachs, N. et al.: Generating efficient code from data-flow programs. In Third *International Symposium on Programming Language Implementation and Logic Programming*, Passau, Germany, August 1991.

Hardesty, L.: Supercomputing on a cell phone, MIT, 2010. http://web.mit.edu/new soffice/2010/supercomputer-smart-phones-0901.html.

Hey, T. et al.: *The Fourth Paradigm: Data-Intensive Scientific Discovery*, 284p., ISBN-10: 0982544200, ISBN-13: 978-0982544204, Microsoft Research, Redmond, WA, October 16, 2009.

Horvitz, E.: Machine learning, reasoning, and intelligence in daily life: Directions and challenges. (Invited talk), *Proceedings of Artificial Intelligence Techniques for Ambient Intelligence*, Hyderabad, India, January 2007.

Kaczmarczyk, L.C.: *Computers and Society: Computing for Good, Chapman & Hall/ CRC Textbooks in Computing*, 305p., CRC Press, Boca Raton, FL, December 14, 2011.

Karsai, M.: Cooperative behavior in complex systems, PhD thesis, *University of Szeged— Department of Theoretical Physics*, Université Joseph Fourier, Hungary, 2009.

Kurzweil, R.: *The Singularity is Near: When Humans Transcend Biology*, 672p., ISBN-10: 0670033847, Viking Adult, New York, 2005.

Lazer, D. et al.: Computational Social Science, *Science*, 323, February 6, 2009.

Lazowska, E. et al.: Report to the President on Computational Science: Ensuring America's competitiveness, June 2005. http://www.nitrd.gov/pitac/reports/20050609_computational/computational.pdf.

Lee, E.A.: Cyber-physical systems—are computing foundations adequate? *Position Paper for NSF Workshop On Cyber-Physical Systems: Research Motivation, Techniques and Roadmap*, Austin, TX, October 2006.

Lee, E.A.: *Disciplined Heterogeneous Modeling*, EECS Department, UC Berkeley, Invited Keynote Talk, MODELS 2010, Norway, 2010.

Lee, E.A. and Sangiovanni-Vincentelli, A.: A framework for comparing models of computation, *IEEE Transactions on Computer-Aided Design of Integrated Circuits and Systems*, 17(12):1217–1229, December 1998.

Levytskyy, A.: Model driven construction and customization of modelling & simulation web applications, *Technische Universiteit Delft, PhD Thesis*, 2008.

Levytskyy, A. et al.: MDE and customization of modeling and simulation web applications, *Simulation Modelling Practice and Theory*, 17(4):408–429, 2009.

MathWorks™, Inc. http://www.mathworks.com/help/simulink/index.html, Using Simulink®, Natick, MA, 2012.

MATLAB and Simulink support package for Arduino. http://www.mathworks.com/academia/arduino-software/arduino-matlab.html

MIT Human Dynamics Lab. http://www.media.mit.edu/research/groups/human-dynamics.

MIT Spectrum, Karaginis, L. interview with Malone T., Collective brainpower, using new technologies to amplify human intelligence, 2010. http://spectrum.mit.edu/articles/normal/collective-brainpower/?utm_campaign=email10su&tr=y&auid=6425928

Moler, C.: *Are We There Yet? Zero Crossing and Event Handling for Differential Equations.* EE Times, pp. 16–17, 1997. Simulink 2 Special Edition.

Mosterman, P.J.: Hybrid Dynamic Systems: *Modeling and Execution, in Handbook of Dynamic System Modeling*, P.A. Fishwick (ed.), Chapter 15, CRC Press, Boca Raton, FL, 2007.

Mosterman, P.J. and Vangheluwe, H.: Computer automated multi-paradigm modeling: An introduction, in *SIMULATION: Transactions of The Society for Modeling and Simulation International*, 80(9), 433–450, Editor Choice Article, 2004.

Mosterman, P.J. and Zander, J., Advancing model-based design by modeling approximations of computational semantics, *Proceedings of the 4th International Workshop on Equation-Based Object-Oriented Modeling Languages and Tools (EOOLT 2011)*, Zurich, Switzerland, 2011.

Mosterman, P.J. et al.: An industrial embedded control system design process, in the Inaugural *CDEN International Conference on Design Education, Innovation, and Practice*, pp. 02B6-1 through 02B6-11, July 29–30, Montreal, Quebec, Canada, 2004.

Mosterman, P.J. et al.: Model-based design for system integration, in the Second *CDEN International Conference on Design Education, Innovation, and Practice*, pp. TB-3-1 through TB-3-10, July 18–20, Kananaskis, Alberta, Canada, 2005.

Mosterman, P.J. et al.: Towards computational hybrid system semantics for time based block diagrams, *In Proceedings of the 3rd IFAC Conference on Analysis and Design of Hybrid Systems (ADHS'09)*, A. Giua, C. Mahulea, M. Silva, and J. Zaytoon (ed.), pp. 376–385, Plenary Paper, Zaragoza, Spain, September 16–18, 2009.

Mosterman, P.J. et al.: A computational model of time for stiff hybrid systems applied to control synthesis, control engineering practice journal (CEP), Elsevier, in *Control Engineering Practice*, 20(1), 2–13, January 2012.

Mun, M. et al.: PEIR, the personal environmental impact report, as a platform for participatory sensing systems research, *In Proceedings of the 7th International Conference on Mobile Systems, Applications, and Services*, Krakow, Poland, pp. 55–68, 2009.

National Science Foundation (NSF): *Advisory Committee for Cyberinfrastructure Task Force on Grand Challenges, Final Report*, March 2011.

National Research Council: *Report of a Workshop on the Scope and Nature of Computational Thinking, Committee for the Workshops on Computational Thinking*; National Research Council 126p., ISBN-10: 0-309-14957-6, 2010. http://books.nap.edu/openbook.php?record_id=12840&page=2.

Nicolescu, G. and Mosterman, P.J. (Eds.): *Model-Based Design for Embedded Systems*, CRC Press, Boca Raton, FL, ISBN 9781420067842, November 2009.

NIST: Report of the Steering Committee for Foundations in Innovation for Cyber-physical Systems, *Foundations for Innovation: Strategic R&D Opportunities for 21st Century Cyber-physical Systems—Connecting computer and information systems with the physical world*, 2013.

PCAST, President's Council of Advisors on Science and Technology (PCAST): Leadership under challenge: Information Technology R&D in a Competitive World. An Assessment of the Federal Networking and Information Technology R&D Program, August 2007. http://www.nitrd.gov/pcast/reports/PCAST-NIT-FINAL.pdf

Picard, R.W.: *Affective Computing*, MIT Press, Cambridge, MA, 1997.

Pollak, B. (ed.): *Ultra-Large-Scale Systems*, 150p., ISBN:0-9786956-0-7, June 2006.

Quinn, A.J. and Bederson B.B. et al.: *Human Computation: A Survey and Taxonomy of a Growing Field, Association for Computing Machinery*, ACM Press, New York, May 7, 2011.

Simonite, T.: *A Look Back at Predictive Assistants and Lurching Giants*, December 2012. http://www.technologyreview.com/news/508971/a-look-back-at-predictive-assistants-and-lurching-giants/?utm_campaign = newsletters&utm_source = newsletter-weekly-computing&utm_medium = email&utm_content = 20121227

Surowiecki, J.: *The Wisdom of Crowds: Why the Many Are Smarter Than the Few and How Collective Wisdom Shapes Business, Economies, Societies and Nations*, 320p., ISBN-10: 0385503865, Doubleday, New York, 2004.

Van Kranenburg, R.: *The Internet of Things, Institute of Network Cultures*, ISBN/EAN: 978-90-78146-06-3, Amsterdam, the Netherlands, 2008.

von Ahn, L.: *Duolingo: The Next Chapter in Human Computation*, 2011, TEDxCMU. http://tedxtalks.ted.com/video/TEDxCMU-Luis-von-Ahn-Duolingo-T, Accessed in April 2013.

von Ahn, L. et al.: *reCAPTCHA: Human-Based Character Recognition via Web Security Measures*, Science. pp. 1465–1468. DOI:10.1126/science.1160379, September 2008.

Wightman, D.: Crowdsourcing human-based computation, *Proceedings of the 6th Nordic Conference on Human-Computer Interaction: Extending Boundaries*, pp. 551–560, ACM New York, NY, 2010. http://www.hml.queensu.ca/files/CHCv27%20%281%29.pdf

Wing, J.: Computational thinking, March 2006/Vol. 49, No. 3, Communications of the ACM, 2006. http://www.cs.cmu.edu/afs/cs/usr/wing/www/publications/Wing06.pdf

Wolfram, S.: *A New Kind of Science*, 1192 p., ISBN-10: 1579550088, 1st edn., Wolfram Media Inc., Champaign, IL, May 14, 2002.

Zander, J.: A vision on collaborative computation of things for personalized analyses, *Proceedings of the 3rd IEEE Track on Collaborative Modeling & Simulation—CoMetS'12 in WETICE 2012, 21st IEEE International Conference on Collaboration Technologies and Infrastructures*, Toulouse, France, June 25–27, 2012.

Zander, J. and Mosterman, P.J.: Computation of things for sustainable development, *Homeland Security Workshop on Grand Challenges in Modeling, Simulation, and Analysis*, Washington DC, March 2010.

Zander, J. and Mosterman, P.J.: *Technical Engine for Democratization of Modeling, Simulations, and Predictions*, WinterSim 2012, Berlin, Germany, 2012.

Zander, J. et al.: Computation of things for human protection and fulfillment, *Proceedings of the IEEE International Conference on Technologies for Homeland Security 2010 (IEEE-HST 2010)*, Waltham, MA, November 8–10, 2010.

Zander, J. et al.: On the structure of time in computational semantics of a variable-step solver for hybrid behavior analysis, *Proceedings of 18th World Congress of the International Federation of Automatic Control (IFAC) 2011*, Milan, Italy, 2011a.

Zander, J. et al.: *A Taxonomy of Model-Based Testing for Embedded Systems from Multiple Industry Domains, Model-Based Testing for Embedded Systems*, J. Zander, I. Schieferdecker, and P. J. Mosterman (eds.), ISBN 9781439818459, CRC Press, Boca Raton, FL, September 2011b.

Part II

Computing for a Sustainable Society

3

Sustainability through Computation

Azam Khan and Kasper Hornbæk

CONTENTS

3.1 Overview

This chapter describes how computation can support individuals, groups, and societies in living sustainably. By sustainability, we refer primarily to the Earth's capacity to support human, animal, and plant life while sustaining our planet's existing climate and biodiversity. We argue that measuring a system's sustainability by characterizing status quo, quantifying the effects of an intervention, and comparing decision alternatives will lead to greater overall sustainability. The chapter first discusses why it is difficult to obtain measures about sustainability. On the one hand, sustainability is a complex issue, drawing on multiple disciplines. On the other hand, sustainability may be described at varying levels of scale, each with a very large number of variables, assumptions, and interactions. Next we turn to computation and discuss its use in analyzing sensed data and simulating sustainability.

We describe strategies for computing sustainability and for handling multiple models, simulations, and parameters. Finally, we present several visions for using computing and simulation to improve sustainability. Pursuing these visions will help us effectively deal with our exponential population growth and also, we argue, help improve sustainability.

3.2 Introduction

Sustainability is a problem of scale. As the human population has increased exponentially in a short period of time, we face new challenges. Suddenly, it has become important to calculate Earth's total *biocapacity*: the biological capacity of an area of land or water to generate resources and to absorb waste. Biocapacity can be weighed against our *ecological footprint*: the measure of the demand that human activity puts on the biosphere. But even if these limitations can be measured [1], the question of how to develop insights, plan actions, and mitigate damage to our shared life support system remains. Meanwhile, our economic system promotes unbounded growth under the assumption of a limitless supply of resources to be consumed independent of any consequences. Furthermore, our infrastructure is designed to separate production from usage obfuscating true environmental costs and leading to a personal sense of limitless abundance.

Sustainability is also a problem of complexity. Environmental systems and subsystems have a level of complexity that makes causal relations difficult or impossible to find. Moreover, many of the variables change dynamically and many are yet to be discovered. The more we learn about the world and our place in it, the more complexity we find at any spatial and temporal scale. Exploring this unfolding web of complexity results in an unprecedented level of uncertainty about the mechanisms and relations in the organic and inorganic physical processes that support our existence. This uncertainty has prompted many questions. At a personal level, we may wonder what natural, mechanical, and business processes are required to supply the water we drink, the food we eat, and the air we breathe. At a global level, we may ask to what limits our planet can be pushed, beyond which natural systems can no longer provide what we desire from them. Fundamentally, our uncertainty about the stability or fragility of Earth's ability to support life has resulted in a global discourse on a strategy of sustainability. The concept of sustainability is the commonsense response to great uncertainty: when an outcome is impossible to predict because of complexity, one may hope that reducing interference with natural systems may increase the overall stability of the system, based on the assumption that natural systems are maximally stable from millions of years of natural optimization. If we can reduce uncertainty by reducing human impact on the environment while we continue to learn and understand more about the systems involved, we hope to be able to extend our time frame for survival and even prosperity.

This chapter argues that an important tool to address scale and to study complexity is *computational modeling*: mathematical models representing the behavior of a complex system by computer simulation. To test a number of hypotheses, a simulation can be created for each scenario. The outcomes of these simulation experiments can support decision making. The unique challenge of modeling the sustainability problem is the multiscale and multidisciplinary nature of the global environmental and economic processes involved. Quantifying sustainability issues, and tracking them over time, will help to calibrate the simulations further developing the models that generate emergent complexity in the effort to improve the resemblance to observed behavior. Next we explain why sustainability must be studied at many scales and where the complexity of measuring and modeling sustainability lies. Then we discuss how computational modeling may be applied to sustainability and present some visions for how to make modeling and simulation impact, thinking about and acting toward sustainability.

3.3 Sustainability

The use of physical resources is central to the topic of sustainability. As resources are extracted from the environment, transformed into products and services, and consumed, a number of effects and side effects occur that impact the environment in a myriad of ways. We discuss efforts to measure these impacts at a global level as well as at the level of an individual product. We describe how these vast differences in scale contribute to the complexity of determining causes from their effects and we provide examples of multi-scale issues.

To address the great complexity of large natural systems, a sustainability strategy can be adopted to minimize the impact. The impact of human intervention on the planet has often been expressed with the metaphor of a *footprint*. The properties of a footprint include the type and amount of (1) original resources together with the process of extraction of those resources, (2) production resources (consumed and wasted) together with the process of production, (3) resources used for distribution together with the process of distribution, (4) resources required during the use period, and (5) wasted/recyclable resources together with the process of disposal and/or reuse [2] (see Figure 3.1).

By understanding and measuring our environmental and ecological footprint, we can work toward reducing it thereby reducing uncertainty of whether or not we are permanently damaging our current or future

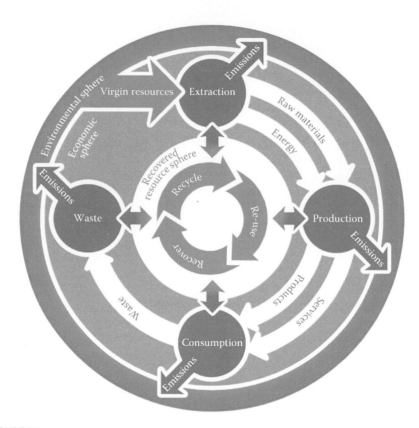

FIGURE 3.1
Resource transfers and cycles from the environmental sphere into the economic sphere. Extending the conservation of resources in the inner recovered resource sphere will minimize the need for additional virgin resources. (Adapted from European Environment Agency, *The European Environment—State and Outlook 2010: Synthesis*, EEA, Copenhagen, Denmark, Figure 4.1, p. 70, 2010.)

environment. To characterize the size and nature of various ecological footprints, the full product lifecycle must be considered. Both products and by-products, that is, resources used in production as well as side effects produced such as waste and emissions, contribute to these footprints.

To better account for resources used, a simple taxonomy of types and sources of resources is helpful. Natural resources may be considered to be *renewable* or *non-renewable*. Renewable resources can be *replenished* at the same rate as they are consumed. For example, water, trees, and other plants are generally considered to be renewable resources as they can be replaced by equivalent natural resources within a human timescale. Also water is considered to be renewable since water table levels can be maintained. However, care must be taken in these generalizations as conservation and resource management must control the *rate* of consumption to ensure renewal. That is, if a forest is completely cut down, important parts of the ecosystem including animals and other plants will be destroyed. Planting thousands of trees after a clear-cutting may not be possible if the ground properties cannot sustain such a large simultaneous resource drain. Also even if a forest could be regrown, it would not be considered to be renewable if it would take longer than a human lifetime to restore the forest.

Some natural resources are often included in the category of renewables but are essentially infinite and could be called *constant* resources: the sun, the earth, and the moon, which can be utilized for solar energy, geothermal energy, wind power, wave power, and tidal energy. *Non-renewable resources* cannot be replenished at the same rate as they are consumed. So-called fossil fuels such as oil, coal, and natural gas were created by natural processes over geological timescales, and so once they have been consumed, they are effectively gone. Unfortunately, the consumption of these non-renewable resources creates harmful emissions in large quantities.

In addition to the categorization of resources as renewable or non-renewable, it is helpful to differentiate *recovered* resources from *virgin* resources. When natural resources are first extracted from the environment, they are considered to be virgin. But when resources (natural or man-made) are recovered after they have been used and can be reused in production instead of virgin resources, they are considered to be recovered resources (see Figure 3.1).

Humanity's rate of consumption of virgin resources can be reduced by the reuse of objects and materials, and by the recovery of materials and the recycling of those recovered resources into new useful objects. Once virgin resources have been removed from the environmental sphere and have entered the economic sphere, our level of sustainability will be increased if we delay additional virgin resource extraction by increasing the duration that these materials are maintained within the inner cycle of use, recovery, and reuse (see Figure 3.1).

While the quantity of resources used will fluctuate, the desired trends would move from destructive to sustainable to restorative levels of consumption (see Figure 3.2). A destructive trend is defined as a growing footprint of consumption of virgin and non-renewable resources. Currently, since we are in a destructive trend, our collective challenge is to both decrease our use of virgin and non-renewable resources and increase our use of recovered and renewable resources. A sustainable trend would start only when recovered, renewable, and constant resources are being consumed at renewable rates. Finally, a restorative trend would be defined as a negative footprint where absolute quantities of recovered and renewable resources are being reduced and the remaining quantity could be returned to natural systems in a usable form. Therefore, we define *sustainability* as the reduction of our footprint toward a sustainable trend of resource usage.

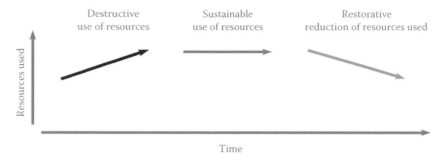

FIGURE 3.2
(left to right) Destructive trend where more resources are used than are returned in a reusable form. Sustainment trend where the quantity of resources used for production and consumption is fixed. Restoration trend where more resources are returned to the environment, as usable input, than are consumed.

3.3.1 Environmental Footprint

Footprints include both products (the amount of resources consumed) and pollution (the amount of waste and emissions produced). Atmospheric, land-based, and water-based pollution (see Figure 3.3) all affect each other but are often discussed independently. Atmospheric pollution is the most popular topic of public discourse and is often casually referred to as a *carbon footprint* but is intended to describe carbon dioxide (CO_2) and other harmful emissions reported in CO_2-equivalent (CO_2e) units. Land-use footprints are not yet generally discussed but are as important as a carbon footprint as these factors are all related. For example, the destruction of virgin rainforest, to grow cattle for instance, is a large and serious problematic use of land that permanently damages biodiversity as well as natural carbon dioxide consumption and oxygen production. *Water footprints* have also been studied (www.waterfootprint.org). Of course, natural water cycles are also an integral part of carbon and land-use footprints. As with the atmosphere, water

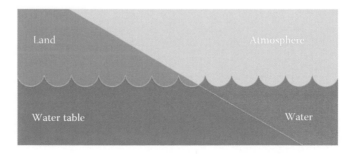

FIGURE 3.3
Human impact on the environment can be classified into a number of interacting areas: land, the water table, (open) water, and the atmosphere.

may seem to be an infinite resource, but the renewal rate of ground water can be anywhere from days to centuries, so careful conservation is required to ensure local ecosystem preservation.

Problematic human impact on natural systems has been examined by the United Nations Environment Programme (UNEP) Sustainable Consumption and Production Branch [3]. This group works to define metrics to better understand the current global resource imbalance between consumption and production and its potential cost to environmental stability. Three dangers are specifically called out as primary factors to consider: global warming potential, land-use competition, and human toxicity. Within these factors, 10 categories of resource use are defined: plastics, coal, natural gas, crude oil, biomass, animal products, crops, iron and steel, other metals, and minerals. While each of the three factors has its own dominant problem category, a combined value reflects how these factors affect the environment. To help support decision making, a weight is chosen for each category based on its overall estimated environmental impact, resulting in one general *environmental footprint*, expressed as an environmentally weighted material consumption indicator (right-hand column in Figure 3.4).

Our environmental footprint, caused by material consumption expressed by this indicator, is led by animal products (34.5%), crops (18.6%), and coal (14.8%). Animal products stand out as the primary global environmental problem, larger than crops and coal combined, because of the massive amounts of resources that are consumed in producing the 294.7 million metric tons of meat from (roughly 58 billion) land and aquatic animals that are killed each year for food [4]. This consumes a large portion of the global water and food supply.

As shown in Figure 3.5, unexpected water demands for everyday food products show how the consumption of water resources is not inherently obvious [5]. For example, the production of 1 kg of beef consumes 16,726 L of water while the same amount of potatoes uses only 132 L. These numbers show how practices, such as factory farming, can destabilize the water table for an entire region. To better measure the water footprint and to capture these hidden environmental costs, the concept of *virtual water* was introduced by Tony Allen in 1993 [5]. This term describes the amount of water consumed in the production process of an agricultural or industrial product. This can also be an important political issue when trade considers the export or import of water-intensive products. For example, the production of a 32 megabyte computer chip of only 2 g requires 32 kg of water [5]. A general lack of transparency in resource consumption in the supply chain of products adds significantly to the complexity discussed earlier.

In terms of global warming potential, greenhouse gas (GHG) emission reports from the U.S. Department of Energy [6] and the Livestock's Long Shadow report from the United Nations Food and Agriculture Organization [7] find that the three primary causes of anthropogenic

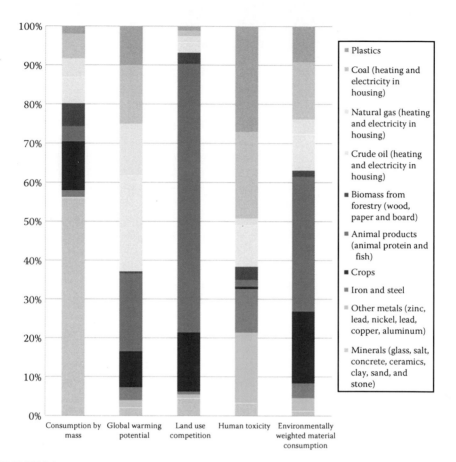

FIGURE 3.4

Environmental damage indicators as reported by the United Nations Environment Programme (UNEP) Sustainable Consumption and Production Branch. (From Hertwich, E. et al., Assessing the environmental impacts of consumption and production: Priority products and materials, A report of the working group on the environmental impacts of products and materials to the international panel for sustainable resource management, United Nations Environment Programme, Nairobi, Kenya, 2010.)

GHGs are buildings (48%), animal products (18%), and transport (14%). Typically, transport is believed to be the climate change culprit even though buildings—or rather the energy utilized for heating, cooling, and lighting buildings—are the dominant air pollution problem (see Figure 3.6). This could be thought of as a multiscale problem as buildings are often the largest man-made structures on Earth, yet directly and indirectly emit gases that cause nanoscale chemical reactions that destroy our atmosphere.

GHGs include water vapor (H_2O), carbon dioxide (CO_2), methane (CH_4), nitrous oxide (N_2O), ozone (O_3), and chlorofluorocarbons (CFCs). These

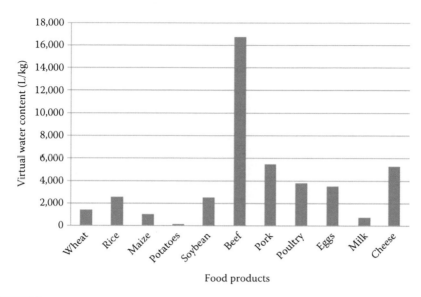

FIGURE 3.5
Comparative water demand from everyday food products (L/kg) (Adapted from Table 4.1). (From Hoekstra, A.Y., ed., Virtual water trade, *Proceedings of the International Expert Meeting on Virtual Water Trade*, Value of Water Research Series No. 12, UNESCO-IHE, Delft, the Netherlands, 2003.)

FIGURE 3.6
The power generation currently utilized to supply heating, cooling, and lighting to buildings causes more GHG pollution of the atmosphere than any other category. Photo of Shanghai. (Courtesy of Azam Khan, 2010.)

substances interact in ways that are difficult to predict. The most dangerous of these is CFCs as they destroy ozone. Unfortunately, when exposed to
ultraviolet light, CFCs release a free Chlorine (Cl) atom that reacts with O_3,
breaking it into O and O_2. This starts a chain reaction where a single Cl atom
can break down 100,000 O_3 molecules (see Figure 3.7) [8]. Although ozone
itself is a GHG, it also acts as a protective layer around the Earth preventing
ultraviolet radiation from damaging life on the surface of the planet. The
ozone layer is relatively thin and is sensitive to disturbances.

To make matters worse, the lifetime of some GHGs can be quite significant. For example, methane (CH_4) has a lifetime of about 10 years [9], N_2O
has a lifetime of 114 years, and some CFCs can last from 45 years to 1700
years [10].

The discovery that ozone destruction would continue for some time,
despite the total cessation of CFC emissions into the atmosphere, motivated
a global resolution to ban the production of CFCs worldwide at the 1987
Montreal Protocol on Substances That Deplete the Ozone Layer. The hope
was that the ozone layer would recover in about 60 years by 2050. However,
in 2011, a massive ozone hole was still observed over the southern hemisphere [11] (see Figure 3.8).

Methane is about 20 times more effective than CO_2 at trapping heat in
the atmosphere. It primarily stems from enteric fermentation in ruminant
animals (37%), such as cows and sheep [7]. Nitrous oxide is considered to
be 296 times more dangerous than CO_2 and, as with methane, the main
global source of N_2O is animals (65%), such as cows and pigs [7]. The animal waste (220 billion gallons of waste each year in the United States, over
830 billion liters [12]) from factory farms is typically collected in *lagoons*
(see Figure 3.9).

The lining of these lagoons often break causing *Escherichia. coli* bacteria
from the feces to poison local water systems affecting downstream farms
[13]. The media often report *E. coli* or *Giardia* contamination of certain veg-
etables and the health risks of consuming them, but the factory lagoon
that was the source of the problem is typically not mentioned or identified [13]. In New York State alone, 3000 impaired river and stream miles
are contaminated by factory farm animal feces. Lagoon spills are also a
top contributor to the impairment of lakes and reservoirs, affecting over
300,000 lake acres in the state [13].

Industrial buildings, including livestock factories, are responsible for more
than half of the GHG emissions from buildings while the remainder is from
residential buildings. Together, they account for 48% of global GHGs, moti-
vating the architectural community to work toward the development of very
low-energy buildings [15].

In summary, unexpected sustainability problems can be found in systems
from the nanoscale of chemical reactions, to the global scale of resource
consumption and transformation. Furthermore, this complexity is affecting
another highly complex system: the global climate (see Figure 3.10).

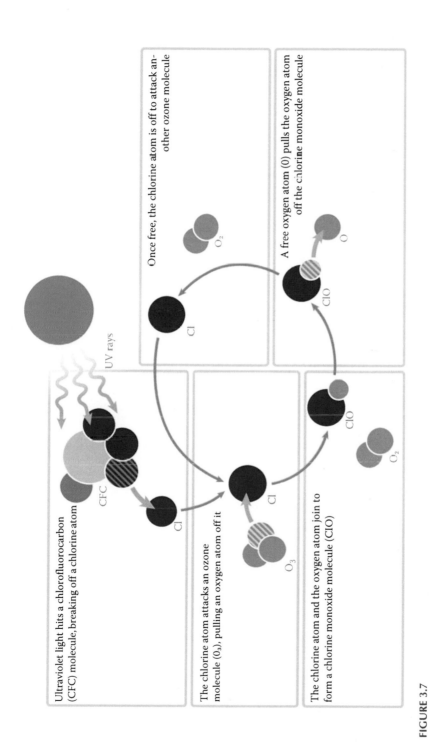

FIGURE 3.7
The ongoing destruction of ozone in the stratosphere by chlorine atoms derived from the breakdown of CFCs in the atmosphere. (Courtesy of Michael Glueck.)

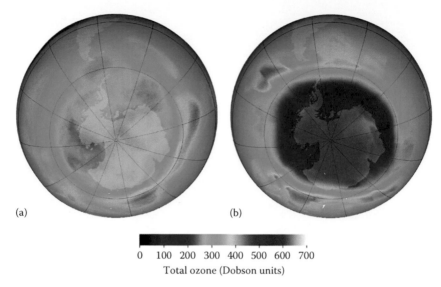

(a)   (b)

```
0    100  200  300  400  500  600  700
        Total ozone (Dobson units)
```

FIGURE 3.8
Massive ozone hole (blue region) seen over southern hemisphere on August 31, 2011 (b), compared with earlier image of the same region on August 31, 1982 (a). (From NASA, Ozone Watch, http://ozonewatch.gsfc.nasa.gov/)

FIGURE 3.9
State-of-the-art lagoon waste management system for a 900 head hog farm. The facility is completely automated and temperature controlled—United States 2002. (Photo courtesy of USDA NRCS/Jeff Vanuga.) (From Steinfeld, H. et al., Livestock's long shadow: Environmental issues and options, Food and Agriculture Organization of the United Nations, Rome, Italy, 2006.)

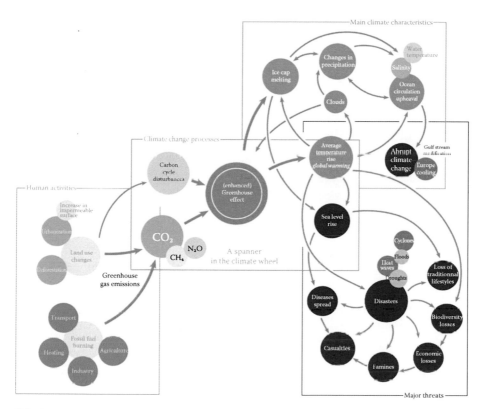

FIGURE 3.10

Climate change: processes, characteristics, and threats, 2005, Philippe ReKacewicz, in UNEP/ GRID-Arendal Maps and Graphics Library, Retrieved June 12, 2013, from http://www.grida. no/graphicslib/detail/climate_change_processes_characteristics-and-threats15a8.

We have described the direct and indirect sources of ecological footprint contributors and have provided a taxonomy of resources to help define a sustainability trend. The taxonomy described virgin or recovered (replenishable or constant) renewable and non-renewable resources. We have shown how sustainability is a problem of both the scale and the complexity and have described the leading environmental issues with respect to land, water, and atmosphere pollution. However, solutions are possible, in terms of human diet change [14] and low-energy building design [15] that could have an enormous positive impact on minimizing the dangers of global warming potential, land-use competition, and human toxicity.

Having discussed the major properties of our ecological footprint, we next contend that two key drivers of the growth of our ecological footprint are economics and infrastructure. These two areas both introduce a level of indirection in resource use that obfuscates causal relations. We propose that transparency in the lifecycle of a resource will improve sustainability.

3.3.2 Ecological Economics

The consumption of products and services result in pollution in the form of waste and emissions. While recycling programs have been running for decades to minimize waste, atmospheric emissions are inherently more difficult to capture and reuse in production processes. The current business approach to emission reduction is to introduce cap-and-trade schemes. While emission trading schemes have been established for CO_2e (equivalent amounts of GHGs to CO_2 with respect to global warming potential), some countries have found taxation to be more effective.

Direct taxation supports cause-and-effect understanding while trades and subsides (funded by indirect taxation) obfuscate which behavior creates which outcome. For natural resources, the complexities of different layers of subsidies, taxation, tariffs, and pricing for different types of users in specific regions have resulted in a tangled web of causes and effects. These indirections render it impossible to properly calculate true cost pricing of a product or service. Standard economic theory states that any voluntary transaction is mutually beneficial for the buyer and seller. However, an exchange of goods or services can cause additional effects on third parties, called *externalities* (see Figure 3.11) [16]. True cost economics is an economic model that includes the cost of negative externalities directly into the pricing of goods and services. The goal is for products and activities that directly or indirectly cause harmful consequences to living beings or to the environment to be accordingly taxed so that their price includes hidden cost. A leader in

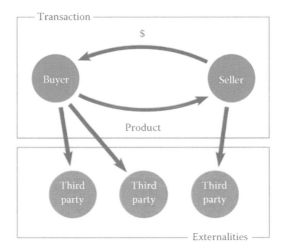

FIGURE 3.11
Transactions that are mutually beneficial to buyer and seller may have adverse effects on third parties, such as people or the environment, that are not accounted for and are passed on to society at large.

green tax reform, shifting taxation from labor to environment and energy, is Denmark:

> Denmark has become much greener during the last two or three decades. Most environmental indicators support this conclusion. The environmental improvements are of course the results of various kinds of regulations. During the last three decades the polluter-pays principle has been implemented widely in the Danish economy. In a Danish context this means that consumers or polluters pay not only the direct costs connected to supplying for instance water and energy, disposing of waste and waste water etc. They also pay the externalities connected to this consumption through a wide range of green taxes. Green taxes have thus played an important role in fully implementing the polluter-pays principle.
>
> Following the introduction of the tax on pesticides used by agriculture in 1996 the consumption has fallen by 40%. Likewise the introduction of the tax on drinking water has contributed to reduce the consumption of water by 25%. A reduction in the water consumption has the added effect of reducing the emissions of nutrients to the water environment from sewage treatment plants and industry. The tax levied directly on the waste water has had the same effect. And together with the direct regulation of sewage treatment the waste tax levied on phosphor, nitrogen and organic material has reduced the emissions of these substances by 75–85% from 1989 to 2008. [17]

While the implementation of green taxes must be carefully managed, as it is still a somewhat indirect system for the reduction of resource consumption, it can clearly have a significant impact. Eliminating externalities would allow products and services to be priced according to their true cost. This would greatly help individual and corporate consumers to make informed choices as they could compare options from vendors in a meaningful way. For example, a large supplier of industrial flooring and carpeting recycles old carpeting into new carpeting and, so, understands the costs and benefits compared to purchasing raw materials [18].

Unfortunately, for many products, especially electronics, the design process does not take into account the end-of-life recovery of materials making them practically impossible to recycle. Recently, however, product design software is integrating lifecycle assessment features encouraging designers and engineers to consider both the manufacturing processes and the recycling processes necessitated by their design choices [19].

3.3.3 Infrastructure Trap

Another indirection that obfuscates resource consumption is the use of infrastructure services such as electric power, natural gas, and water. The term *infrastructure* can be defined in several ways. Infrastructure can be thought of as a spatial term. For example, common services for a number of buildings, such as water or energy, could be supplied by remote infrastructure instead of being locally generated. The term can also be considered as applying to large numbers of users, as opposed to services that cater only to a few individuals. Well-water is an example of a local water supply serving a single family while municipal water systems can serve millions of people. Finally, infrastructure could also be measured in terms of implementation cost or efforts, making these systems seem permanent once a large investment has been made. An example would be the deployment of telephone landlines to every individual home and building in a city versus the installation of cellular phone transmission towers across the same area. Given these dimensions, we can see that great benefits can be had from the implementation of infrastructure. The development of reliable infrastructure services, notably in the first half of the twentieth century, was an enabling technology propelling cities and some countries to economic dominance. Thus, the modern city assumes the existence of highly reliable common infrastructure services. The massive scale and cost of urban infrastructure projects, together with their ubiquitous nature, make them essentially invisible, and in a sense, this is the goal of infrastructure so that daily life is not burdened with finding clean water, food, and other basic necessities.

However, infrastructure is inherently an indirect method to supply services. When we turn on a computer, we do not see a truckload of coal being dumped into a distant furnace. Yet, coal-burning is the primary method for electricity generation in the United States [6]. These stations exhaust the resulting smoke (CO_2, SO_2, NO_x, and CO) into the atmosphere. As these GHG emissions are permitted at no cost to the power station, electricity can seem to be quite inexpensive, but the cost of dealing with this pollution is indirectly passed on to society at large as an externality. Furthermore, the significant overhead in starting or stopping these stations means that they cannot dynamically respond to large changes in demand. For example, coal power stations can take 12–120 h (0.5–5 days) to start, depending on their current temperature [20]. After the major power outage in the northeastern United States and Canada in 2003, nine nuclear power plants took 6 h to 11 days to return to service [21]. This cycle-time overhead leads to a strategy of *oversupply* to ensure reliability in delivering electricity. To minimize the oversupply, a load-balancing approach is being implemented in some regions. By varying the pricing depending on the demand at the time, the hope is to distribute the demand as evenly as possible. The concept of a *smart grid* was introduced that can vary pricing dynamically, as well as redirect a number of power sources to locations of power needs in an efficient manner. A smart grid system would also ideally be designed to accept more variable energy sources such as solar or wind power.

Urban infrastructure creates benefits and costs, especially as it relates to sustainability. Strategies, such as energy demand–response management, reflect scaling problems of infrastructure. But even if oversupply is minimized, this minimizes only *extra* waste. The amount of pollution created by generating the power required to meet the actual demand is unchanged. So while smart grids should be developed to minimize extra waste and to enable alternate energy sources to participate more easily in the power grid, a smart grid is an enabling technology but is, itself, not a sustainability solution.

Alternatively, local supply of power and water removes the need and cost of infrastructure development, improvement, and oversupply strategies. A local supply strategy is also scalable and, as a direct method of supply, helps users to develop a sense of their usage as well as an inherent sense of scarcity rather than abundance. However, a local supply of resources does not provide the freedom to consume them at any rate as natural replenishment rates still apply.

If individuals become more self-reliant, dependence on infrastructure is reduced, leading to an overall increase in resilience to large-scale disturbances (see Figure 3.12). However, off-grid structures typically require more land (for solar, rainwater, or wind collection as well as septic beds), making it difficult to achieve the population density of typical cities. In this sense, the development of high-density cities entraps citizens to be dependent on infrastructure that indirectly supplies services creating sustainability problems of both urban scale and human behavior.

In some cases, the level of the community is an appropriate scale for the consideration of resource production and consumption. For example, an early experiment toward a 100% renewable energy community was the Danish island of Samsø in 1997. The Danish Ministry of Energy chose Samsø to "study how high a percentage of renewable energy a well-defined area could achieve using available technology." The island is 114 km² in area with a population of approximately 4200 people. By 2007, on-land and off-shore windmills and three new district heating systems were built, successfully

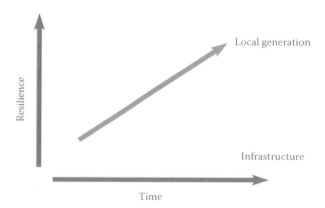

FIGURE 3.12
The resilience of individual buildings to withstand service disruptions increases when the need for infrastructure is reduced.

generating all of their electricity and offsetting all carbon emissions. The Samsø community is now attempting to broaden their focus beyond energy to include the more comprehensive topic of sustainability [22].

3.3.4 Summary

Once we have a sense of our impact on the Earth in measured terms, together with an understanding that, to some degree, this is caused by our economic and infrastructure systems, we can develop sustainability strategies with measured results. In other words, if we can measure our environmental footprint, as well as measure the outcome of specific strategies to reduce our footprint, we will be able to learn which approaches work and which do not.

We propose that, as sustainability is a global problem, automated measurement will be required to make progress. These large data sets will require automated processing to support our understanding and decision-making processes. They will greatly improve the transparency of the resource lifecycle, and more specifically, these data sets can inform and calibrate our computational models to improve our cumulative knowledge about sustainability.

3.4 Computation

We have briefly touched on some key problems affecting environmental sustainability including animal consumption, energy consumption in buildings, economic models, and infrastructure scalability. Because of the overall problem scale, automation is necessary to measure materials and processes to help develop specific sustainability strategies, as well as to monitor results and progress. Humans monitor many systems to collect data, but these efforts are limited by manual labor, and often these values are not networked together to support aggregation. Automated monitoring, through digital sensors, be they real or virtual, satellites or thermometers, submeters or light switches, sales figures or odometers, can collect data in a more scalable manner that is also more amenable to aggregation. The process of aggregating the data will expose areas where additional sensors would help paint a more complete picture of human activity and environmental responses.

Automation, through computational instrumentation and computational modeling, can create the opportunity to perform multiscale analysis, develop and test hypotheses, and measure whether outcomes perform as expected when changes are implemented. We describe a method to combine instrumentation and modeling in support of decision making and then differentiate decisions that primarily effect the decision-maker from decisions that primarily affect others. For both types of decisions, we discuss examples of computational support.

3.4.1 Global Symbiotic Simulation

The combination of computational instrumentation and modeling is known as *symbiotic simulation* where the computational system interacts with the physical system in a mutually beneficial manner. Sensor-data serves as real-time input to the simulation that can execute a number of what-if experiments, trying different strategies, to generate alternatives. Selected scenarios that are put into practice can then be validated by monitoring the outcomes, using the same sensors, to help the system learn about the effectiveness of that decision. In turn, the physical system may benefit from the effects of decisions made by the simulation system [23,24].

The interaction could be direct or indirect. For certain real-time applications, the simulation could directly control the physical system, also called a symbiotic simulation control system (SSCS). For example, the heating ventilation air conditioning system in a building could be augmented with an SSCS by executing simulations of various expected occupancy and weather patterns. For a smart grid, an SSCS could be utilized to help direct energy flows in real time from sensor data in a demand–response system, based on what-if simulations that attempt to predict various outcomes. However, for longer-term decision-making applications, such as policy making, the interaction between the simulation and the physical system would be indirect, through visualizations and other decision-support mechanisms such as expert systems, supplied to human teams. Such *human-in-the-loop* systems are referred to as symbiotic simulation decision support systems (SSDSS) (see Figure 3.13).

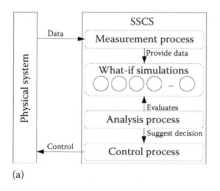 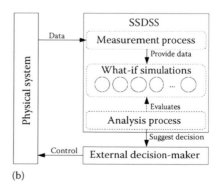

(a) (b)

FIGURE 3.13
(a) Overview of symbiotic simulation control system (SSCS). (b) Overview of symbiotic simulation decision support system (SSDSS). (From Aydt, H. et al., Symbiotic simulation systems: An extended definition motivated by symbiosis in biology, *Proceedings of the 22nd Workshop on Principles of Advanced and Distributed Simulation*, Los Alamitos, CA, pp. 109–116, 2008.)

In the context of sustainability, we propose that a global symbiotic simulation (GSS) could provide a high level of automation for monitoring and analysis. Such a system would be continuously running, generating alternatives that would become increasingly more helpful as additional sensor data streams are supplied to the system. It would combine SSCS and SSDSS at low and high levels of abstraction, respectively. Decisions proposed and adopted would be tracked to study their actual impact and the difference from the predicted result.

A number of global simulation systems currently exist but are not symbiotic with the physical systems they examine. For example, global climate model simulations are the primary systems that are global in nature [25], but they are solely for forecasting purposes. As an attempt to go further, the Group on Earth Observations is developing a Global Earth Observation System of Systems (GEOSS) specifically to "support decision making in an increasingly complex and environmentally stressed world" [26]. While not intended as a modeling and simulation platform, the GEOSS mission is to coordinate multidisciplinary scientific data in support of policy and decision making and, so, could function as the measurement process component of a GSS. The MIT Integrated Global System Model adds a Human Activity Model to improve estimates of economic drivers of resource consumption (http://globalchange.mit.edu).

A GSS, as described earlier, would require a multiscale multidisciplinary approach, including the modeling of environmental and economic activities. Known policies and adherence to them could be included as well to see if their impact can be detected. Undertaking the development of a GSS would necessitate a kaizen (continuous improvement) methodology so that the system can function despite knowing that the underlying model will always be incomplete.

The process flow for a GSS includes measurement, what-if simulation, analysis, and decision making (see Figure 3.13). The key value of creating such a system would come from the quality of what-if simulations, which could be generated or manually created, in support of decision making.

Decision making is often considered in isolation. However, in context, the decision-making process can be viewed as a lifecycle (see Figure 3.14). Guided by a strategy, well-informed decision-makers can make effective decisions to take specific actions [27]. By monitoring the action and its outcomes, learning can guide new and improved strategies. Decisions have a fixed lifetime, and only monitoring the outcomes will reveal when a new decision will be necessary [28]. For example, at a large scale, urban planners may decide to build a road connecting two regions that previously only had a railroad. By monitoring road traffic, planners can determine if the connection was successful and if additional lanes or additional trains would better support sustainable growth. At a smaller scale, by monitoring power usage and the availability of new technologies, consumers may decide whether a newer more efficient appliance would be better to purchase than continuing to operate an older machine.

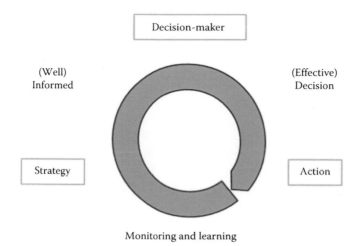

FIGURE 3.14
Decision making shown as a lifecycle in support of continuous improvement.

3.4.2 Sustainability Strategies

Strategies help guide the decision-making process. The types of strategies required will be determined by the types of causes of resource consumption. The primary causes can be classified as being either *elective* or *imposed*. For example, an animal-based diet versus a vegetable-based diet is an individual choice, while the amount of energy consumed to heat or cool a building is not an individual choice and is imposed by the design of the building and its systems (see Figure 3.15). Elective causes are chosen on an ongoing basis, so opportunities to choose differently occur daily. However, imposed causes are inherently more difficult to change, and these causes and their impact often remain in place for long periods of time.

Imposed causes could also be called infrastructural to distinguish when and where changes would be necessary. This classification can help us to

FIGURE 3.15
Elective causes of resource consumption are taken by individuals. However, individuals, such as architects and engineers, may also impose certain consumption rates on others through their design choices.

understand when human behavior changes are necessary to solve a problem and when infrastructure improvements must be developed and applied. While not directly a cause, an implicit factor to consider is scale. That is, there are so many people and buildings involved that automation would be required to be an integral part of any solution if improvements are to outpace the normal growth of continued unsustainable practices. Unfortunately, sustainability is not the normal course of action. Worse still, process and economic complexity confounds efforts to be more sustainable. Complexity must therefore be addressed and plans must be laid, followed, monitored, measured, and managed, in order to achieve sustainability. Finally, lessons learned must be included in further planning.

3.4.3 Computation for Elective Consumption

Consumers elect which resources to use every day. By necessity, people make food choices daily that can greatly vary the amount of resources that are required. Beyond this, people may choose to purchase clothing, electronics, furniture, etc. Because these consumption patterns are routine, people may not reflect on the resource demands that their choices incur. While some people collect receipts to track their spending, it is considerably more difficult to monitor and track resource consumption. Ideally, resource consumption would be directly related to cost so that consumers would know that products that cost more clearly consumed more resources in their creation. However, as mentioned earlier, externalities are not normally accounted for in pricing so it would require in-depth investigation into the specific goods and services and their histories and origins to determine the true costs to support meaningful comparative shopping and decision making.

Even companies that provide goods and services would be challenged to provide resource consumption information and to accurately report such data. In fact, it would be necessary to include the entire supply chain in the reporting of any previous step in production. The importance of having detailed information about a supply chain was demonstrated during the mad cow disease crisis in the 1990s prompting farm-to-table traceability of all beef sold in Europe and Japan. To this end, www.sourcemap.com [29] supports the authoring of high-level supply chain maps (see Figure 3.16). Again, complexity calls for computerized automation to collect resource history and processes. Instrumenting supply chains with a number of sensor types would provide live data that could contribute to more automated, detailed, and accurate cost and resource estimates. Simulations could also make use of these data to help optimize supply chains in terms of both cost and resource reduction.

Once aggregate resource consumption information can be delivered to consumers, computation can be utilized for resource tracking, goal setting, measurement, reporting, and persuasive encouragement for behavior change. Some of these approaches have been explored in the areas of

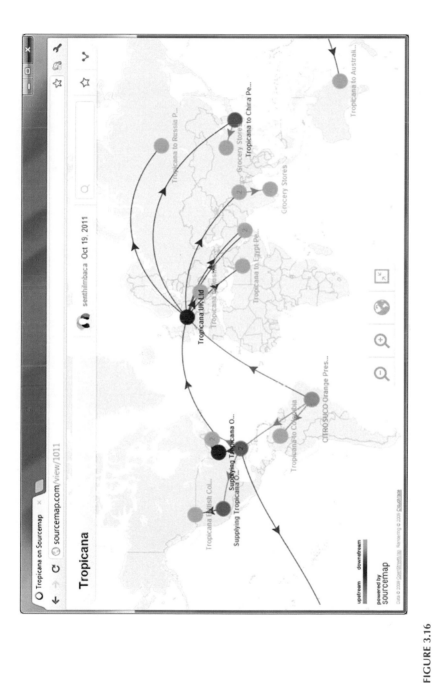

FIGURE 3.16

Supply chain mapping and visualization for a single product. (From Bonanni, L. et al., Small business applications of sourcemap: A web tool for sustainable design and supply chain transparency, *Proceedings of the 28th International Conference on Human Factors in Computing Systems—CHI'10*, Atlanta, GA, pp. 937–946, 2010.)

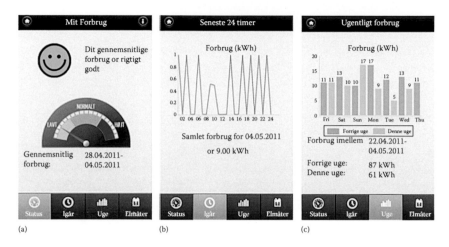

(a)   (b)   (c)

FIGURE 3.17
Mobile computing to assist users in understanding home energy consumption. (a) Total weekly household power usage (kWh) compared to similar household in the same geographic region showing a low reading. (b) Graph of household consumption (kWh) over the last 24 h. (c) Chart of consumption per day for the last week compared to the week before.

ubiquitous computing and human–computer interaction. In particular, the area of *persuasive technology* seeks to determine how best to influence users to transition and adhere to improved practices [30] in terms of health, resource consumption, and purchasing. Mobile computing has been used to enhance resource understanding for home energy consumption (see Figure 3.17) [31] and grocery purchasing (see Figure 3.18) [32,33].

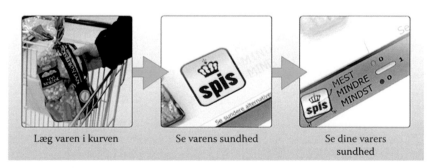

Læg varen i kurven   Se varens sundhed   Se dine varers sundhed

FIGURE 3.18
Mobile computing to assist users in making healthy purchasing choices while grocery shopping. First, place item in cart (left). The cart displays the item's classification as "eat most," "eat less," "eat least" (middle). Finally, (right) the summary for all items in the cart is updated for each of the three categories. (From Kallehave, O. et al., Persuasion in-situ: Shopping for healthy food in supermarkets, *Proceedings of the 2nd International Workshop on Persuasion, Nudge, Influence, and Coercion through Mobile Devices, CHI'11*. ACM, Vancouver, BC, Canada, pp. 7–10, 2011.)

When efficiency efforts have been successful, a rebound effect has been observed where gains are lost because of new increased consumption. However, many treatments of this topic are oversimplified, and changes in consumption must be considered together with co-benefits, negative side effects, and spillover effects [34].

The relationship between choices and consequences is difficult to determine without significant effort in measurement and tracking. However, computational support from the resource supply chain and during decision making can help users consume less or consume differently, adding more renewable resources and more efficient usage of all resources. Still, while consumers may intend to act more sustainably, many of their resource usage decisions have been made by others directly and indirectly. Next, we discuss consumption that is imposed upon consumers and citizens in general.

3.4.4 Computation for Imposed Consumption

To varying degrees, people have resource consumption imposed on them every day. Furthermore, people such as product designers, architects, and engineers make choices that impose consumption levels on others for the lifetime of their designs. These impositions can be beneficial or detrimental. Unfortunately, many critical design decisions are made early in the design process that implicitly dictate detrimental resource consumption during extraction, manufacturing, packaging, shipping, and usage. However, if the consequences of these choices were revealed to the designer before such decisions were made, significant improvements would be possible whose positive impact would exist throughout the artifact's lifetime.

Recently, design software has attempted to provide decision support through consequence visualization [19], indicating a number of related factors such as material attributes, CO_2 footprint, embedded energy and water, estimated raw cost of materials, Restriction on Hazardous Substances (RoHS) and Waste, Electrical and Electronic Equipment (WEEE) compliance, toxicity, and end-of-life regulations (see Figure 3.19). Given these estimates while still developing the computational model of the product, designers can reduce resources required during manufacture and usage, and improve the reuse and recycling of objects.

While this is beneficial *before* usage, other systems are necessary *during* usage to help users understand what quantities of resources are used during the operation of a device or the consumption of the product. For consumer products and homes, mobile computing devices may be helpful in curbing resource consumption (see Figures 3.16 through 3.18), but for larger structures, complexity is greatly increased, and specialized systems may be necessary. For example, buildings contain several types of mechanical equipment that include sensors as part of their normal function. Sensor types include thermostats for ambient temperature and humidity, motion detectors to

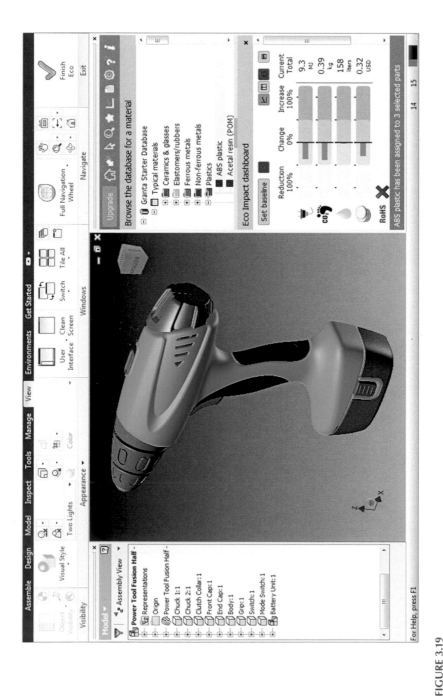

FIGURE 3.19
Computational modeling software for product design (Autodesk inventor), with a sustainability panel (right-hand side) presenting material properties to support the decision-making process.

FIGURE 3.20
Sensors for measuring electricity, light, temperature and humidity, motion, and CO_2.

control lighting, or meters and submeters for reporting and billing of power consumption (see Figure 3.20). However, despite this instrumentation of buildings in support of building automation, many buildings are manually controlled because of the complexity of appropriately programming the automation system for a specific building.

At COP15, the *UN Climate Change Conference* 2009 held in Copenhagen, a building on the campus of the University of Copenhagen, called the Green Lighthouse, was opened as the first public net-zero energy building in Denmark (see Figure 3.21). So-called net-zero energy buildings, which generate as much power as they themselves require, have been developed in the past few years indicating that there is enormous room for improvement to reduce or even eliminate buildings as the primary cause of atmospheric CO_2.

> Generally, a net-zero building strategy starts with dramatic efficiency gains of 75% through design and technology, with the remaining 25% of power demand being generated locally with sources such as solar or wind power. Within the technological efficiency gains is the digitization of the building control system and its device and sensor network. This so-called "smart building" paradigm shift in building systems could form the basis of a reliable platform for ubiquitous computing for sustainability [35].

As shown in Figure 3.21, buildings have many interacting subsystems that must be efficient individually but must also operate effectively together to achieve a net-zero energy footprint. The figure indicates both the complexity that causes so much energy use in buildings in the first place and the complexity that will be inherent to any solutions that are designed, constructed, and occupied. Generally, this complexity is *designed* by the architects and engineers. In modern projects, Building Information Modeling (BIM) is often adopted in the design and construction phases of a building project to help capture and manage this complexity. BIM is essentially a computational model of a building using a database approach for managing the objects and relations between objects in an architectural model [36]. Standard data models have been proposed that also include mechanical, electrical, and piping data, construction planning, and personnel information. In following the principles of an SSCS outlined earlier, combining the BIM with the

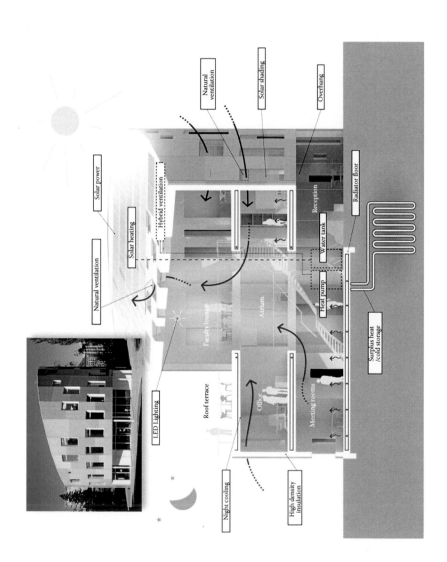

FIGURE 3.21
Green lighthouse diagram of low-energy techniques, such as natural ventilation, heating, and cooling, together with depiction of occupants and environmental conditions. (Courtesy of Christensen & Co., Arkitekter A/S, Green Lighthouse, University of Copenhagen, 2009.)

buildings sensor network [37] to create an improved building automation system will enable a significant portion of a GSS. In a fully instrumented building, effects and properties shown simply as static artist interpretations in Figure 3.21 can be shown in the context of a detailed computational model as dynamic values updating in real time, as in Figure 3.22.

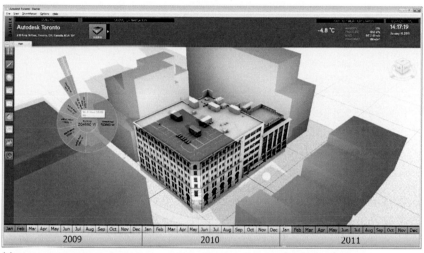

(a)

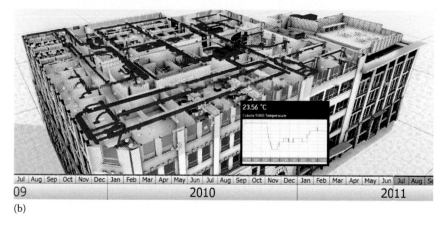

(b)

FIGURE 3.22
Visualization software for data aggregation and animation over time of building usage and responses based on collected sensor, meter, and submeter data streams. The software superimposes the visualizations of the physical sensor data over top of the computational building model. (a) Exterior view of building with hierarchical breakdown of electricity usage in sunburst visualization. (b) Interior cut-away view showing sensor locations (blue), ventilation system (red), lighting system (peach), temperature shading on walls and floors, and tooltip for a specific sensor with temperature graph. (From Attar, R. et al., Sensor-enabled cubicles for occupant-centric capture of building performance data, *2011 Conference Proceedings: ASHRAE Annual Conference. ML-11-C053*, Montreal, QC, Canada, p. 8, 2011.)

Many opportunities exist to greatly reduce imposed consumption in both product design and creation as well as in the design and operation of architecture. Decision support software in design and operations tools with life-cycle monitoring, together with advancements in simulation for architecture and urban design, could enable many aspects of a GSS.

3.4.5 Ways Forward

We propose that each aspect of the GSS problem be encoded into a large collaborative cyberphysical simulation framework that can support multiple disciplines and that can be continuously refined and improved. While such a system calls for both theoretical and pragmatic advancements in dynamic distributed simulation, knowledge representation, human–computer interaction, and software engineering, work is progressing in all of these areas as part of the general advancement of the scientific method to include collaborative computational modeling and simulation. In particular, the Symposium on Simulation for Architecture and Urban Design (SimAUD) has begun to take this on as its mission. The core approach is the discrete event system specification (DEVS) formalism, first proposed by Ziegler [38]. Strict adherence to a simulation formalism such as DEVS will likely be necessary to ensure scalability of the system with automated model checking. The SimAUD research community is currently developing a DEVS-specific integrated development environment (IDE), called DesignDEVS, as a customized interface to a large distributed collaborative simulation framework (see Figure 3.23). The intent is to leverage the modular and hierarchical nature of DEVS computational modeling to support contributions from industry, academia, and government. Interacting with such a large complex system is also a fundamental challenge, for both authors and users, with respect to elective and imposed consumption. In the field of human–computer interaction research, a sustainability community has been formed to examine these issues at the SIGCHI conference [39]. Ultimately, research communities in most areas of computer science will be necessary to participate in this global challenge.

3.5 Conclusion

We have presented environmental and economic processes, characteristics, and threats as they relate to sustainability. These issues are further burdened by scale and complexity. To both understand and address sustainability, we have proposed computational approaches to measurement and modeling to improve the simulation of sustainability.

The multidisciplinary nature of a GSS of sustainability calls for a collaborative solution. A challenge in the development of such a system is

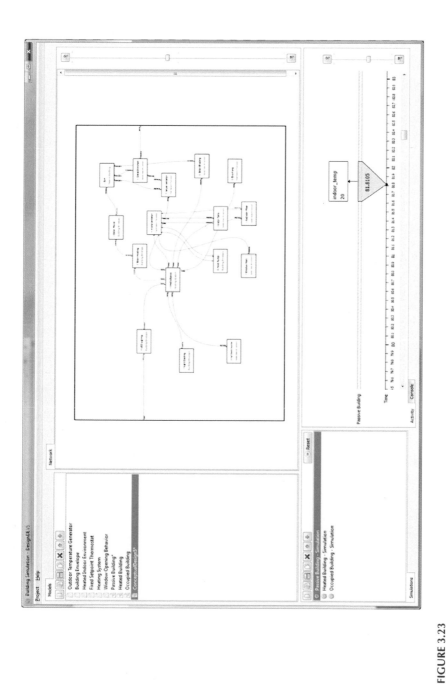

FIGURE 3.23
DesignDEVS IDE under development to explore the design of a large collaborative computational simulation framework. (Courtesy of Rhys Goldstein and Simon Breslav.)

the coordination of disparate types of measurements and models required to represent actual systems at several levels of detail. Moreover, the ability for a system to support continuous improvement during continuous operation will be critical. Decision making and policy making supported by simulation are further areas of research that will be necessary, involving human–computer interaction and visualization research.

Although measuring and modeling sustainability are a grand challenge for the computing community, we are optimistic that progress can quickly be made to better inform and support individuals, groups, and societies in living sustainably.

Acknowledgment

Thanks to Emilie Møllenbach for helpful discussion and review of this chapter.

References

1. Ewing, B., Moore, D., Goldfinger, S., Oursler, A., Reed, A., and Wackernagel, M. (2010). *The Ecological Footprint Atlas 2010*. Global Footprint Network, Oakland, CA.
2. European Environment Agency. (2010). *The European Environment—State and Outlook 2010: Synthesis*. EEA, Copenhagen, Denmark, Figure 4.1, p. 70.
3. Hertwich, E., van der Voet, E., Suh, S., Tukker, A., Huijbregts, M., Kazmierczyk, P., Lenzen, M., McNeely, J., and Moriguchi, Y. (2010). Assessing the environmental impacts of consumption and production: Priority products and materials. A report of the working group on the environmental impacts of products and materials to the international panel for sustainable resource management. United Nations Environment Programme, Nairobi, Kenya.
4. United Nations Food and Agriculture Organization. (2011). Global information and early warning system on food and agriculture, *Food Outlook*, November 2011, p. 8.
5. Hoekstra, A.Y. (ed.) (2003). Virtual water trade, *Proceedings of the International Expert Meeting on Virtual Water Trade*, Value of Water Research Series No. 12, UNESCO-IHE, Delft, the Netherlands.
6. U.S. Energy Information Administration. (2008). *Assumptions to the Annual Energy Outlook*. U.S. EIA, Washington, DC.
7. Steinfeld, H., Gerber, P., Wassenaar, T., Castel, V., Rosales, M., and de Haan, C. (2006). *Livestock's Long Shadow: Environmental Issues and Options*. Food and Agriculture Organization of the United Nations, Rome, Italy.

8. Molina, M.J. and Rowland, F.S. (1974). Stratospheric sink for chlorofluorometh-anes: Chlorine atom-catalyzed destruction of ozone. *Nature* 249:810–812.
9. Boucher, O., Friedlingstein, P., Collins, B., and Shine, K.P. (2009). The indirect global warming potential and global temperature change potential due to meth-ane oxidation. *Environmental Research Letters* 4(4):1–5.
10. IPCC. (2007). Climate change 2007: The physical science basis. *Contribution of Working Group I to the Fourth Assessment, Report of the Intergovernmental Panel on Climate Change*, Solomon, S., D. Qin, M. Manning, Z. Chen, M. Marquis, K.B. Averyt, M. Tignor, and H.L. Miller (eds.). Cambridge University Press, Cambridge, U.K., 996pp., Table 2.14, p. 212.
11. NASA. Ozone Watch, http://ozonewatch.gsfc.nasa.gov/. Accessed April 1, 2012.
12. Marks, R. (2001). Cesspools of shame: How factory farm lagoons and sprayfields threaten environmental and public health. Natural Resources Defense Council and the Clean Water Network, Washington, DC, p. 4.
13. Schade, M. (2001). The wasting of rural New York state—Factory farms and public health. Citizens' Environmental Coalition & Sierra Club, Ottawa, ON, Canada, p. 33.
14. Risku-Norja, H., Kurppa, S., and Helenius, J. (2009). Dietary choices and greenhouse gas emissions—Assessment of impact of vegetarian and organic options at national scale. *Progress in Industrial Ecology—An International Journal* 6(4):340–354.
15. Harvey, L.D.D. (2009). Reducing energy use in the buildings sector: Measures, costs, and examples. *Energy Efficiency* 2:136–163.
16. Koomey, J. and Krause, F. (1997). *Introduction to Environmental Externality Costs*. Lawrence Berkeley Laboratory, CRC Handbook on Energy Efficiency, CRC Press, Inc., Boca Raton, FL, p. 16.
17. Larsen, T. (2011). Greening the Danish tax system. Federale Overheidsdienst Financiën—België, No. 2, pp. 91–111.
18. Khan, A. and Marsh, A. (2011). Simulation and the future of design tools for ecological research. *Architectural Design (AD)* 81(5):82–91.
19. Khan, A. (2011). Swimming upstream in sustainable design. *Interactions* 18(4):12–14, ACM.
20. Lefton, S. and Besuner, P. (2006). The cost of cycling coal fired power plants. *Coal Power Magazine*, Winter 2006, pp. 16–20.
21. Liscouski, B. and Elliott, W., Eds. (2004). Final report on the August 14, 2003 blackout in the United States and Canada: Causes and recommendations. U.S.–Canada Power System Outage Task Force, Ottawa, Canada, p. 112.
22. Jørgensen, P.J., Hermansen, S., Johnsen, A., Nielsen, J.P., Jantzen, J., and Lundén, M. (2007). Samsø—A renewable energy island: 10 years of development and evaluation. Chronografisk, Århus, Denmark, p. 7.
23. Aydt, H., Turner, S.J., Cai, W., and Low, M.Y.H. (2009). Research issues in symbiotic simulation. *Winter Simulation 2009 Proceedings*, Austin, Texas, pp. 1213–1222.
24. Aydt, H., Turner, S.J., Cai, W., and Low, M.Y.H. (2008). Symbiotic simulation systems: An extended definition motivated by symbiosis in biology. *Proceedings of the 22nd Workshop on Principles of Advanced and Distributed Simulation*, Los Alamitos, CA, pp. 109–116.

25. Ohfuchi, W., Nakamura, H., Yoshioka, M.K., Enomoto, T., Takaya, K., Peng, X., Yamane, S., Nishimura, T., Kurihara, Y., and Ninomiya, K. (2004). 10-km mesh meso-scale resolving simulations of the global atmosphere on the earth simulator. *Journal of the Earth Simulator* 1:8–34.
26. Williams, M. and Achache, J., Eds. (2010). *Crafting Geoinformation: The Art and Science of Earth Observation*. Banson Production, Cambridge, U.K., p. 2.
27. Struss, P. (2011). A conceptualization and general architecture of intelligent decision support systems. *MODSIM2011: 19th International Congress on Modelling and Simulation*. Modelling and Simulation Society of Australia and New Zealand, Perth, Western Australia, pp. 2282–2288.
28. Drucker, P.F., Hammond, J., and Keeney, R. (2001). *Harvard Business Review On Decision Making*. Harvard Business School Press, Allston, MA, pp. 12–18.
29. Bonanni, L., Hockenberry, M., Zwarg, D., Csikszentmihalyi, C., and Ishii, H. (2010). Small business applications of sourcemap: A web tool for sustainable design and supply chain transparency. *Proceedings of the 28th International Conference on Human Factors in Computing Systems—CHI'10*, Atlanta, GA, pp. 937–946.
30. Fogg, B.J. (2002). *Persuasive Technology: Using Computers to Change What We Think and Do*. Morgan Kaufmann, Burlington, MA.
31. Kjeldskov, J., Skov, M.B., Paay, J., and Pathmanathan, R. (2012). Using mobile phones to support sustainability: A field study of residential electricity consumption. *CHI'12*. ACM, Austin, TX, pp. 2347–2356.
32. Kallehave, O., Skov, M.B., and Tiainen, N. (2011). Persuasion in-situ: Shopping for healthy food in supermarkets. *Proceedings of the 2nd International Workshop on Persuasion, Nudge, Influence, and Coercion through Mobile Devices, CHI'11*. ACM. Vancouver, BC, Canada, pp. 7–10.
33. Linehan, C., Ryan, J., Doughty, M., Kirman, B., and Lawson, S. (2010). Designing mobile technology to promote sustainable food choices. *MobileHCI 2010*, Lisbon, Portugal.
34. Hertwich, E.G. (2005). Consumption and the rebound effect: An industrial ecology perspective. *Journal of Industrial Ecology* 9:85–98.
35. Khan, A. and Hornbæk, K. (2011). Big data from the built environment. *Proceedings of the 2nd International Workshop on Research in the Large (LARGE '11)*. ACM, Beijing, China, pp. 29–32.
36. Eastman, C., Teicholz, P., Sacks, R., and Liston, K. (2008). *BIM Handbook: A Guide to Building Information Modelling*. John Wiley & Sons, Hoboken, NJ, 2008.
37. Attar, R., Hailemariam, E., Breslav, S., Khan, A., and Kurtenbach, G. (2011). Sensor-enabled cubicles for occupant-centric capture of building performance data. *2011 Conference Proceedings: ASHRAE Annual Conference*. ML-11-C053, Montreal, QC, Canada, p. 8.
38. Zeigler, B.P., Praehofer, H., and Kim, T.G. (2000). *Theory of Modeling and Simulation*, 2nd edn. Academic Press, San Diego, CA.
39. Khan, A., Bartram, L., Blevis, E., DiSalvo, C., Froehlich, J., and Kurtenbach, G. (2011). *CHI 2011 Sustainability Community Invited Panel: Challenges Ahead. CHI EA '11*. ACM, Vancouver, Canada, pp. 73–76.

4

Computer as a Tool for Creative Adaptation: Biologically Inspired Simulation for Architecture and Urban Design

Taro Narahara

CONTENTS

4.1 Introduction

In today's design methodologies in architecture and urban design, we normally attempt to anticipate all current and future design requirements and potential changes for buildings prior to construction and endeavor to resolve all issues in a single (relatively static) solution. However, this solution may not always be able to respond to ongoing radical population growth and environmental changes. Moreover, the physical scale of buildings and the complexity involved in building programs have been increasing to unprecedented levels. In such conditions, increased use of process-based four-dimensional design strategies (space + time) can be anticipated.

This four-dimensional design thinking is not only promising for developing more flexible and adaptable architecture but also for expanding the territory of design to systems issues. This chapter investigates the potential of architecture and urban design to produce systems that can grow over time and introduces computer-based approaches in architecture with particular focus on computational emulation of natural processes in design through conceptual models. The main focus is not on providing immediate solution methods to resolve any specific professional problems in architecture but rather on investigating the emergent characteristics of these simple models' methods that can potentially evolve new design solutions over time and on showing how tools employing the methods can be used beyond passive evaluation and analysis tools for given architectural instances.

In recent years, many scientists have started to gain the advantages of self-organizing systems in nature through their computational models in areas such as telecommunication networks and robotics. The main advantages of such systems are robustness, flexibility, adaptability, concurrency, and distributedness. With limited space, the author has chosen to write about how this specific type of computation can influence architectural design methods. After a brief overview of computer-based methods in architecture, the feasibility and application area of extensible systems in architecture will be reviewed, and the chapter speculates as to the possibilities of open frameworks for design using computational methods through relatively simple yet explicit model examples.

4.2 Early Applications of Computation in Architecture and Urban Design

Early in the 1970s, one of the pioneers in the field, William Mitchell of MIT, foresaw the possibilities and applicability of computers to architectural design and set the agenda for computer-aided design (CAD) and its education (Mitchell 1977). Around the same time, Nicholas Negroponte founded MIT's Architecture Machine Group, which was later expanded into the MIT Media Lab. The group's original focus was to create an architecture machine to help users design buildings without architects (Negroponte 1970). This unique approach set the primary motive and foundation for the current development in architectural computing, often referred as *generative design*. Negroponte's group conceived automated architectural plan generator software based on parameter-inputs by users.

However, research in the field remained largely theoretical until the drastic increase in desktop computational power in the mid-1980s allowed for practical application by architects. In the 1970s, it was hard to draw a line in a computer; by the 1980s, many architecture firms were starting to adopt CAD software (such as AutoCAD) for the documentation of drawings from manual drafting. By the late 1990s in the United States, most of the drawings by architects, including construction documents and conceptual renderings, had been digitized. Since then, beyond merely being a representation and visualization tool, computation tools have been actively used for the evaluation of given instances in architecture in order to analyze their performance. Finite Element Analysis software has been widely used by engineers and even architects for structural analysis, and environmental analysis software is available for simulation and evaluation of day lighting, solar radiation, thermal performance, energy analysis, and so on (Autodesk Ecotect Analysis 2012).

More contemporary application areas of computer-based methods are building information technology (BIM) and generative design. A BIM is a process involving the generation and management of digital representation of a facility (Eastman et al. 2008; NBIMSP 2012), and BIM tools provide a comprehensive 3-D model that is parametrically adjustable based on attributes such as construction cost, time, materials, and product-specific information from manufactures, without users having to reconstruct models every time a change in design occurs. This type of software is often also referred to as Parametric Modeling software among architects. In principle, 2-D construction drawings such as plans and sections can be instantly acquired by slicing the 3-D model, and the use of the latest crowd computing technologies starts to allow data sharing of building information. On the other hand, generative design has been widely used in the schematic phase of architectural design since the 2000s, after scripting capabilities in common CAD software matured (AutoLisp 2012; MEL 2012; Rhinoscript 2012). The method generates possible design iterations based on a set of rules or an algorithm, normally using a computer program, and one of the pioneering works by Stiny and Gips (1972) has developed into a Shape Grammar, which is a widely accepted approach within this area.

Today, many architects write their own project-based custom code for geometrical operations using scripting capabilities. The prime example of this new attitude—the architect simultaneously being a designer and toolmaker—is Frank Gehry, who developed his own technical consulting firm, Gehry Technologies (2012), and software, Digital Project (2012), to resolve complex issues on geometry and fabrication for his projects such as the Guggenheim Museum Bilbao (Shelden 2002). The firm and the tool contributed to rationalizing advanced geometry and construction issues for projects by others such as Beijing National Stadium, known as the Bird's Nest Stadium, by Herzog & deMeuron. In terms of application in a more social context, the Dutch firm MVRDV (1999, 2005) has actively incorporated computer programs to solve political and economic issues such as zoning, and their unique computational interfaces include a form of digital game with architects and mayors as its players. In academia, the Centre for Advanced Spatial Analysis (CASA) has done extensive research on city planning, policy, and architecture using digital technologies in geography and the built environment. The Space Syntax group (2012) has developed human-focused techniques for the analysis of spatial configurations using the connectivity of urban spaces based on cognitive factors such as visibility (Hillier 1999) and has served as consultant for important firms' projects, most recently the London Olympic Legacy design proposal. Lastly, much effort has been made to lay a scientific foundation for the field by scholars from various interdisciplinary areas, including Gero (2012).

In terms of more conceptual implications, recent theories of form in architecture have focused on computational methods of formal exploration and expression, and prominent works include Novak's "Computational Compositions" (1988), Mitchell's *Logic of Architecture* (1990), Eisenman's "Visions Unfolding" (1992), Frazer's *An Evolutionary Architecture* (1995), Lynn's *Animate Form* (1999), and Kostas Terzidis's *Algorithmic Architecture* (2006).

4.3 Complexity, Time, and Adaptation in Architecture and Urban Planning

One of the main concerns in architecture today is the increasing quantity of information to be processed during design and the level of complexity involved in most building projects. As globalization and economic development increase, large-scale urban development has become ever more essential. Complex threads of relationships among buildings and urban infrastructures are intertwined to produce inseparable connections. Dependencies among these structures are extremely intense, not only pragmatically but also aesthetically. Nearly half a century ago, Alexander (1964) already foresaw these conditions and stated the following:

> In any case, the culture that once was slow-moving, and allowed ample time for adaptation, cannot keep up with it. No sooner is adjustment of one kind begun than the culture takes a further turn and forces the adjustment in a new direction. No adjustment is ever finished. And the essential condition on the process—that it should in fact have time to reach its equilibrium—is violated.

Today most buildings and infrastructure designs require dynamic and collaborative engagements by multiple professionals, and a conventional knowledge-based approach alone may not be able to respond to emerging building types. For example, housing projects for thousands of people have been emerging in urban areas, and demands for planning and design of buildings with multiple occupancies and complex programs are becoming a challenge. The social impact of such buildings can completely alter

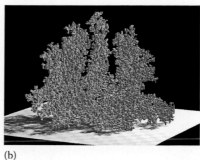

(a)   (b)

FIGURE 4.1
In these high-rise residential towers in Hong Kong, housing thousands of people, identical floors are simply stacked one on top of another (a). Algorithmically optimized 3-D clusters with maximized opening areas (b). (From Narahara, T., Self-organizing computation: A framework for generative approaches to architectural design, Doctor of Design dissertation, Harvard University Graduate School of Design, Cambridge, MA, 2010.)

the behavioral dynamics and physical conditions of local environments and often leads to redefinitions of transportation systems and infrastructures at far greater scales. Moreover, lack of flexibility and adaptability to ever-changing environments has resulted in the necessity to carry out expensive and invasive operations of demolition and reconstruction.

For instance, conventional solutions for multifunctional large-scale complexes, high-rise office towers, and housing complexes are often to design a plan resolving all the problems within a single floor and simply stack one on top of another (Figure 4.1). This approach may satisfy a number of initial requirements on a temporary basis. However, there are always other potential spatial configurations worthy of investigation. Selecting optimal solutions that can accommodate unpredictable additions and renovations for future adaptation has become a difficult task for architects. Our ultimate goals for successful design have shifted to seeking solutions for satisfying more long-term needs in a flexible manner.

On the other hand, some buildings do not include a comprehensive solution from the outset for all the potential scenarios of the future. Instead, some of those buildings possess systems that allow them to adapt to future changes over time by altering their designs spontaneously based on simultaneous feedback from a number of simple entities (or agents) inside the system. These feedback systems can be effectively distributed to formulate globally satisfactory working solutions as a collective result. These methods do not always guarantee the best solution in a deterministic sense; however, they may prove effective where there is no deterministic and analytical means to derive solutions. As a natural consequence of adapting to radical population growth, sometimes these characteristics can be seen in low-cost housing developments in less regulated zones with no supervision by professionals (Figure 4.2). These developments from human designs do not provide

(a)　　　　　　　　　　　　　　(b)

FIGURE 4.2

Extreme results of bottom-up design. (a) Favela in Rio today called Rochina. (Photo by Ciaran O'Neil, http://irishabroad.blogspot.com) (b) Kowloon Walled City in 1994 before its demolition. (Photo by Dan Jacobson, Kowloon Walled City, Hong Kong, 1989, available at http://commons.wikimedia.org/wiki/File:Kowloon_Walled_City.jpg)

completely positive results, however, characteristics of dynamic adaptations seen in these examples suggest ideas for future computational implementations. This type of design approach is often referred to as *bottom-up* and is found in many natural systems.

As a promising means to respond to the aforementioned complexities in recent architecture culture, this chapter explores the active adoption of computational methods inspired by biological systems in the next section. Several pioneering works are reviewed with particular emphasis on computational emulation of natural processes in design, and the chapter, more specifically, investigates advantageous use of computation inspired by *self-organizing* systems.

4.4 Computational Emulation of Natural Processes in Design

Self-organization is a characteristic that can be found in systems of many natural organisms: flocking of birds, pigmentation of cells in animal skin patterns, and collective building behaviors by social animals and insects such as termites. These behaviors are often referred to as "emergent" behaviors, and *emergence* refers to "the way complex systems and patterns arise out of a multiplicity of relatively simple interactions" (Camazine et al. 2002). Original theories of self-organization, developed in the context of physics and chemistry, are defined as the emergence of macroscopic patterns out of processes and interactions defined at the microscopic level (Nicolis and Prigogine 1977; Haken 1983). Such systems' behaviors display many characteristics that are similar to the bottom-up approach in some artificial systems (Figure 4.3).

(a) (b)

FIGURE 4.3

(a) Collective constructions by social insects (termites). (Photo taken and supplied by Brian Voon Yee Yap, Cathedral Termite Mounds in the Northern Territory, 2005; this image is from Dan Jacobson, available at //commons.wikimedia.org/wiki/File:Termite_Cathedral_DSC03570.jpg under the creative commons cc-by-sa 3.0 license.) (b) Experiment on construction automation by the author at Design Robotics Group at Harvard.

Not only in natural systems have we witnessed self-organizing growth processes, but also in some artificial systems. Beyond the scale of buildings, we have witnessed self-organizing growth processes in many formations of cities on large scales over long spans of time. Although results of these processes are not always successful in all aspects of design, the processes display heuristic and almost trial-and-error types of approaches that are robust and flexible enough to dynamically adapt to ever-changing environments. As the cities deliberately created by designers and planners rarely display the level of flexibility seen in these spontaneous city growth patterns, it is worth investigating their characteristics. In this area, Batty (2006) and Batty and Longley (1994) at CASA have published seminal work on urban growth modeling using algorithms such as fractals. Further rigorous computational reinterpretation and application of the principles underlying artificial self-organizing phenomena will possibly enhance and elaborate the advantages of these systems to a more practical level.

In recent years, many scientists have started to obtain the advantages of self-organizing systems in nature through their computational models (Bonabeau et al. 1999). The main advantages of such systems are robustness, flexibility, adaptability, concurrency, and multiplicity. In the bio-inspired computation field, application of ants' foraging behaviors to telecommunication networks (Schoonderwoerd et al. 1996; Di Caro and Dorigo 1997) and control of multiple robots (swarm robotics; Lipson et al. 2005; Murata 2006) are a few examples of applications that use distributed controls to gain more flexibility and robustness in their systems. *Self-organizing computation* is a computational approach that brings out the strengths of the dynamic mechanisms of self-organizing systems: *structures appear at the global level of a system from interactions among its lower-level components.* In order to computationally implement the mechanisms, the system's constituent units (*subunits*) and the rules that define their interactions (*behaviors*) must be described. The system expects emergence of global-scale spatial structures from the locally defined interactions of its own components. Development of adaptive design systems may benefit from active implementations of self-organizing logic by gaining its characteristics, such as flexibility, adaptability, and tolerance for growth.

In architecture, following and synthesizing natural principles as a basis of building design date back to the work of Otto et al. (1985) in Stuttgart, Germany, in the 1970s, and much of the effort has been spent on translating principles found in that decade into our contemporary professional practice that requires computational methods for the complete process of design. Most recently in this area, Hensel et al. (2010) and Weinstock (2010) have reported their theories and experiments regarding biomimetics and emergence in architecture.

There were several inspirational projects built by architects in the past, and obviously, the development of flexible and adaptable architecture has been a perennial theme among practitioners. During the 1960s in Japan, metabolists introduced megastructures that could constantly grow and adapt by plugging prefabricated pods onto the infrastructural core. Similar reconfigurable systems have been proposed by others including Archigram (Cook 1999). However, original visions of metabolic growth and adaptation were rarely realized physically, as the sizes and weights of the pods were practically very difficult to reconfigure. In the 1990s, construction automation by general construction companies in Japan shed light on the concept of self-reproduction in architecture: architecture that can produce architecture (Shiokawa et al. 2000). However, there was still a clear division between assembler and assemble relationships. Mechanical components that could produce buildings were far from actual livable architectural spaces. Thus they could only repeat, producing an identical or similar building at a time, and no future adaptation was available. These precedents indicate the difficulties of designing universal subunits that could tolerate technological, environmental, and circumstantial changes associated with structures and the differences in scale between artificial and natural systems (Figure 4.4).

Most artificial modular systems' building configurations are predetermined by designers and tend to require longer periods to actively organize themselves into a meaningful form or to require assistance from other assembler systems such as cranes. Reconfigurable swarm robotics systems devised by computer scientists are a more advanced concept that affords active behaviors to their subunits. In order to fulfill this gap between natural and artificial systems, it may be inferred that subunits of artificial systems need to gain more active dynamic behaviors while maintaining a fine granular size relative to global systems. Recent interests in nano/micro robotics all point in a biomimetic engineering direction, and development of such nano devices may be one way to assimilate emergence. The graph in Figure 4.5

(a)   (b)

FIGURE 4.4
(a) Nakagin capsule tower by Kurokawa, Tokyo, 1971. (Photo by Tomio Ohashi, courtesy of Kisho Kurokawa Architect & Associates.) (b) Habitat 67 by Moshe Safdie, Montreal, Canada, 1967. (Photo by Wladyslaw, Montral:Habitat 67, 2008; image is from Wladyslaw, available at // commons.wikimedia.org/wiki/File:Montreal_QC_Habitat67.jpg.)

shows the relative size of a subunit over the entire system on the x-axis and the lifetime or time for a growth period (in years) on the y-axis. Regardless of the sizes of entire systems, the natural systems tend to show smaller ratios for the relative size in the x-axis compared to those of many artificial systems, especially ones that were deliberately designed by professionals. Termite nests are composed of varied debris transported by termites, and their sizes are much finer than replaceable pods proposed by metabolists. When we get reduction in this relative size of a subunit, systems start to display more prominent characteristics of emergent behaviors.

However, recent advances in technologies may allow architects to envision more actively responsive structures in the near future. For example, new construction materials including ultra-high-strength concrete can span far greater lengths and expand architectural scope and possibilities. Moreover, recent advancements in sensor technologies have opened possibilities for buildings to have active adaptable mechanisms, and series of kinetic structures by Calatrava (2012) and Hoberman Associates (2012) foreshadow artistic integration of sensor technologies, robotics, and architecture. Application of robotics in architecture has already become a trend in academia, and Gramazio and Kohler at ETH Zurich (2012) and Design Robotics Group at Harvard (2012) have actively used industrial robotic arms for their design experiments since 2005 (Figure 4.3a).

In line with Alexander's prediction half a century ago (Alexander 1964) about the rapidly changing culture of his day, the author speculates that

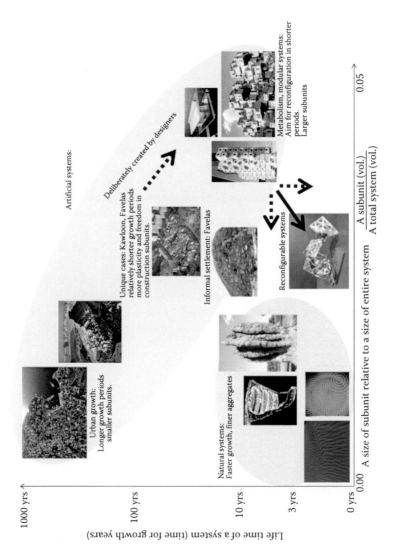

FIGURE 4.5
Various systems' lifetime and their subunit scales.

there will be more demands for buildings to be able to adapt to newly emerging needs for different qualities and quantities of architectural and urban-scale components. Within the limited allowable growth areas in dense urban settings, the desire to be able to accommodate new needs will increase demand for more adaptable buildings that can avoid future demolitions or drastic reconstructions. The hypothesis of this chapter is that the computational implementation of self-organizing principles is one potentially valuable strategy to improve the design of systems that are required to change over time.

4.5 Emergent Formations: Examples of Self-Organizing Computation

The following section presents relatively simple yet explicit examples of self-organizing computation in order to clarify the preceding more conceptual and abstract discussions about design and self-organization. The main focus is not on providing immediate solution methods to resolve any specific professional problems in architecture, but rather on investigating the emergent characteristics of these simple models' methods that can potentially evolve new design solutions over time, and on showing how tools employing the methods can be used beyond passive evaluation and analysis tools for given architectural instances.

Lane formation is a fascinating emergent phenomenon we can observe from simple agent-based pedestrian simulation (Figure 4.6). In crowds of oppositely walking pedestrians, the gradual formation of varying lanes of pedestrians moving in the same directions are observed. This is an empirically observed collective phenomenon and has been recorded in many real-life locations such as crowded pedestrian streets crossing in the city of Tokyo (Katoh et al. 1980). The emergence of this spatiotemporal pattern is a result of nonlinear interactions among pedestrians, and groups of pedestrians can find efficient walking formations solely from locally embedded individual behaviors without imposing any global geometry. Each pedestrian's embedded behavior is simply avoiding others, blocking his or her way in local neighborhood conditions, yet the global collective behavior that emerges from interactions of simple behaviors shows self-organized characteristics of a crowd.

Pedestrian agents as in Figure 4.7 are further developed (Narahara 2007), displaying more cognitive behaviors in reaction to spatial characteristics such as transparent surface, opaque surface, and furniture, in a system called the Space Re-Actor (Figure 4.7). Each agent is assigned a psychological profile with a different degree of sociability and reacts to proximity and visibility of others in the same space. Behaviors of agents are stochastically applied

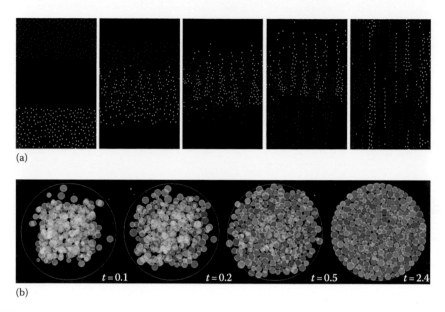

(a)

(b)

FIGURE 4.6
Pedestrian simulation by the author and emergence of lanes (a). Circle packing using a bubble mesh method (simulation by the author) (b).

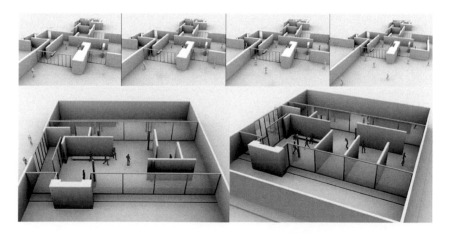

FIGURE 4.7
The Space Re-Actor: Simulation and visualization of human behavior. (From Narahara, T., The space re-actor: Walking a synthetic man through architectural space, MS thesis, Massachusetts Institute of Technology, Cambridge, MA, 2007.)

based on statistical information about human behaviors under certain circumstances, which implies that every run of the simulation can be slightly different, and normally, analyses based on stochastic simulations require multiple trials to verify their results. Though this model does not directly create new design, its real-time editing feature for spatial plans allows users to select spatial configurations with desired pedestrian flows and behaviors.

Another example of self-organizing computation using local interactions by agents is close packing of circles within a circle. Circle packing is one typical case where simulation using dynamics excels the performance of any analytical means, and it can be implemented by relatively straightforward code. Simple, locally implemented physical motions of bubbles—pushing and squeezing against each other—can eventually lead a group of bubbles to form a globally cohesive packing layout.

A physical implementation of previous examples in a more architectural context is the following robotic device with locally embedded sensors and microcontrollers (Figure 4.8). Unlike pluggable pods by metabolists being inert objects, subunits here are scaled architectural components that can reconfigure themselves into globally functional configurations based on feedback from locally distributed intelligence. Bottom-up control strategies allow the device to optimize its orientation with respect to a light source, independent of how and where the unit is placed using a multidimensional optimization algorithm. The project aims to demonstrate a design system that can respond to a dynamically changing environment over time without imposing a static blueprint of the structure in a top-down manner from the outset of design processes (Narahara 2010b).

In architecture, difficulties of developing computational design systems include the multiple objectives typical of an architectural problem and the considerable size of the search space that contains possible formal and programmatic architectural solutions. In the case of adaptive growth models, objectives are also ever-changing dynamic properties of the models, and the computational design systems of such models are expected to possess certain solution-seeking behaviors that can maneuver through the vast dynamic "solution-scape."

The earlier examples have yet to yield reliable tools for practice, though they display generative characteristics. It is worth noting that all the preceding methods rely on a heuristic. A heuristic method is a solving of a problem by iterative processes of trial and error and is intended to find optimal solutions rather than to find a single deterministic solution. As can be seen in the last project, calculations of dynamic reconfigurations can be fairly extensive if you search all one by one (i.e., brute-force search). This fact implies that the deterministic analytical means are less adequate where we need concurrent solutions for dynamically changing conditions, and we may need to rely on heuristic search as the complexity of the project increases.

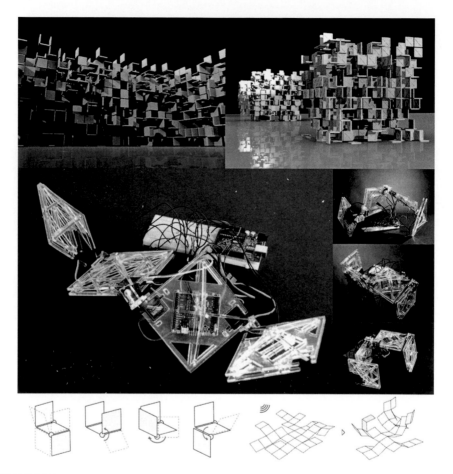

FIGURE 4.8
Robot tries to find better configurations to receive more light exposure on its four panels using Nelder–Mead multidimensional optimization. (From Narahara, T., *Int. J. Archit. Comput.*, 8, 1, 29, 2010a.)

4.6 Program for Simulating the Origins of Urban Form

In this section, the chapter further explores a generative aspect of agent-based models on an urban scale and proposes a computational method that simulates growth processes of settlement patterns.

There is an earlier work that introduced methods to produce spatial patterns using the behavior of spatial agents. Helbing et al. (2001) have done computational simulations using their "active walker model" and simulated trail formations observed within a human trail system. Their agents use a marking behavior that leaves modifications of the ground that make it more comfortable to walk on. This implementation is similar to formation of trail systems by certain ant species using chemotaxis, and an algorithmic implementation of chemotaxis was also introduced by Dorigo (1992). In architectural research, Schaur (1991) investigated empirical precedents of a human trail system, and these precedents indicate that many computational simulation results are able to obtain characteristics of pattern formations in a human trail system.

The system proposed in the following text is a computer program written in C# and is applicable to any terrain found in Google Earth by exporting geographical data. The system's three primary components are terrain, agents, and buildings. The system assumes that the development starts from unoccupied empty terrain, and behaviors of wandering settlers are simulated by the computational agents. The terrain surface is subdivided into a grid of triangulated patches that can store dynamic local information about traffic intensity of agents (settlers) and local acuteness of a terrain (slope). The buildings and streets (trails) will be gradually generated by the agents as a part of their behaviors. The primary behavioral characteristics of agents include physical mobility such as hill-climbing ability, their attraction toward environmental conditions, their selection of paths based on local traffic density, and presence or absence of global destinations.

These primary factors that govern the heading direction vectors of agents are named as attraction to gentle slope, attraction to traffic intensity, and attraction to destinations. The earlier three attractions can coexist simultaneously in many real-life scenarios. A normalized sum of three vectors weighted by three factors, slope-factor (*s*-factor), traffic-intensity-factor (*t*-factor), and destination-factor (*d*-factor), produces the heading vector of an agent. By manipulating the proportions of these three weights, the behavior of agents can be directed. These values are also dynamically changeable parameters based on changing environmental potentials.

$$Vagent = s * Vslope + t * Vtraffic + d * Vdest$$

$$(s + t + d = 1.0; \ 0 \le s, t, d \le 1.0)$$

The system advances every finite step of time. The value for attraction—intensity of traffic—is directly related to the visibility of the trails. Traces of trails by agents are perceptible by others in the region at adjacent patches and attract them. Trails that are frequently used by agents become attractors for agents in their neighborhoods. Consequently, the frequency of these trails' usage is amplified. In contrast less used trails will eventually fade away, and this is implemented by applying a reduction rate to the traffic intensity value of all subdivision surface areas every step of a simulation. The positive and negative feedback work as a pair to produce organic gradual transformations of street patterns such as branching, bundling, and emergence of hierarchy among them (Figure 4.9).

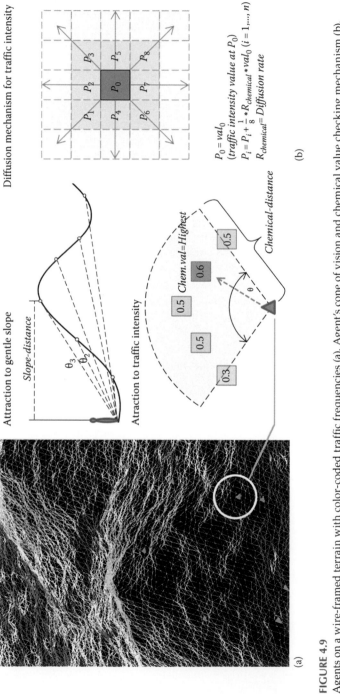

FIGURE 4.9
Agents on a wire-framed terrain with color-coded traffic frequencies (a). Agent's cone of vision and chemical value checking mechanism (b).

As a preliminary scenario with fixed behaviors of agents throughout the run of simulations, the results are categorized in roughly three emergent patterns, *direct paths*, *minimal ways*, and *detours*, and other in-between patterns. The high value for slope-factor induces agents to find a comfortable walking path. By avoiding climbing or descending a steep hill, agents produce *detours*. In this simulation, the given terrain has two steep hilltops at the south and north, and detours around these hilltops are recognized from the results. The high value for destination-factor induces agents to find the shortest path to form a *direct path* system. The high value for traffic-intensity-factor stimulates agents to minimize overall length of circulation. When agents have more frequent trips between destinations, shorter overall length of the system is beneficial because of its lower construction costs of roads: *minimal ways*. These three factors can be applied in various different proportions to find compromise solutions among three different motivations (Figure 4.10).

One of the biggest merits of the system is that purely microscopic behaviors can produce a global configuration without requiring any macroscopic information to be input into the system. When numbers of cities are small, resulting configurations are rather predictable. However, deriving a street configuration from a large number of cities at arbitrary locations on irregular terrain geometry is a challenge. Multiagent simulation is not necessarily the fastest

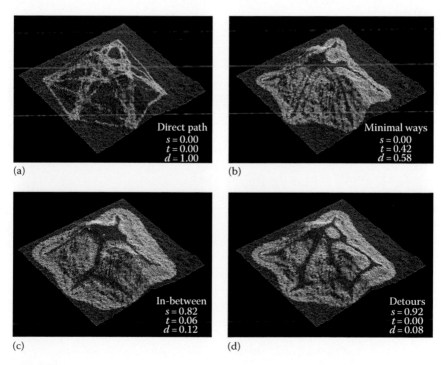

(a) Direct path
$s = 0.00$
$t = 0.00$
$d = 1.00$

(b) Minimal ways
$s = 0.00$
$t = 0.42$
$d = 0.58$

(c) In-between
$s = 0.82$
$t = 0.06$
$d = 0.12$

(d) Detours
$s = 0.92$
$t = 0.00$
$d = 0.08$

FIGURE 4.10
Eight cities on uneven terrain: results with various values for *s*, *t*, and *d* factors.

method of derivation, but it is a reasonably robust method for deriving street configuration as it requires only microscopic behaviors as input information.

4.7 Growth Simulation

The system can change behaviors of subunits over time based on stimuli from the changing environment. In the following section, shifts in behaviors of agents stimulated by environmental changes are considered. Once the activity level of an entire environment reaches a certain maturity, agents start to rely on information that already exists in environments. Agents start to follow higher traffic frequency areas instead of trails that no one has been taking (i.e., t-value increases). The earlier behaviors were implemented by setting a threshold value to shift behaviors of agents. As the number of terrain patches that possess traffic intensity value exceeds the threshold, agents start to check the traffic frequency around them to make decisions about their heading directions (Figure 4.11).

After the emergence of street networks, some of the intersections of several arteries become population concentration areas, and these areas have the potential to grow into cities. The algorithm finds peak areas of traffic intensity value above a certain threshold value and finds places where these peaks are forming clusters. Traffic-intensive patches that are within a certain distance from each other are read as one island. If these islands are larger than a certain minimum size, they are considered as city areas. After sufficient careful trials, numbers that produce results that represent a natural scale of development relative to the scale of terrain are empirically adopted. Once cities are registered by the system, all agents will travel around these cities.

Emergence of buildings is dependent on environmental potentials that are mainly frequencies of traffic and topographical conditions of a terrain. The frequencies of traffic at each patch of the terrain are considered as an indication of population density. When this value is higher than a certain threshold value, the site becomes a potential location for buildings under a certain probability. The sizes and heights of the buildings are defined by a negative exponential function of traffic intensity value to reflect upper-bound heights found in existing settlements.

Slopes of the site are another criterion for building sites. If the maximum slope of the site is above a certain degree, it is quite likely that no settler is willing to build any structures at such a steep slope. This maximum value can be dependent on regional tectonic cultures, available materials and technologies, and climatic conditions. Buildings are modeled to have an ability to align themselves to the heading directions of agents and shift of buildings locations to avoid direct collisions. Buildings also calculate the average rotation angle of others within a certain distance away and attempt to align them

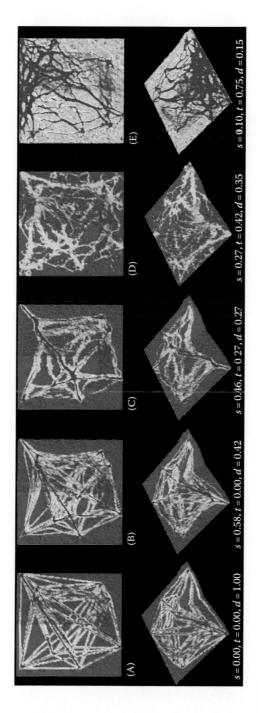

FIGURE 4.11
Results of emerging patterns on a terrain with eight predefined stationary destination points and two hills with various values for *s*, *t*, and *d* factors.

to the average angle. This is a self-organizing behavior often seen among a flock of birds. Synchronization among the rotation angles of buildings is eventually expected to produce natural arrays of buildings.

Once the overall environmental potential grows above the aforementioned minima, cities will have emerged. At this stage, settlers are no longer randomly wandering migrants seeking temporary shelter. Instead, they act as heterogeneous self-driven agents based on clear objectives of their own and travel through these newly emerging cities. Spatiotemporal developments of the site are organized by the behaviors of settlers. However, the environmental changes induced by the settlers also simultaneously influence their behaviors.

4.8 Results of Simulation of an Existing Site

Figure 4.12 shows results obtained from the proposed system tested on topographic data from an existing site, San Miniato in the region of Tuscany in Italy. The site was employed due to its unique correlations between its landform and urban settlements. The original environment of the system is completely vacant: unoccupied. The early phase of the experiments has captured patterns seen in some existing settlements (Figure 4.12, top). The formal results of the simulation are induced by a repertoire of a few behaviors implemented in agents, and some of the behaviors spontaneously emerged from the growth of a system itself. This inductive characteristic of the system implies that some tendencies found in human design activities are, to some extent, captured or approximated by a few repertoires of major behaviors by agents.

Figure 4.12 (bottom) shows the hypothetical results where incoming population and traffic intensities of the site area continue to increase. The current site, the San Miniato area, has a moderately settled, relatively rural condition, yet some of the results from later phases show developments close to the typical density of metropolitan-class cities. Various parameters that govern the local behaviors of agents changed the later patterns of the developments of the cities. Schemes from five different parameter settings display various extreme urbanistic future scenarios for San Miniato in a speculative domain. One can decide which schemes to select by finding which schemes induce global behavioral patterns to emerge that one prefers agents to enact.

Of course, configurations based on decisions by groups of humans are often results of complex negotiations and superimpositions of multiple results over time, and complete descriptions of these processes may not easily be reduced to a simple set of parameters. The system has yet to be equipped with social, political, and economic considerations. However, in principle, dynamics associated with such considerations can be computationally described by implementing more complex behaviors and contextual constraints, in addition to the three primary behaviors, in order to move toward a more practical application.

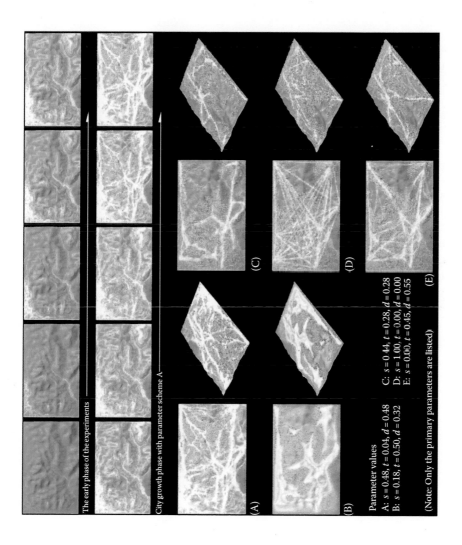

FIGURE 4.12

City growth phase with scheme A: gradual growth of paths and buildings (top). City growth phase with five different parameter settings (bottom).

In this system, the behaviors of agents are updated accordingly as new paths and buildings are generated. This coevolutionary process between agents and environments is known to exist in many self-organizing systems. One obvious advantage is its ability to *represent and generate growth processes over time*. This means that the approach can design a system in transition, and it is a necessary feature for simulating decentralized dynamics of settlements.

Environmental growth of the system and behavioral changes of agents lead the entire system to gradually adapt itself to emerging states of the system, such as "building allocations that can maintain functional traffic patterns" or "developing shortest path patterns for agents to travel around cities." These objectives are ex post interpretations of the results and are not provided directly as *globally* defined initial requirements or goals for the system. During the course of the system's run, these changes toward satisfaction of certain objectives emerge as a result of the system's ability to adjust its response to stimuli according to the state of the environment, solely from *locally* defined individual behaviors of agents. This characteristic of the system leads us to speculate regarding the possibility of creating a growth system that can discover new unknown objectives (in a sense, the emergence of new architectural programs) by further developing this computational approach, and such a system would be potentially applicable to planning and strategic development of architectural projects in earlier stages.

Finally, one of the unique characteristics of self-organizing computation is its *nonreliance on any external knowledge.* Unlike some existing city generation software, the system does not impose specific design templates such as grid, radial, or branching patterns. Reliance on a preexisting template may preclude the possibility of discovering what original inputs naturally turn into. With a few primary inputs related to the site's geographical information as initial conditions, the system can spontaneously produce all design components using self-organizing computation.

In this manner, the inherent characteristics of the resulting configurations are traceable back to several parameter values that govern the behavior of the system, and certain sets of parameters that lead to characteristics similar to existing urban phenomena can be studied. The goal of the experiment is to derive forms from behaviors instead of supplying a formal knowledge of design patterns at the outset.

4.9 Conclusion

Computational design application tools in architecture are on the brink of transition from being mere analytical tools to becoming more creative tools that can induce emergence of new solutions, or at least serve as *coevolvers* of design solutions for humans. This transition in the role of computational tools will have a big impact on our design communities and will pose a question about what the actual roles of human design experts in the field are. Further, the rise of collective design interface platforms may change our current value systems in architecture and design. This chapter has investigated computational strategies that could advance this transition and proposed generative approaches through conceptual experiments. An obvious next step is to find real-life scenarios to which these approaches are applicable and develop feasible systems that go beyond conceptual desktop experiments to become practical off-the-shelf solutions.

Acknowledgments

Firstly, I would like to thank Dr. Justyna Zander and Dr. Pieter J. Mosterman for this great opportunity. I would also like to sincerely thank my former academic advisers, Professor Martin Bechthod and Professor Kostas Terzidis at Harvard University, and Professor Takehiko Nagakura at the Massachusetts Institute of Technology, for their insightful guidance and constant support. Finally, I would like to thank my current employers, Dean Urs Gauchat and Professor Glenn Goldman at New Jersey Institute of Technology for their generous academic support.

References

Alexander, C. (1964). *Notes on the Synthesis of Form*, Harvard University Press, Cambridge, MA.

Autodesk Ecotect Analysis. (2012). *Autodesk Ecotect Analysis, Sustainable Building Design Software*, http://usa.autodesk.com/ecotect-analysis/ (accessed: August 8, 2012).

AutoLisp. (2012). AutoLISP developer's guide, http://docs.autodesk.com/ACDMAC/2013/ENU/PDFs/acdmac_2013_autolisp_developers_guide.pdf (accessed May 31, 2013).

Batty, M. (2006). *Cities and Complexity: Understanding Cities with Cellular Automata, Agent based models, and Fractals*, MIT Press, Cambridge, MA.

Batty, M. and Longley, P. (1994). *Fractal Cities: A Geometry of Form and Function*, Academic Press, San Diego, CA.

Bonabeau, E. et al. (1999). *Swarm Intelligence: From Natural to Artificial Intelligence*. Oxford University Press, New York.

Camazine, D., Franks, S., and Theraulaz, B. (2002). *Self-Organization in Biological Systems*, Princeton University Press, Princeton, NJ.

Centre for Advanced Spatial Analysis (CASA), University College London. (2012). *The Bartlett Center for Advanced Spatial Analysis*, http://www.bartlett. ucl.ac.uk/casa (accessed: August 8, 2012).

Cook, P. (1999). *Archigram*, Princeton Architectural Press, New York.

Design Robotics Group. Material processes and systems research, Graduate School of Design, Harvard University, (2012). http://research.gsd.harvard.edu/drg/ (accessed: August 8, 2012).

Di Caro, G. and Dorigo, M. (1997). AntNet: A mobile agents approach to adaptive routing, Technical Report IRIDIA/97-12. Universite Libre de Bruxelles, Belgium.

Digital Project (2012). Digital project: Overview, http://www.gehrytechnologies. com/node/4 (accessed: August 8, 2012).

Dorigo, M. (1992). Optimization, learning and natural algorithms, PhD thesis, Politecnico di Milan, Italy.

Eastman, C. et al. (2008). *BIM Handbook: A Guide to Building Information Modeling for Owners, Managers, Designers*, Wiley, Hoboken, NJ.

Eisenman, P. (1992). Visions unfolding: Architecture in the age of electronics media, *Domus*, 734, 17–24.

Fanelli, G. (1990). *Toscana*, Francesco Trivisonno Firenze, Cantini, Florence, Italy.

Frazer, J. (1995). *An Evolutionary Architecture*, Architectural Association, London, U.K.

Gehry Technologies. (2012). http://www.gehrytechnologies.com/ (accessed: August 8, 2012).

Gero J. S. (2012). http://mason.gmu.edu/~jgero/ (accessed: August 8, 2012).

Gramazio and Kohler, Architecture and digital fabrication, ETH Zurich Department Architecture. (2012), *Gramazio and Kohler, Architectur und Digitale Fabrication, ETH Zürich Departement Architectur*, http://www.dfab.arch.ethz.ch/ (accessed: August 8, 2012).

Haken, H. (1983). *Synergetics,* Springer-Verlag, Berlin, Germnay.

Helbing, D., Farkas, I. J., and Bolay, K. (2001). Environment and Planning B, 2001, Self-organizing pedestrian movement. *Environment and Planning B: Planning and Design*, 28, 361–383.

Hensel, M., Menges, A., and Weinstock, M (2010). *Emergent Technology and Design— Towards a Biological Paradigm for Architecture*, Routledge, London, U.K.

Hillier B. (1999). *Space Is the Machine: A Configurational Theory of Architecture*, Cambridge University Press, Cambridge, U.K.

Hoberman Associates, Inc. (2012). *HOBERMAN, Transformable design, profile work insights*, http://www.hoberman.com/home.html (accessed: August 8, 2012).

Katoh, I. et al. (1980). Characteristics of lane formation, *Nihon Kenchiku-gakkai Ronbun houkoku-shu*, 289, 119–129.

Lipson, H. et al. (2005). Self-reproducing machines, *Nature*, 435, 7038, 163–164.

Lambot, I. and Girard, G. (1999). *City of Darkness: Life in Kowloon City*, Watermark Publishing, England, U.K.

Lynn, G. (1999). *Animate Form*, Princeton Architecture Press, New York.

MEL (Maya Embedded Language). (2012). Autodesk Maya 2013, MEL and expressions, http://download.autodesk.com/global/docs/maya2013/en_us/index.html?url=files/Running_MEL_Run_MEL_commands.htm,topicNumber=d30e700724 (accessed May 31, 2013).

Mitchell, W. (1977). *Computer-Aided Architectural Design*, Petrocelli/Charter, New York.

Mitchell, W. (1990). *The Logic of Architecture*, MIT Press, London, U.K.

Murata, S. (2006). *Modular Structure Assembly Using Blackboard Path Planning System*. *ISARC2006*, Tokyo Institute of Technology, Interdisciplinary Graduate School of Science and Engineering, Tokyo, Japan.

MVRDV. (1999). *FARMAX*, 010 Publishers, Rotterdam, the Netherlands.

MVRDV. (2005). *Spacefighter: The Evolutionary City Game*, Actar, Barcelona, Spain.

Narahara, T. (2007). The space re-actor: Walking a synthetic man through architectural space, MS thesis, Massachusetts Institute of Technology, Cambridge, MA.

Narahara, T. (2010a). Design for constant change: Adaptable growth model for architecture, *International Journal of Architectural Computing (IJAC)*, 8, 1, 29–40.

Narahara, T. (2010b). Self-organizing computation: A framework for generative approaches to architectural design, Doctor of Design dissertation, Harvard University Graduate School of Design, Cambridge, MA.

NBIMSP (National Building Information Model Standard Project Committee). (2012). *National BIM Standard-United States Version 2, an initiative of the National Institute of Building Sciences building SMART alliance*, http://www.buildingsmartalliance.org/index.php/nbims/faq/ (accessed: March 2, 2012).

Negroponte, N. (1970). *The Architecture Machine: Towards a More Human Environment*, MIT Press, Cambridge, MA.

Nicolis, G. and Prigogine, I. (1977). *Self-Organization in Non-Equilibrium Systems*. Wiley & Sun, New York.

Novak, M. (1988). Computational compositions, *Computing in Design Education*, *ACADIA '88 Workshop Proceedings*, ed. P. J. Bancroft, University of Michigan, Ann Arbor, MI, pp. 5–30.

Otto, F. et al. (1985). Natürliche Konstruktionen: Formen und Strukturen in Natur und Technik und Prozesse ihre Entstehung (with a contribution by F. Otto) Deutsche Verlags-Anstalt, Stuttgart, Germany.

Rhinoscript. (2012). RhinoScript Wiki in McNeel Wiki, http://wiki.mcneel.com/developer/rhinoscript (accessed May 31, 2013).

Santiago Calatrava LLC. (2012), *SANTIAGO CALATRAVA*, http://www.calatrava.com/ (accessed: August 8, 2012).

Schaur, E. (1991). Non-planned settlements: Characteristics features—Path system, surface subdivision, ed. F. Otto, Vol. 39 of *Institute for Lightweight Structures* (IL.), University of Stuttgart, Stuttgart, Germany.

Schoonderwoerd, R., Holland, O., Bruten, J., and Rothkrants. L. (1996). Ant-based load balancing in telecommunications networks, *Adaptive Behavior*, 5, 169–207.

Shelden, D. R. (2002). Digital surface representation and the constructability of Gehry's architecture, PhD thesis, MIT, Cambridge, U.K.

Shiokawa, T. et al. (2000). Automated construction system for high-rise reinforced concrete buildings, *Automation in Construction*, 9, 229–250.

Stiny, G. and Gips, J. (1972). Shape grammars and the generative specification of painting and sculpture, in *Information Processing Freiman, C.V. (Ed)*, 71, pp. 1460–1465. North-Holland Publishing Company, Amsterdam, the Netherlands.

The Space Syntax group. (2012). Space syntax, trusted expertise in urban planning, building design & spatial economics, http://www.spacesyntax.com/ (accessed May 31, 2013).

Terzidis, K. (2006). *Algorithmic Architecture*, Architectural Press, Burlington, MA.

Weinstock, M. (2010). *The Architecture of Emergence—The Evolution of Form in Nature and Civilization*, Wiley Academy, London, U.K.

Yatsuka, H. and Yoshimatsu, H. (1997). *Metabolism: senkyūhyaku rokujū-nendai Nihon no kenchiku avangyarudo*, Inakkusu Shuppan, Tokyo, Japan.

5

Advanced Structural Health Monitoring Based on Multiagent Technology

Kay Smarsly and Kincho H. Law

CONTENTS

5.1 Introduction

Civil structures such as bridges, buildings, dams, or wind turbines are complex systems that are vital to the well-being of our society. Safety and durability of civil structures play a very important role to ensure society's economic and industrial prosperity. Unfortunately, many of our ageing civil structures are deteriorating because of continuous loading, harsh environmental conditions, and, often, inadequate maintenance. For example, in the

United States more than 150,000 bridges—about 25% of the U.S. bridges—are considered structurally deficient [1]. In Germany, as another example, more than 80% of the federal highway bridges show signs of deteriorations, with required repair and maintenance costs estimated at more than €6.8 billion [2]. Generally, all structures deteriorate over time. Therefore, an important task is to reliably assess the structural conditions and to ensure that safety standards are met throughout the operational live of a structure. It is also beneficial for the owners and facility operators of a structure to know the extent of deteriorations and to estimate the remaining life of a structure.

The conventional approach to assess the condition of a structure is to periodically perform visual inspections. Visual inspections, however, are limited to identify damages that are visible on the surface of a structure. Moreover, visual inspections are labor intensive and highly subjective since the results depend on the experience and the judgment of the inspector [3]. Besides visual inspections, nondestructive evaluation (NDE) technologies, such as ultrasonic, radiographic, magnetic particle, or eddy current testing techniques have been used to detect structural damage, but they must be applied to localized areas. Thus, NDE technologies require a priori knowledge of probable damage locations.

There has been a technological push in developing structural health monitoring (SHM) systems, as a supplement to visual inspections and NDE, to assess the safety and durability of civil structures [4]. As opposed to time-based inspections with visual and NDE techniques, an SHM system is installed permanently on a structure to monitor its conditions on a continuous basis. SHM systems, in principle, consist of sensors (e.g., accelerometers, temperature sensors, displacement transducers) that are installed in the structure to collect response measurements caused by some external or internal stimuli. The measurements are then transmitted to a centralized computer, which holds the responsibility for storing and processing the data collected from the sensors. Once being stored in the centralized computer, the data can be analyzed automatically by software programs or manually by human experts.

SHM systems can be greatly enhanced by taking advantage of the advances in computing and communication technologies. For instance, instead of bundling all monitoring activities and functions in a centralized computer, software modules can be developed in a distributive fashion such that the modules reside on different machines remotely connected to the centralized computer [5]. This decentralized software paradigm allows geo-spatially distributed users and software modules to participate in the monitoring activities, to access and manage the sensor data, and to work collaboratively in assessing the conditions of the monitored structure. Moreover, with the advancements in wireless communication and embedded computing technologies, wireless sensor nodes, incorporated with powerful microcontrollers, are now available. It has been shown that wireless sensor nodes are capable of not only collecting but also analyzing and condensing the sensor data using embedded software at the sensor node, as well as communicating and collaborating with other sensor nodes in a wireless SHM network [6,7].

As SHM systems evolve from a spoke-hub architecture, where all data are centrally stored and processed, to a decentralized configuration with software modules spatially distributed on different machines or sensor nodes, multi-agent technology can be beneficially deployed in the implementation of SHM systems to enhance their modularity, robustness, accuracy, and reliability. A multi-agent system is composed of self-contained cooperating software entities, referred to as "software agents," that can be adopted to perform the monitoring tasks in an autonomous and flexible manner—without any direct intervention by humans. As will be illustrated in this chapter, software agents are implemented for SHM applications both in a decentralized software framework and in the form of *mobile software agents* in a wireless sensor network. After a brief introduction to multi-agent technology, an agent-oriented system development process is presented using a simple data monitoring scenario as an example. Thereupon, the implementation and validation of multi-agent systems for two different SHM applications are described. In the first application, the deployment of a decentralized agent-based software system for long-term monitoring of a wind turbine is presented. The multi-agent system is designed to ensure the robustness of wind turbine monitoring in that any system (such as network) malfunctions as well as component (such as sensor) malfunctions are detected. In the second demonstrative application, mobile software agents are introduced in a wireless sensor network to facilitate the timely detection of changes of a structural specimen in a laboratory setting.

5.2 Multi-Agent Technology

Multi-agent systems are an active research area of increasing practical relevance. Multi-agent technology has been successfully applied to a broad range of applications, including transportation and logistics [8,9], traffic management [10], vehicle control [11], military applications [12], power engineering applications [13,14], water resource management [15], manufacturing systems [16,17], and building control systems [18,19], as well as process control, air traffic control, electronic commerce, business process management, and health care [20]; the experiences gained in several decades of research in multi-agent systems are summarized and discussed in [21]. As multi-agent technology evolves, the term *software agent* has been overused both as a technical concept and as a metaphor [22]. Many efforts have been attempted to define the properties and characteristics that constitute a *software agent* [23–25]. Following the definition by Wooldridge and Jennings, who echo the definitions proposed in the seminal work of Maes [26] and Russell and Norvig [27], a software agent is referred to as a piece of software or device "situated in some environment that is capable of [performing] *flexible autonomous* action in order to meet its design objectives" [23]. A software agent can act *autonomously* without any direct intervention of humans or other software agents and has full control over its internal state. Unlike a software object that, upon being invoked by another object, has no control over the execution of its methods, a software agent decides for itself whether or not to perform an action upon request from another agent or human user. Additionally, a software agent possesses the *flexible* features of reactivity, pro-activeness, and social ability [28]:

- *Reactivity*: Software agents perceive their environment and trigger in a timely manner appropriate actions in response to external events and changes that occur in the environment.

- *Pro-activeness*: Instead of solely acting in response to their environment, software agents are capable of taking initiatives and apply self-contained, goal-directed actions.

- *Social ability*: Software agents are capable of interacting via cooperation, coordination, and negotiation with other agents (and possibly with humans) to perform specific tasks in order to achieve their design objectives.

A multi-agent system comprises of a collection of software agents working collaboratively in an environment, which, in the broadest sense, can be anything external to the agents that the agents can sense and interact with [29].

5.3 Development of Multi-Agent Systems for Structural Health Monitoring Applications

The development of a multi-agent system for SHM applications presented in this chapter follows the design methodology as described by Smarsly and Hartmann [30]. The design methodology adopts well-established agent-oriented design approaches and specifications, such as Gaia [31], ASEME [32,33], SPEM [34], Prometheus [35], and Tropos [36]. The development of an agent-oriented system consists of five basic phases:

1. Requirements analysis
2. Agent-based system analysis
3. Agent-based system design
4. Object-oriented implementation
5. System validation

Using a simple scenario, the first three phases of the development process will be illustrated in this section. The implementation and validation phase will be discussed in subsequent sections where a distributed SHM software system for a wind turbine and a mobile multi-agent SHM system for a wireless sensing network are presented.

5.3.1 Requirements Analysis

The purpose of the requirements analysis phase is to define the functionalities needed in the multi-agent system. First, a *scenario model*, which defines distinctive sets of functions for scenario activities, is constructed. Actors participating in the scenario model are then identified and categorized in an *actor category model*. The requirements are expressed in a *computer-independent model* using actor diagrams that show the identified actors and their individual goals.

Figure 5.1 shows an example of an actor diagram illustrating a scenario of a user's request for sensor measurement data. A (human) actor, labeled *user*, interested in analyzing the condition of the observed structure requests measurement values from a database system. Through the artificial (computational) actor *personal assistant*, the request is constructed and forwarded to a *database wrapper* actor responsible for providing database access, which handles the request and extracts the requested measurement data from the database system.

5.3.2 Agent-Based System Analysis

The purpose of the system analysis phase is to further refine the requirement models and to define the actors' roles and their interactions. The actor diagrams are transformed into use case diagrams defining the agent capabilities. Figure 5.2 shows the use case diagram that corresponds to the actor diagram depicted in Figure 5.1. Specifically, the actors are transformed into *roles*, and the goals of the actors are transformed into *use cases*. Derived from the *personal assistant* actor, a specific *technician assistant* role is identified and designed to proactively support its counterpart (a human technician)—in this case requesting measurement data. The use case diagram also shows the *technician* role derived from the actor *user*, the *database wrapper*

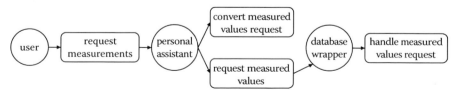

FIGURE 5.1

An actor diagram showing a scenario of a user's request for measurement data.

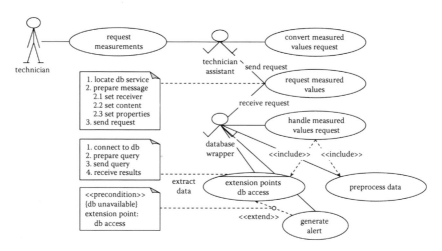

FIGURE 5.2
Use case diagram for obtaining measurement data.

role, and the interdependencies between the roles necessary to obtain the measurement data.

Following the Gaia methodology for agent-oriented analysis and design, a *roles model* and an *interaction model* are generated [31]. Represented as roles schemata, the roles model defines the functions to be carried out by a role. The interaction model, represented as a set of interaction protocols, defines the interaction patterns between the roles. Figure 5.3 shows the interaction protocols for the goals *request measured values* and *generate alert* as depicted in the scenario shown in Figure 5.2. All participating roles, parameters, and interaction

Protocol	Request measured values	Generate alert
Initiator(s)	Technician assistant	Database wrapper
Receiver(s)	Database wrapper	Technician assistant
Purpose, parameters	Specific measurements from a database system are requested Input: specification of request Output: extracted data	Alerts are generated in case of a database exception (e.g., database not available) Input: database message Output: generated alert
Process	RequestMeasuredValues = locateDBService. prepareMVRMessage. sendMVRequest. receiveMVResults PrepareMVRMessage = setMVRReceiver. setMVRContent.setMVRProperties	GenerateAlert = prepareDBAMessage. sendDBAMessage PrepareDBAMessage = setDBAReceiver. setDBAContent.setDBAProperties

FIGURE 5.3
Interaction Protocols Associated with the *Database Wrapper* Role.

processes are defined in the interaction protocols. An interaction process consists of a set of activities (shown underlined in Figure 5.3), to be performed by an agent without involving interactions with other agents, and protocols, which are activities that require interaction between two or more agents.

Figure 5.4 shows the roles schema of the *database wrapper* where the protocols and activities that the role participates in are included. It should be pointed out that the roles model and the interaction model, because of their strong interdependencies, are necessarily developed together (and often iteratively) to ensure accuracy. Once all the roles and protocols are developed, a functionality table is defined, which associates each activity of the roles schemata with a technology or tool (e.g., a database or an engineering application) that is to be employed to efficiently implement the roles.

5.3.3 Agent-Based System Design

The design phase aims to develop low-level details required to implement the multi-agent systems using traditional object-oriented techniques.

Role schema: database wrapper

Description:

 This role integrates database systems into the monitoring system

Protocols and activities:

   connectToDB, disconnectFromDB, extractData, generateAlert, handleAnalysisRequest,
      handleDBRequest, handleMeasuredValuesRequest, prepareDBAMessage, prepareResults,
      preprocessData, receiveDBRequest, receiveDBResults, sendDBAMessage, sendDBQuery,
      sendResults, setDBAContent, setDBAProperties, setDBAReceiver

Permissions:

 reads: *database, database metadata // e.g. database URL, drivers, operations, etc.*

 writes: *database*

 changes: *database*

 generates: *result, alert // result of a request, alert in case of db exception*

Responsibilities:

 Liveness:

  database wrapper = (receiveDBRequest.handleDBRequest.sendResults)$^\omega$

  handleDBRequest = handleMeasuredValuesRequest|handleAnalysisRequest

  handleMeasuredValuesRequest = extractData.[preprocessData].prepareResults

  handleAnalysisRequest = handleMeasuredValuesRequest.addInternalAnalysisResults

  extractData = connectToDB.(generateAlert|(sendDBQuery.receiveDBResults.
     disconnectFromDB))

  generateAlert = prepareDBAMessage.sendDBAMessage

  prepareDBAMessage = setDBAReceiver.setDBAContent.setDBAProperties

 Safety: *database available = true*

FIGURE 5.4
Database Wrapper Roles Schema.

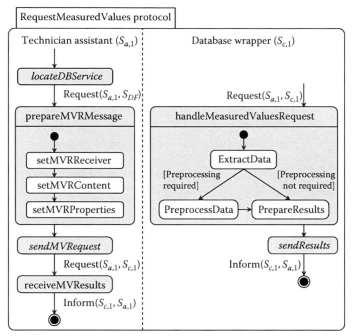

Participants:
technician assistant $(S_{a,1})$, database wrapper $(S_{c,1})$, directory facilitator $(S_{DF})^*$

*Regularly part of common agent platforms.

FIGURE 5.5
The *requestMeasuredValues* protocol.

Inter-agent control and intra-agent control are the main aspects to be considered in the design phase. The objective is to define modular, reusable, and platform-independent models.

As an example, Figure 5.5 shows a state diagram for the *requestMeasured-Values* protocol depicted in Figure 5.3. In the table, activities are represented as monitoring states. The activities that require agent interaction, such as *send Results,* are shown in italics. Furthermore, interagent message exchange is depicted by arrows where information about the senders, the receivers, and the messages is provided. The design models, which are platform independent, are defined in compliance with the specifications issued by the IEEE Foundation for Intelligent Physical Agents (FIPA) that promotes agent-based technology, interoperability of agent applications, and the interoperability with other technologies.

The design models provide sufficient details that allow agent types and their corresponding class objects to be created. Two SHM systems, implemented using the agent-oriented design methodology, are described as follows. The first system is a multi-agent system for the monitoring of a wind turbine structure; the second system is a mobile multi-agent system employed in a wireless sensor network. The implementation of specific agents in each system will be described and their functions validated.

5.4 Agent-Based Structural Health Monitoring of a Wind Turbine

An SHM system has been implemented to monitor a 500 kW wind turbine in Germany. The purpose of the SHM system is to systematically collect sensor data for monitoring the structural condition of the wind turbine, to detect damages and deteriorations, and to estimate its service life span. During the operation of the SHM system, malfunctions of sensing units have been observed, interrupting the continuous data acquisition and resulting in the loss of valuable sensor data. A multi-agent system is designed and implemented to automatically detect system and component malfunctions and to notify responsible human individuals in a timely manner. This section first provides an overview of the SHM system installed on the wind turbine and then presents the implementation and validation of the multi-agent system for automated data monitoring and malfunction detection.

5.4.1 Architecture of the SHM System

Figure 5.6 shows the basic architecture of the SHM system that is composed of a sensor network and a decentralized software system. Installed in the observed wind turbine, the sensors are connected to data acquisition units (DAUs) controlled by an on-site local computer (on-site server). Remotely connected to the sensor network, a decentralized software system with software modules distributively residing on different computer systems is designed to automatically execute crucial monitoring tasks, such as storing, converting, and analyzing the collected data sets.

The wind turbine is instrumented with accelerometers, temperature sensors, and displacement transducers to collect structural and environmental data necessary for assessing the condition of the wind turbine. In addition, an ultrasonic anemometer is mounted on a telescopic mast near the wind turbine for acquiring wind information. Figure 5.7 depicts the sensor instrumentation in the interior of the wind turbine tower at the level of about 21 m as well as the anemometer installed next to the wind turbine. The sampled and digitized data sets are temporarily stored in the local computer and, using a permanently installed digital subscriber line (DSL) connection, periodically forwarded to the decentralized software system. The DAUs, placed at the ground level in the tower and connected to the on-site server, are shown in Figure 5.8.

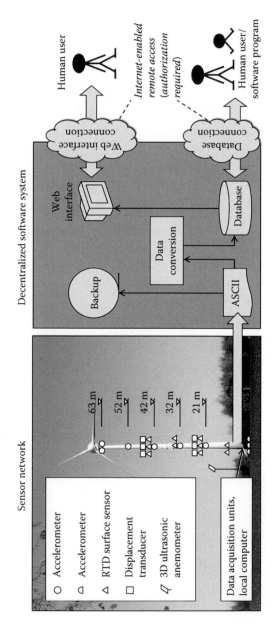

FIGURE 5.6
Architecture of the SHM system.

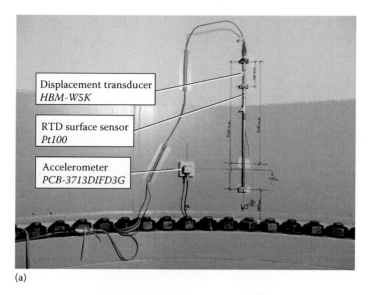

(a)

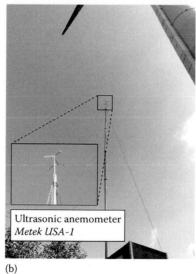

(b)

FIGURE 5.7
Instrumentation of the wind turbine. (a) Instrumentation of the wind turbine tower at the level of 21 m. (b) Anemometer installed on a telescopic mast.

The decentralized software system is designed to persistently store the data sets collected by the sensor network; it also provides remote access to the SHM system for human users as well as software programs. The software modules of the decentralized software system are installed on different computers at the Institute for Computational Engineering (ICE) in

FIGURE 5.8
DAUs and on-site server at the ground level of the wind turbine tower during installation.

Bochum (Germany). They include (1) server systems for online data synchronization, conversion, and transmission, (2) redundant array of independent disks (RAID)-based storage systems for periodic backups, (3) a structured query language (MySQL) database system for persistent data storage, and (4) a web interface for remote access to the data. Figure 5.9a illustrates the web interface used to visualize the measurement data taken from the wind turbine. Specifically, the figure shows the data recorded on April 24, 2011, including wind speed as well as the radial and tangential horizontal accelerations of the tower at 63 m height. Database connections are implemented utilizing java database connectivity (JDBC), an industry standard for database-independent connectivity between Java applications and database systems. As shown in Figure 5.9b, a direct database connection allows authorized users to download the data from the database and supports remote access to the data from external software programs, for example, to automatically execute damage detection algorithms.

To illustrate a typical malfunction of a temperature DAU, Figure 5.10 shows anomalies within the data sets collected by the DAUs. The malfunction is characterized by a large number of identical measurements because of an internal system problem in the DAU at time t_0. The DAU, instead of recording and storing the actual temperature measurements, repeatedly stores the last normal value (sensed right before the occurrence of the malfunction).

5.4.2 Implementation of a Multi-Agent System for Data Monitoring

To autonomously detect malfunctions in the DAUs, a multi-agent system is implemented and integrated into the SHM system. Once a malfunction is observed, the multi-agent system reacts and informs the responsible individuals through e-mail alerts about the defects, so that the affected DAUs can

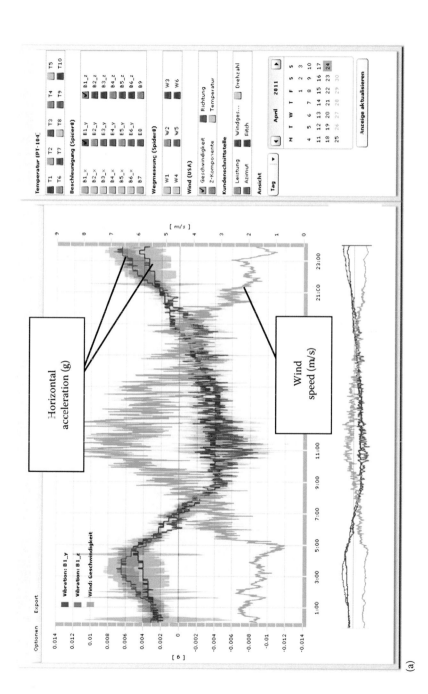

(a)

FIGURE 5.9

Acceleration and wind speed data measured on April 24, 2011, visualized through remote access. (a) Web interface connection.

(continued)

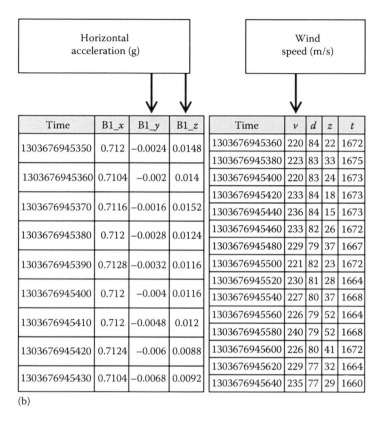

Time	B1_x	B1_y	B1_z
1303676945350	0.712	−0.0024	0.0148
1303676945360	0.7104	−0.002	0.014
1303676945370	0.7116	−0.0016	0.0152
1303676945380	0.712	−0.0028	0.0124
1303676945390	0.7128	−0.0032	0.0116
1303676945400	0.712	−0.004	0.0116
1303676945410	0.712	−0.0048	0.012
1303676945420	0.7124	−0.006	0.0088
1303676945430	0.7104	−0.0068	0.0092

Time	v	d	z	t
1303676945360	220	84	22	1672
1303676945380	223	83	33	1675
1303676945400	220	83	24	1673
1303676945420	233	84	18	1673
1303676945440	236	84	15	1673
1303676945460	233	82	26	1672
1303676945480	229	79	37	1667
1303676945500	221	82	23	1672
1303676945520	230	81	28	1664
1303676945540	227	80	37	1668
1303676945560	226	79	52	1664
1303676945580	240	79	52	1668
1303676945600	226	80	41	1672
1303676945620	229	77	32	1664
1303676945640	235	77	29	1660

(b)

FIGURE 5.9 (continued)
Acceleration and wind speed data measured on April 24, 2011, visualized through remote access. (b) Direct database connection.

be restarted remotely or replaced immediately. Following the agent-oriented design methodology as described in the previous section, the multi-agent system is implemented using the java agent development framework (JADE) that complies with the FIPA standards [37–39].

First, several software agents are developed and installed at the ICE in Bochum, two of which, namely the *Agent Management System (AMS) agent* and the *Directory Facilitator (DF) agent*, are specifically designed for administration purposes and provide an overall description of the multi-agent system. The AMS agent is responsible for the overall management actions, including creating and terminating other agents. The DF agent implements a yellow page service, allowing an agent to register and advertise its services as well as to look up other agents that offer the services necessary to cooperatively accomplish a specific monitoring task. The AMS and DF agents together provide the overall configuration description of the multi-agent system. For the scenario particular to data monitoring, a "data interrogator agent," implemented for analyzing the monitoring data

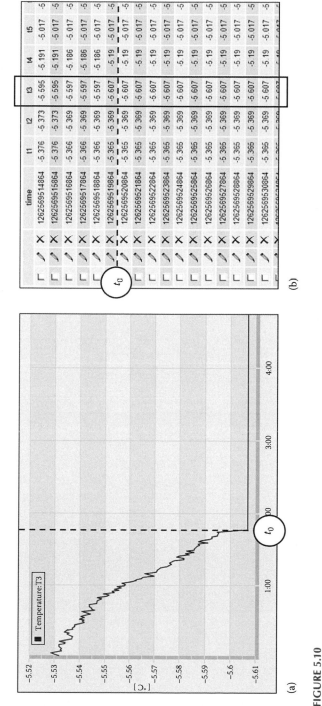

FIGURE 5.10
Visual and numerical representation of a malfunction exemplarily illustrated by temperature sensor T3 through the web interface (a) and the direct database connection (b).

with respect to malfunctions, and a *mail agent* for sending notifications to the responsible individuals are developed.

The data interrogator agent analyzes the sensor data collected by the DAUs to detect malfunctions. As in the example described earlier, one malfunctioning symptom for a temperature DAU is characterized by a large number of identical measurements. To detect such a malfunction, the data interrogator agent extracts and analyzes the temperature data sets collected by the SHM system at certain time intervals. The agent utilizes a set of predefined configuration files specifying interrogation parameters, database URL, database driver, sensor specification, scheduled interrogation intervals, and other administrative information [5]. To access the sensor data, the data interrogator agent interacts with the database of the SHM system through the database connection shown in Figure 5.6 and assumes the role of the database wrapper defined in the roles model as shown in Figure 5.4. The data interrogator agent submits SQL queries, which include simple computational procedures such as calculating the mean values and standard deviations, directly to the MySQL database, and retrieves and analyzes the data. The security of database requests and data transmissions is provided by the database system, which requires password and username as well as secure drivers to be specified by an agent trying to access the database.

Upon detecting a malfunction, responsible individuals are immediately notified. On behalf of the data interrogator agent, notifications are issued by the mail agent via an agent-based e-mail messaging service. First, the mail agent collects all relevant data necessary for composing the e-mails. The e-mail header is automatically created based on the metadata obtained by the mail agent from the configuration files; the e-mail content is created based on the anomaly information received from the data interrogator agent. Once the e-mails are composed, they are sent by the mail agent to the e-mail clients (the individuals responsible for the malfunctioning devices) using the simple mail transfer protocol (SMTP)—an Internet standard for e-mail transmission. Secure e-mail messages are ensured by username- and password-based authentications that the mail agent, similar to a human user, must specify when trying to access the SMTP server.

The actions undertaken by an agent to fulfill a monitoring task require interactions with other agents and the execution of local agent behaviors, which define how the agents perceive, analyze and, in particular, react to various external events (e.g., sensor malfunctions, detected anomalies, or unavailable monitoring data). A local agent behavior includes all relevant information about interaction protocols, communication details, and specifications of languages required to carry out the agent's design objectives. Agent behaviors can be executed either sequentially or concurrently.

To illustrate the interactions between the agents for data monitoring, Figure 5.11 shows three agent behaviors, `CyclicInterrogation`, `MailInteractionInitiator`, and `MailInteractionResponder`, that

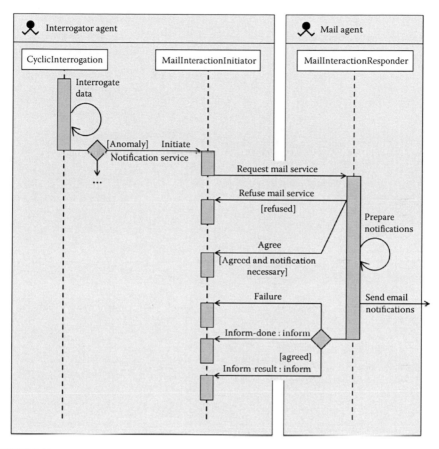

FIGURE 5.11
Monitoring procedure executed in case of a detected malfunction.

are necessary to execute the *malfunction detection* task. The behaviors
CyclicInterrogation and MailInteractionInitiator belong to
the data interrogator agent. The third agent behavior MailInteraction-
Responder is an integral part of the mail agent. As shown in Figure 5.11,
the malfunction detection executed by the data interrogator agent is initiated by its CyclicInterrogation behavior, which periodically requests
and analyzes the monitoring data with respect to anomalies. Upon detecting an anomaly, that is, a possible malfunction, the data interrogator agent
requests the e-mail notification service to notify the responsible human
individuals. To carry out the request, the data interrogator agent uses its
MailInteractionInitiator behavior for contacting the mail agent. The
notification of the human users through e-mail is accomplished by the mail
agent's MailInteractionResponder behavior. As illustrated in Figure
5.11, upon receiving and accepting a notification request, the mail agent
informs the data interrogator agent about its decision on agreeing or refusing

to handle the request. Once agreed, the mail agent processes the request by preparing and sending the e-mail notifications to the users responsible for handling the detected anomaly. If the e-mails are sent successfully, the mail agent informs the data interrogator agent, and the monitoring procedure is finalized. Otherwise, a failure message is sent from the mail agent to the data interrogator agent, which may require further action.

5.4.3 Validation and Operation of the Multi-Agent System

The multi-agent data monitoring system has been installed at Stanford University, United States, and remotely connected to the decentralized software system hosted at ICE in Bochum, Germany. Figure 5.12 shows two malfunctions, detected from the data sets collected by the temperature DAUs in January 2010 (Figure 5.12a) and in July 2010 (Figure 5.12b). As can be seen, these malfunctions are characterized by the large number of identical measurements as previously described, which, as a result of a DAU-internal

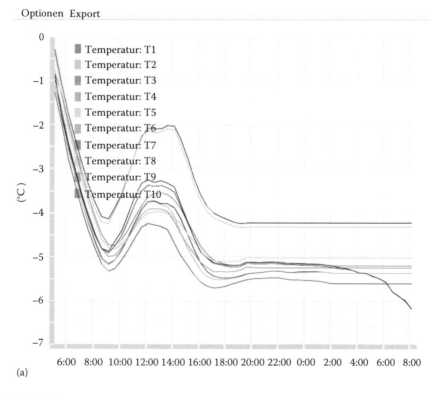

(a)

FIGURE 5.12
Visual representation of two detected malfunctions and e-mail notification generated and sent by the software agents. (a) Malfunction occurred on January 3, 2010 (affected sensors: T1, ... , T8).

system crash, are repeatedly stored by the DAU instead of regular measurements. As shown in Figure 5.12c, the e-mail alert sent in direct consequence of the malfunction includes detailed information on the detected malfunction and is generated automatically by the mail agent using the analysis results provided by the data interrogator agent.

In this application example, a multi-agent system has been presented for the autonomous detection of system malfunctions within an existing wind turbine SHM system. Since its initial deployment various anomalies in the collected data have been automatically detected. The affected DAUs have been remotely restarted by the technicians in a timely fashion. Consequently, considerable amounts of valuable sensor data has been saved and are available for assessing the structural condition of the wind turbine. In total, the long-term deployment of the multi-agent system has demonstrated the performance, the reliability, and the robustness of the multi-agent approach for autonomous SHM.

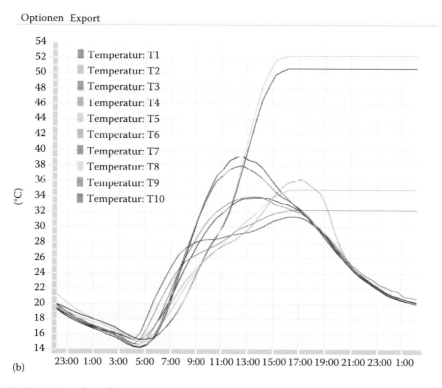

(b)

FIGURE 5.12 (continued)
Visual representation of two detected malfunctions and e-mail notification generated and sent by the software agents. (b) Malfunction occurred on July 19, 2010 (affected sensors: T5, T6, T7, T8).

(continued)

Von: monitoring@inf.bi.ruhr-uni-bochum.de
An: smarsly@stanford.edu
Betreff: [Generated Email] Wind Turbine Safety Report
Datum: 20.07.2010 03:33:20

This email has autonomously been generated
and was sent by a software agent. Do not reply.

This message is sent by the 'InterrogatorAgent'.
Content:

Analysis properties:
Database driver: com mysql.jdbc.Driver
Database table: pt104
Number of values observed per sensor: 86400
Number of significant values: 150
Sensors analyzed: tl;t2;t3;t4;t5;t6;t7;t8;t9;t10;

Anomaly detected:
DAU id: ptl04
Date: Mon July 19 17:57:59 CEST 2010
Sensor: t5
Errornumber: 34.764
(...)

Properties:
IP address: 134.147.216.69
Internet host: tunsql
Hap: tunsql: 1099/JADE
Responsible agent: mailAgent@tunsql: 1099/JADE
Agent container: tunsql: 1099/JADE
Container state: Ready(4)
Agent platform: tunsql: 1099/JADE
Mail host: delta.inf.bi.rub.de
Port: 25
Debugging: True
Protocol: smtp
Sender: monitoring@inf.bi.ruhr-uni-bochum.de
Receivers: 2
Local config file: config\mail. shm

(c)

FIGURE 5.12 (continued)
Visual representation of two detected malfunctions and e-mail notification generated and sent
by the software agents. (c) Extract of the e-mail notification generated and sent in consequence
of malfunction (b).

5.5 An Agent-Based Wireless Sensor Network for Structural Health Monitoring

Eliminating the need for installing communication cables, wireless sensor networks have been proposed as a viable alternative for SHM systems; they reduce material as well as labor costs and allow flexible network configurations. However, because of the limited internal resources of wireless sensor nodes, new and innovative concepts are necessary to implement resource-efficient operation for reliable monitoring. Multi-agent technology, supporting dynamic code migration and automated, cooperative information processing, is explored as a means to achieve resource efficiency and system reliability. Code migration—agents physically migrating from one wireless sensor node to another—enables an SHM system to dynamically adapt to changes and altered conditions of its environment and to reduce network load and latency. Software agents are installed on the wireless sensor nodes continuously executing simple resource-efficient routines to locally process the measurement data. When a sensor or the sensor network detects potential structural anomalies, specialized agents are physically migrating to the relevant sensor nodes; the migrating agents perform more complex—and resource-consuming—algorithms to further analyze the sensor data locally within the sensor nodes. In essence, multi-agent technology is incorporated into a wireless sensor network:

1. To allow the sensor nodes to collaboratively self-assess the condition of the observed structure
2. To enhance the resource efficiency of the sensor nodes with respect to data communication and onboard memory usage

5.5.1 Implementation of an Agent-Based Wireless Sensor Network

As depicted in Figure 5.13, the wireless sensor network consists of a base node and clusters of sensor network systems. Each cluster is composed of a head node that manages the cluster and a number of sensor nodes that collect and locally analyze the data. The base node is connected to a local computer that, in the prototype implementation presented herein, includes a MySQL database system for persistent storage of selective sensor data and an information pool that stores properties of the observed structure (such as modal properties), sensor information, and a catalog of available data analysis algorithms.

For rapid prototype purposes, Oracle SunSPOT devices are deployed for the wireless sensor network. The sensor nodes, built upon the IEEE 802.15.4 standard for wireless networks, comprise a fully capable, CLDC 1.1-compatible Java Platform, Micro Edition (Java ME). The computational

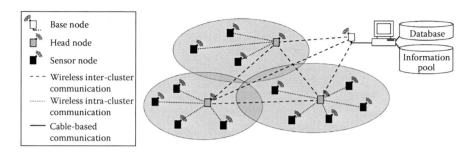

FIGURE 5.13
Hierarchical architecture of the wireless sensor network.

core is a 32 bit Atmel AT91RM9200 executing at 180 MHz maximum internal clock speed. Each wireless sensor node includes a 3-axis linear accelerometer that measures a bandwidth of 4.0 kHz over a scale of ±6 g as well as an integrated temperature sensor, a number of analog inputs, momentary switches, general purpose I/O pins, and high-current output pins [40,41]. For the software implementation, a Squawk Java virtual machine, running without an underlying operating system, ensures a lightweight execution of multiple embedded applications on the sensor nodes [42,43].

Two basic types of software agents are implemented, namely, the *onboard agents,* permanently residing at the head and sensor nodes, and the *migrating agents* located at the head nodes to be sent to the sensor nodes upon request. The onboard agents installed on the sensor nodes are capable of making their own decisions and acting in the wireless sensor network with a high degree of autonomy. They continuously take measurements from the observed structure, perform simple routines for detecting suspected abnormalities, and aggregate the measurement data (e.g., to extract representative features and values from the collected data sets). The aggregated data is then transmitted to the database system installed on the local computer for persistent storage.

As opposed to onboard agents that are permanently residing at the nodes, migrating agents are capable of physically migrating from one node to another in real time. While the onboard agents at the sensor nodes are continuously executing relatively simple yet resource-efficient routines, the migrating agents are designed to carry out more comprehensive data analysis algorithms directly on a sensor node. Acting upon a request by an onboard agent in case of detected or suspected abnormal changes of the monitored structure, a migrating agent is dynamically composed with the most appropriate diagnostic algorithm selected from the information pool for analyzing the detected anomaly. Figure 5.14 shows an abridged UML class diagram illustrating the main classes of the mobile multi-agent system. The following discussion focuses on the design and implementation of the onboard and migrating agents.

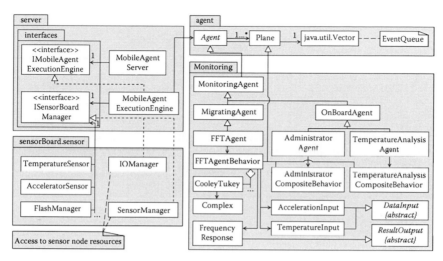

FIGURE 5.14
Abridged UML class diagram of the mobile multi-agent system.

On the sensor nodes, two categories of onboard agents, *administrator agents* and *temperature analysis agents,* are prototypically implemented. An administrator agent is designed to manage the local resources of a sensor node. For example, an administrator agent is responsible for battery level monitoring and for memory management of the sensor node. The temperature analysis agent continuously collects and aggregates the temperature data and detects significant temperature changes in the structure that may signify abnormality. The agent then communicates the aggregated data and the analysis results periodically to the head node, which in turn forwards the data through the base node to the local computer for persistent storage and further processing.

Onboard agents are also installed on the head nodes although they play a very different role than those at the sensor nodes. Unlike onboard agents situated on the sensor nodes, the onboard agents operating on head nodes are primarily administrative agents, which serve as local coordinators for their cluster without being responsible for sensing or analyzing the measurement data. Furthermore, the onboard agents on the head nodes are responsible for performing cluster-internal management and administration tasks, including communication with other clusters and the base node.

Migrating agents implement the different analysis algorithms to be utilized for diagnostic purposes. Technically, the bytecodes of the migrating agents and algorithms are stored on the head nodes. Depending on the symptoms received from the sensor nodes, different types of migrating agents, comprising different analysis algorithms, are assembled during runtime. The corresponding agent objects, together with all relevant agent capabilities and algorithms, are composed on the head node and transmitted to the sensor nodes to carry out the diagnostic functions.

For implementing the wireless agent migration, the characteristics of the Squawk Java virtual machine, as described earlier, are advantageously utilized. Squawk employs an application isolation mechanism that represents each application as an object. Being completely isolated from other objects, objects running on a node, such as migrating agents, can be paused, serialized, and physically transferred (together with agent behaviors, agent states, and required algorithms), all while Squawk instances are running on other nodes. The mobile multi-agent system is implemented based on the *MAPS* agent architecture as proposed by Aiello et al. [44].

5.5.2 Validation of the Agent-Based Wireless Sensor Network

For the proof of concept using multi-agent technology for wireless SHM, validation tests are conducted in the laboratory. As shown in Figures 5.15 and 5.14, the wireless sensor nodes are mounted on an aluminum plate that serves as a test structure. Heat is introduced to the test structure simulating an anomaly to be detected and handled by the wireless SHM system. The test structure, a 900 mm × 540 mm aluminum plate ($t = 0.635$ mm) with one edge being clamped, is instrumented with an array of nine precision temperature sensors, type LM335A manufactured by National Semiconductor. For this experimental test, the wireless monitoring system installed is composed of one cluster—comprising one head node and three sensor nodes— as well as one base node that connects the wireless monitoring system to the local computer. As shown in Figure 5.15, the test structure is divided into three monitoring sections A, B, and C. Each of the three sensor nodes (S_A, S_B, and S_C) is responsible for monitoring one section and is connected to three

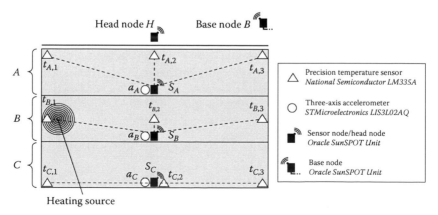

FIGURE 5.15
Overview of the prototype wireless sensor network.

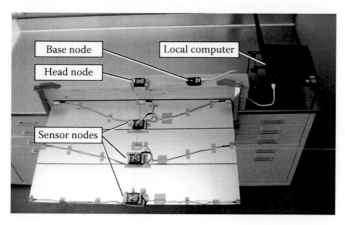

FIGURE 5.16
Laboratory setup: wireless sensor network installed on an aluminum test structure.

temperature sensors that are located within its monitoring section. In addition, each sensor node includes an integrated STMicroelectronics LIS3L02AQ accelerometer. Figure 5.16 shows the fully instrumented test structure. The main objectives of the validation tests are (1) to verify the condition monitoring capabilities of the wireless monitoring system and (2) to obtain system performance data for evaluating the resource efficiency achieved by the newly proposed concepts.

In the laboratory tests, the onboard agents operating on the sensor nodes are continuously sensing temperature measurements using the temperature sensors attached. The collected temperature measurements are locally checked with a very simple and resource-efficient procedure whether they exceed a certain threshold, then forwarded to the local computer, and stored in the database. To simulate structural changes in the test structure, heat is introduced underneath the aluminum plate near monitoring section B. A critical plate temperature $T_{crit} = 60°C$ is predefined as a threshold value indicating that a potential anomaly of the structure may occur and attention by the monitoring system may be required. Figure 5.17 illustrates the temperature distribution calculated from the temperature measurements collected by the onboard agents when T_{crit} is reached at the temperature sensor $t_{B,1}$.

Upon detecting a potential anomaly, the monitoring procedure follows the steps depicted in the sequence diagram shown in Figure 5.18. First, the onboard agents of sensor node S_B, which is responsible for the temperature sensor $t_{B,1}$, notify the head node about the observed situation. A migrating agent is immediately composed and instantiated on the head node to analyze the current condition of the test structure in more detail. First, the information pool installed on the local computer is queried for appropriate actions to be undertaken. In this example, the Cooley–Tukey

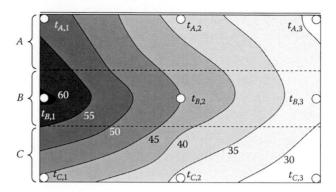

FIGURE 5.17
Temperature distribution (°C) on the upper side of the test structure.

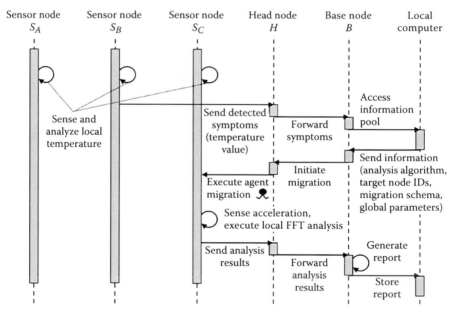

FIGURE 5.18
Monitoring procedure automatically executed in consequence of the potential anomaly.

FFT algorithm, a common algorithm that transforms a function from the time domain into the frequency domain, is selected to analyze the structural condition by determining the modal parameters of the test structure. Furthermore, based on the information provided by the information pool, details on the wireless migration are specified. In this case, sensor node S_C, instead of sensor node S_B where the threshold value has first been exceeded, is defined as the target node for the agent migration. The reason is that sensor node S_C along with its internal accelerometer is installed at the free end of the aluminum test structure and can most likely generate more sensitive results than S_B when acquiring acceleration measurements for analyzing the modal properties of the structure. Together with the selected algorithm and migration path, modal parameters for the original state of the test structure are also passed onto the head node for composing the migrating agent.

Subsequently, the migrating agent composed and initialized on the head node is equipped with a Cooley–Tukey FFT algorithm and with global system information, such as first modal frequency of the test structure in its original state. Once migrated to sensor node S_C, the migrating agent accesses the sensor node's internal accelerometer, collects the acceleration measurements, and calculates the current first modal frequency of the test structure. Figure 5.19 shows the frequency response function computed by the migrating agent based on the vertical accelerations at S_C. The current first modal frequency is calculated to be 1.6 Hz (which is not much different from the first modal frequency of the structure in its original state). The results are sent by the migrated agent to the local computer, where they are stored in the form of a safety report accessible by any responsible individuals.

In summary, the prototype agent-based wireless monitoring system operates (1) resource efficiently and (2) reliably, considering both the local and global structural phenomena. Coupling local information, directly processed on the sensor nodes, with global information, provided by the central information pool, the monitoring system is capable of detecting and handling structural changes in real time. As corroborated by the validation tests conducted, the resource consumption of the prototype system, compared to conventional systems, could significantly be reduced. Owing to the migration-based approach, a reduction of more than 90% of wirelessly communicated data could be achieved as compared with traditional approaches that

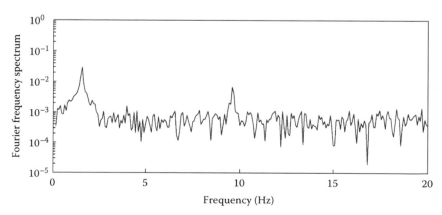

FIGURE 5.19
Frequency response function calculated by the migrating agent.

communicate all collected raw sensor data to a central server. In addition, the memory consumption, as measured in the laboratory experiments, has been reduced by approximately 70 kB per sensor node in comparison with those conventional approaches that perform embedded analyses directly on the sensor nodes without applying dynamic code migration.

5.6 Summary

This chapter describes the deployment of multi-agent techniques in the development of SHM systems. A methodology, from requirements analysis to system design, for developing agent-based monitoring systems has been presented, followed by two application examples elucidating the implementation and validation of the agent-based approach. In the first example, an agent-based decentralized software system for the autonomous detection of SHM system malfunctions has been presented. The agent-based software system has been integrated into an existing decentralized SHM system for real-time monitoring of a wind turbine to detect system and component malfunctions. In the second example, multi-agent technology has been deployed in a wireless sensor network for accurate monitoring and a reliable, timely detection of changes in the structural conditions based on mobile code migration. Specifically, the design, the implementation, and the validation of an agent-based wireless monitoring system, able to autonomously detect structural changes, have been demonstrated. Because of its modularity and extendibility, multi-agent technology is a promising paradigm for the implementation of robust and accurate SHM systems, allowing continuing upgrades and new advancements to be incorporated in an autonomous and distributed manner.

Acknowledgments

This research is partially funded by the German Research Foundation (DFG) under grant SM 281/1-1, awarded to Dr. Kay Smarsly, and by the U.S. National Science Foundation (NSF) under grant CMMI-0824977, awarded to Professor Kincho H. Law. The authors would like to express their gratitude to Professor Dietrich Hartmann of the Ruhr University Bochum, Germany, for his generous support and for his leadership in the development of the SHM system for the wind turbine, which has been instrumented within the DFG-funded research project HA 1463/20-1.

References

1. HSNW, 150,000 U.S. bridges are rated 'deficient', Homeland Security News Wire, Locust Valley, NY, 2010.
2. Smarsly, K., Law, K.H., and König, M., Resource-efficient wireless monitoring based on mobile agent migration, in *Proc. SPIE (Vol. 7984): Health Monitoring of Structural and Biological Systems 2011*, Kundu, T., Ed., 2011, Digital library.
3. Moore, M. et al., Reliability of visual inspection for highway bridges, volume I: Final report, FHWA-RD-01-020, U.S. Federal Highway Administration, McLean, VA, 2001.
4. Wang, Y., Wireless sensing and decentralized control for civil structures: Theory and implementation, Doctoral thesis, Stanford University, Stanford, CA, 2007.
5. Smarsly, K., Law, K.H., and Hartmann, D., A multiagent-based collaborative framework for a self-managing structural health monitoring System, *ASCE Journal of Computing in Civil Engineering*, 26, 76, 2012.
6. Lynch, J.P., Overview of wireless sensors for real-time health monitoring of civil structures, in *Proceedings of the 4th International Workshop on Structural Control and Monitoring*, New York, June 10–11, 2004.
7. Smarsly, K., Law, K.H., and König, M., Autonomous structural condition monitoring based on dynamic code migration and cooperative information processing in wireless sensor networks, in *Proceedings of the 8th International Workshop on Structural Health Monitoring 2011*, Chang, F.-K., Ed., DEStech Publications, Inc., Lancaster, PA, 2011, 1996.
8. Perugini, D. et al., Agents in logistics planning—Experiences with the coalition agents experiment project, in *Proceedings of the 2nd International Joint Conference on Autonomous Agents and Multiagent Systems (AAMAS 2003)*, Melbourne, VIC, Australia, July 14–18, 2003.
9. Graudina, V. and Grundspenkis, J., Technologies and multi-agent system architectures for transportation and logistics support: An overview, in *Proceedings of the International Conference on Computer Systems and Technologies—CompSysTech' 2005*, Varna, Bulgaria, June 16–17, 2005.
10. Burmeister, B., Haddadi, A., and Matylis, G., Application of multi-agent systems in traffic and transportation, *IEE Proceedings Software Engineering*, 144, 51, 1997.
11. Bazzan, A.L.C. and Klügl, F., *Multi-Agent Systems for Traffic and Transportation Engineering*, 1st edn., Information Science Reference, Hershey, PA, 2009.
12. Don Perugini et al., Agents for military logistic planning, in *Proceedings of the 15th European Conference on Artificial Intelligence (ECAI-2002)*, Lyon, France, July 21–26, 2002.
13. McArthur, S.D.J. et al., Multi-agent systems for power engineering applications— Part I: Concepts, approaches and technical challenges, *IEEE Transactions on Power Systems*, 22, 1743, 2007.
14. McArthur, S.D.J. et al., Multi-agent systems for power engineering applications— Part II: Concepts, approaches and technical challenges, *IEEE Transactions on Power Systems*, 22, 1753, 2007.
15. Gunkel, A., The application of multi-agent systems for water resources research—Possibilities and limits, Diploma thesis, Institut für Hydrologie der Albert-Ludwigs-Universität Freiburg i.Br., Freiburg, Germany, November 2005.

16. Shen, W. et al., Applications of agent-based systems in intelligent manufacturing: An updated review, *Advanced Engineering Informatics*, 20, 415, 2006.
17. Maturana, F. et al., MetaMorph: An adaptive agent-based architecture for intelligent manufacturing, *International Journal of Production Research*, 37, 2159, 1999.
18. Fuhrmann, T. and Neuhofer, B., Multi-agent systems for environmental control and intelligent buildings, Report, Department of Computer Science, University of Salzburg, Salzburg, Austria, 2006.
19. Qiao, B., Liu, K., and Guy, C., A multi-agent system for building control, in *Proceedings of the IEEE/WIC/ACM International Conference on Intelligent Agent Technology (IAT'06)*, Hong Kong, China, December 18–22, 2006.
20. Jennings, N.R. and Wooldridge, M.J., Applications of intelligent agents, in *Agent Technology: Foundations, Applications and Markets*, Jennings, N.R. and Wooldridge, M.J., Eds., Springer, Berlin, Germany, 1998, p. 3.
21. Uhrmacher, A.M., Weyns, D., and Mosterman, P.J., Eds., *Multi-Agent Systems. Simulation and Applications (Computational Analysis, Synthesis, and Design of Dynamic Models Series)*, CRC Press Inc., Taylor & Francis Group, Boca Raton, FL, London, U.K., 2009.
22. Nwana, H.S. and Ndumu, D.T., An introduction to agent technology, in *Proceedings of Software Agents and Soft Computing: Towards Enhancing Machine Intelligence, Concepts and Applications*, Nwana, H.S. and Azarmi, N., Eds., Springer, London, U.K., 1997, p. 3.
23. Wooldridge, M.J. and Jennings, N.R., Intelligent agents: Theory and practice, *The Knowledge Engineering Review*, 10, 115, 1995.
24. Nwana, H.S., Software agents: An overview, *The Knowledge Engineering Review*, 11, 205, 1996.
25. Moraitakis, N., Intelligent software agents application and classification, in *Surveys and Presentations in Information Systems Engineering*, Imperial College London, London, U.K., 1997.
26. Maes, P., Artificial life meets entertainment: Life-like autonomous agents, *Communications of the ACM*, 38, 108, 1995.
27. Russell, S. and Norvig, P., *Artificial Intelligence: A Modern Approach*, 1st edn., Prentice-Hall Inc., Upper Saddle River, NJ, 1995.
28. Wooldridge, M.J., *An Introduction to MultiAgent Systems*, John Wiley & Sons, Ltd., West Sussex, U.K., 2002.
29. Wooldridge, M.J. and Lomuscio, A., Multi-agent VSK logic, *Proceedings of the 7th European Workshop on Logics in Artificial Intelligence (JELIAI-2000)*, Springer, Berlin, Germany, 2000.
30. Smarsly, K. and Hartmann, D., Agent-oriented development of hybrid wind turbine monitoring systems, in *Proceedings of the International Conference on Computing in Civil and Building Engineering*, Tizani, W., Ed., Nottingham University Press, Nottingham, U.K., 2010.
31. Wooldridge, M.J., Jennings, N.R., and Kinny, D., The Gaia methodology for agent-oriented analysis and design, *Journal of Autonomous Agents and Multi-Agent Systems*, 3, 285, 2000.
32. Moraitis, P. and Spanoudakis, N.I., The agent systems methodology (ASEME): A preliminary report, in *Proceedings of the 5th European Workshop on Multi-Agent Systems EUMAS 07*, Hammamet, Tunisia, December 13–14, 2007.
33. Moraitis, P. and Spanoudakis, N.I., The Gaia2Jade process for MAS development, *Applied AI*, 20, 251, 2006.

34. Object Management Group (OMG), Software and systems process engineering meta-model specification, Version 2, OMG document number: Formal/2008-04-01, online: http://www.omg.org/spec/SPEM/2.0/PDF (accessed January 17, 2012), Object Management Group, 2008.
35. Padgham, L. and Winikoff, M., Prometheus: A methodology for developing intelligent agents, in *Proceedings of the 17th Conference on Object-Oriented Programme, Systems, Languages, and Applications*, Seattle, WA, November 4–8, 2002.
36. Susi, A. et al., The Tropos metamodel and its use, *Informatica*, 29, 401, 2005.
37. Bellifemine, F. et al., JADE—A white paper, *EXP in Search of Innovation*, 3, 6, 2003.
38. Bellifemine, F., Caire, G., and Greenwood, D., *Developing Multi-Agent Systems with JADE*, 1st edn., John Wiley & Sons, Ltd., West Sussex, U.K., 2007.
39. Bellifemine, F. et al., Jade Programmer's guide, online: http://jade.tilab.com/doc/programmersguide.pdf (accessed January 17, 2012) Telecom Italia S.p.A., Milan, Italy, 2010.
40. Oracle Corporation, SunSPOT programmer's manual—Release v6.0, Oracle Corporation, Redwood Shores, CA, 2010.
41. Oracle Corporation, Main board technical datasheet rev 8.0, Oracle Corporation, Redwood Shores, CA, 2010.
42. Smith, R.B., Cifuentes, C., and Simon, D., Enabling Java TM for small wireless devices with Squawk and SpotWorld, in *Proceedings of the 2nd Workshop on Building Software for Pervasive Computing*, San Diego, CA, October 16, 2005.
43. Simon, D. et al., Java on the bare metal of wireless sensor devices: The squawk Java Virtual Machine, in *Proceedings of the 2nd International Conference on Virtual Execution Environments*, Ottawa, ON, Canada, June 14–16, 2006.
44. Aiello, F. et al., A Java-based agent platform for programming wireless sensor networks, *The Computer Journal*, 54, 439, 2011.

6

ICT for Smart Cities: Innovative Solutions in the Public Space

Ina Schieferdecker and Walter Mattauch

CONTENTS

6.1 Introduction

Research and development for *smart cities* has been ongoing for almost a decade now. However, only now its rise can be seen in close relation to the newest developments in information and communication technologies (ICTs) for mobile broadband communication, sensor technologies, machine-to-machine (M2M) communication, and efficient data and computing resources.

However, what makes a technology smart? According to [7], smart technology comes up to human cognition when it senses and recognizes patterns as humans do. A smart technology recognizes the semantics of patterns it perceives, be it text, speech, or video stream and it proactively associates and correlates data based on grasped meaning. It leverages real-world and real-time background information to assimilate new data, reasons from what it has just learned, provides common sense recommendations and predictions, and finally decides and acts on behalf of humans.

Smart cities have, in particular, the ability to acquire components and models thereof through experience and knowledge, and they use these components productively to solve novel problems and deal successfully with unanticipated circumstances (e.g., natural disasters). However, there is not a uniform, homogeneous understanding of a smart city. In fact, the following definitions demonstrate that in the worldwide discussion, diverse key aspects of a smart city are set as follows:

- According to the California Institute for Smart Communities [1], *Smart Communities* are making conscious efforts to use information technology to transform life and work within its region in significant and fundamental, rather than incremental, ways.

- A Korean vision of the *ubiquitous city* (U-city) [2] focuses on various types of computers/sensors and information systems that are built into houses, office buildings, and streets. The prior objective is to get them connected and provide various services (residential, medical, business, governmental) available in a ubiquitous and integrated manner.

- Nikos Komninos, an Athens-based professor of Urban Development and Innovation Policy [3], defines *intelligent cities* as territories with knowledge-intensive activities and embedded routines of social cooperation. They are characterized by a developed communication infrastructure, digital spaces, and knowledge/innovation management tools and a proven ability to innovate, manage, and resolve problems that appear for the first time. All intelligent cities are digital cities as well, but not every digital city is necessarily intelligent. While digital cities basically provide services via digital communication, intelligent cities use these services for their problem-solving capabilities.

At the same time, these definitions all establish a procedural understanding of smart cities (see also [4-6]): smartness of cities relates to urban management and the involvement of the different stakeholders. Smart technologies are used for issue detection, for resolving those issues in an intelligent way, and for continuously improving urban processes to advance the quality of life and the attractiveness of the city.

6.2 Vision of a Smart City

By managing and constantly improving urban processes, a smart city operates as a service provider for citizens, for enterprises of the region, and for tourists and visitors of the city. A smart city is to become a self-sustaining and sustainable ecological urban environment for an increased urban quality of life and city attractiveness. This can be fostered and achieved by interplay of a number of different technical components.

As a basic prerequisite, networked and integrated urban resources are needed that provide data access to and intelligent control of physical urban infrastructures. These technologies enable the necessary information management, information transport, coordination and control, etc. Like in a living organism, the ICT-enriched infrastructures form a nervous system for the smart city that orchestrates the flow of information between its entities.

Smart cities have, in particular, the ability to acquire, through experience, knowledge, and models of itself, its components, and its environment and to use them productively to analyze and solve novel problems and deal successfully with unanticipated circumstances.

A smart city platform contributes to sustainable growth and employment in knowledge-intensive industries, and it fosters a technology base in future Internet-relevant technology and service industries. Next to that, the platform provides an environment that empowers providers and citizens to develop and deploy services, and it accelerates the development of new network-based services for citizens and consumers.

The future information and communication networks of a smart city will evolve from the networks as we know them today. They will follow a pattern of unbounded growth that leads to even further penetration. But apart from the integration of continuous technological advances, this coevolution will occur in a complex response to requirements from innovative emerging services and applications. Major technological drivers for these future networks derive from

- General mobility and wired/wireless network integration.
- New solutions for M2M communication will result from the progressive instrumentation of the living environment with sensors and actuators.
- Ubiquity of network connectivity deeply embedded into our living environments following the always best connected (ABC) paradigm of services and the Internet of things.
- Technological heterogeneity with diversity and ubiquitous collaboration instead of a one-size-fits-all approach.

The future networks create an underlying technological substrate penetrating and transforming all aspects of society. Networking connectivity is turning into a generic utility and becoming part of everyday life much like machines, electricity, and means of transportation did in the wake of the industrial revolution. But along with that role it becomes a critical infrastructure whose uninterrupted availability is a prerequisite for the proper functioning of the whole of society (e.g., cloud computing, e-Health, e-Government, and emergency communication). Complete large-scale outages would lead to disastrous consequences; hence, it is of paramount importance that the communication infrastructure is organized as an autonomous resilient system capable of remaining operational on its own through intrinsic self-configuration and self-healing properties.

Societal, economic, and political demands will shape characteristics, deployment, and usage of the future ICT. Hence, research in that area must be firmly embedded into a multidisciplinary context anticipating existing trends and demands that go beyond pure technological engineering. In addition, ICT research must respond to generic strategic challenges for the development of the information society.

6.3 State of the Art

Research and development for smart cities has been ongoing for almost a decade; however, it is only now that its rise can be seen in close relation to the newest developments in ICT for mobile broadband communication, sensor technologies, M2M communication, and efficient data and computing resources.

Different to the urban trends in developing countries, Europe is facing specific challenges that include the following:

- Demographic developments: negative growth, aging, differentiation of population
- Weaker economic growth: economic consolidation, diverging work places, financial limitations of public authorities, subsidy reduction
- Postulate for sustainability: economic development in relation to ecological and social dimensions
- Social change: singling, small families, single parents
- Technological progress: permanency of technical innovation in every sector
- Quest for future: migration disposition and necessity

Furthermore, in Europe special attention must be given to existing urban infrastructures—green field developments for smart cities are possible in rare cases only. For example, according to London's environmental advisor to the Mayor, Isabel Dedring, 80% of the buildings that will make up the city in 2050 do already exist today [9]. Another speciality of Europe is that 40% of inhabitants live in medium-size cities with populations of approximately 500,000. Given that, research results for small-size and medium-size cities will be used within Europe, while instead megacity research results will be relevant for outside Europe.

More and more cities engage in research projects as a partner providing its infrastructures or cities as a *living lab* for technology development. Cities give consideration to the fact that participating in innovative research projects can leverage their status quo and boost their attractiveness and competitiveness as a municipality. Therefore, all over the world, smart city initiatives are gaining momentum. Subsequently, a selection of exemplary initiatives is given:

- Initiatives in selected European cities: Recognizing the importance of going smart, mainly driven by the insight that urban life cannot be sustained in the near future with current approaches, a few selected cities in Europe have taken the initiative of implementing smart technologies for a more sustainable future. Barcelona, for example, is Wi-Fied by a coverage-complete network owned by the city hall with the aim of offering local services to its administration as well as third-party services. Barcelona also commences with smart parking and noise monitoring, and the city advocates the use of the Internet to stimulate virtual communities and to promote citizen participation in municipal government. More examples can, among others, be found in Aarhus, Amsterdam, Berlin, London, Lyon, Malaga, Oulu, Paris, Santander, Stockholm, Trento, Vienna, or Zurich.

- Selected multicity initiatives in Europe: SmartCity [12] is a European initiative that loosely connects eight European cities (i.e., Rome, Wroclaw, Cluj, Brussels, Istanbul, Novi Sad, Maribor, and Amsterdam) to promote artistic and technology activities aiming to improve living conditions in urban areas. Another growing initiative is the Living Labs [13], a *citizen's laboratory* with the aim to overcome the digital gap older citizens suffer, to include elders in the codesign of innovative solutions for their well-being, and to create new professional profiles and jobs for citizens over 55 years of age. At more of a political level, the EUROCITIES initiative [14] is worth mentioning, which unites local governments of more than 140 large cities in over 30 European countries with the aim to influence and work with the European Union (EU) institutions to respond to common issues that impact the day-to-day lives of Europeans. Finally, EIT ICT Labs [15] was established in 2009 as a new EU instrument for promoting innovation in Europe. It is a consortium of businesses, universities of technology, and research centers in Europe that will use ICT as the key enabler driving the transformation toward the Future Information and Communication Society, with the ultimate goal of transforming ICT innovation into quality of life. Among several thematic areas targeted, the digital cities' domain aims at creating intelligent and sustainable cities enhanced with innovative and disruptive service-based applications for the citizens of Europe and beyond. In 2011, acatech—the German National Academy of Engineering and Sciences—formulated recommendations for actions on smart cities [16].

- Selected initiatives outside Europe: While at first *smart cities* were an initiative predominantly driven by European stakeholders, there is a growing amount of smart city initiatives worldwide. The Intelligent Community Forum was founded in 2004 in the United States [19] to study economic and social development of the twenty-first-century community and to share best practices by communities. The International Telecommunication Union's (ITU's) digital cities initiative [17] acts as an information portal to various but disconnected city initiatives worldwide; it is constantly updated with the latest news and developments. Furthermore, the Massachusetts Institute of Technology (MIT) has launched an initiative on smart cities [18] that offers a range of technology advances to aid urban living. In addition, Australia and Japan advocate the *post fossil living* [20]. Seoul has gone one step further by engaging its citizens in giving them a direct role in running the city by means of contributions to city policies and direct discussions with city officials [21]. China is developing concepts to sense their country, environment, and society ubiquitously [22].

- Commercial initiatives: Although many companies such as Siemens [10], Deutsche Telekom [11], Microsoft [45] and SAP [54] invest in smart city concepts, the only notable developments in smart cities at a coherent, international company level are operated by IBM [8, 23]. It unites commercial interests as well as interests of utilities, city halls, and citizens themselves. IBM works with a large amount of large, medium, and small companies—thus constituting an integrator of integrators to facilitate smarter living.

No matter which emphasis a smart city initiative is taking, ICT plays a central role in any development toward or evolution of a smart city. ICT does not make a city smart by itself; however, there is no smart city without a strategic positioning of ICT and substantial investment into it.

The assumption that ICT is evolving toward the fundamental backbone of a sustainable and livable city, however, inevitably leads to a number of additional questions that are subject matters of current and future *smart city* research, which we would like to sketch in the following chapters: What should the technical architecture of a city ICT look like? Who are the stakeholders that drive or should drive the smart city evolution? How can we measure and benchmark the ICT-related smartness of a city? Who/what organization is in charge of the network and data security? How can a city's ICT be monitored while at the same time protecting personal data and other privacy issues?

6.4 ICT of a Smart City

Urbanization, globalization, and demographic and climate changes are putting ever-increasing demands on our cities. At the same time, however, the opportunities to efficiently and sustainably manage energy, material resources, and people are increased through the use of ICT. It must be an objective of a smart city to capitalize on these opportunities. An important basis for this process is the integration of information and communication systems into the various technical systems and infrastructures of a city. Such integration enables the flexible control of supply and disposal networks, especially for electricity, water, and gas as well as goods. Furthermore, integration enables novel solutions for mobility, administration, and public safety in the city. An optimally networked city not only renders everyday life easier in all aspects but also provides environmentally sustainable solutions. A smart city is an informed, networked, mobile, safe, and sustainable city.

As a consequence, urban ICT is more and more evolving to become the backbone of smart cities: potentially everything becomes addressable, communicates, and can be monitored, controlled, and optimized by people with the help of computers. Urban ICT is realized as a system of systems that provides information exchange to urban users (public employees, local businesses, citizens, tourists, and others) at any request and need, at any place, at any time, and at any preference. The systems collect, exchange, aggregate, analyze, and provide data and information of various kinds.

ICT-based systems are used in order to govern the city by automating the control, remote maintenance, and optimization of physical infrastructures (e.g., grids, streets, parks, buildings, lamps) that are directly relevant for areas such as energy supply, mobility, logistics, safety, and security and by generating and providing information about these infrastructures. Data are automatically generated via sensor networks or interconnected machines and servers. City-related objects (vehicles, street lamps, wastebins, etc.) are enhanced to become smart objects. Next to their autonomous behavior and functioning, smart infrastructures/objects provide specific data on their own status and their environment. These data are transferred via reliable and secured (wired or wireless) networks and kept in municipal data stores (public or private). Where needed and possible, data are provided in real time and continuously as streams. Data are visualized on various end devices including personal phones, computers, tablets, but also devices in public areas such as traffic information screens or information billboards.

Furthermore, ICT systems are used by the city to optimize communication and governance processes and to facilitate, for example, the municipal administration (e-Government), but also the enterprises (e-Business) or the health care system (e-Health) within a city or a metropolitan area. ICT systems are also increasingly used to enhance citizen engagement and political codetermination (e-Transparency, e-Participation). The users in a city are in various ways like consumers and providers/producers (*prosumers*) of information and data—they act as information and services consumers (e.g., via smartphones, tablets, laptops, or desktop computers in private life and as professionals, but as well via public terminals), as data producers (e.g., via smart home sensors), or even as ICT providers (e.g., via shared wireless networks) (Figure 6.1).

In a smart city, mechanical processes as well as many human activities are continuously based upon, managed, or at least influenced by data and information. Value-added services are derived from the mash-up of open, commercial, and personal data. Multimodal information is distributed through various channels according to specific user needs and access rights as well as according to specific situations and devices. Public services increasingly allow for bidirectional communication between the municipal administration and other city stakeholders such as citizens, businesses, and nongovernmental organizations (NGOs).

Data that are generated in specific urban contexts can be provided in raw or in conditioned form via the Internet. The current worldwide open data initiatives (e.g., [29,30,32]) aim at making even more data available to anyone, any purpose, and any use by public authorities but also by companies, NGOs, or the like. If data can be directly accessed by machines as static artifacts (e.g., files) or as dynamic services (e.g., application programming interfaces (APIs)), new kinds of services can be developed and provided to the public and to city professionals.

Thus, ICT has become a critical urban infrastructure—critical in a multiple sense: ICT does not only enhance urban supply, logistics, and disposal infrastructures to become smart and energy-efficient infrastructures—many safety-critical processes depend on ICT systems today. In hazards, accidents, and catastrophes, ICT is an important prerequisite for emergency forces, police, and health-care organizations to efficiently save lives and minimize damages. In addition, with the use of personal data, there is a specific responsibility and need to secure these data against all kinds of abuse to a high extent. And finally, the ICT infrastructure itself has become that important for the functioning of any urban society that ICT architects and designers have to turn special attention to the protection and the resilience of the systems against damage, destruction, pollution, or direct attacks.

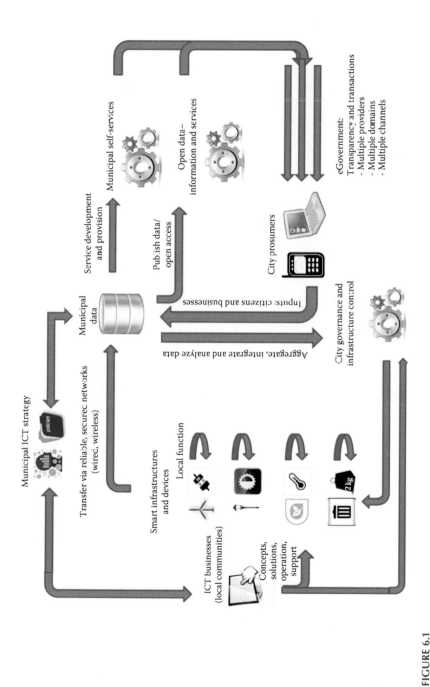

FIGURE 6.1
Municipal ICT as a system of systems facilitating governance, infrastructure control, and information exchange.

ICT systems, at the same time, are a much more dynamic field than most other parts of an urban infrastructure. The quality of the ICT system is a clear advantage in competition concerning the economic performance but also the quality of life within a city. Therefore, a community of local ICT providers is needed to continuously refine and enhance the ICT infrastructure in a smart city and to provide solutions to the city management. Because of this, urban ICT systems cannot be planned like canalization or rail systems with stable components lasting for decades. ICT systems are systems in constant evolution that demand mechanisms of data integration and constant technical improvement as new technologies are coming to the market. A direct municipal strategy for ICT development is essential to set priorities and—not only technically but as well politically—form a *smart city*.

6.5 ICT Architecture of a Smart City and Its Elements

Despite the heterogeneity of ICT solutions in urban management, there are basic concepts, layers, and interfaces that are required to enable access to and communication of required data and information where needed; to overcome silos of data, information, and communication in different application domains; and to provide value-added services and applications to the stakeholders in a city. This multilayer architecture is explained in the following. It has been prototypically used in different reference scenarios that are described at the Fraunhofer FOKUS ICT for smart cities research portal [24]. Its deployment at larger scale, however, is still subject to research and development work, standardization, and adoption.

6.5.1 Overview

In a smart city, the technical communication infrastructure is the root that provides everlasting data flow. Hardware and software systems convert data into useful information so that it becomes comprehensible and visible to the citizen of the smart city. The urban ICT infrastructure is, often hidden, the basis for the collection and transfer of city data into the municipal data platform (Figure 6.2). Flexible, safe, and robust access provides a framework for seamless data availability and use. In addition, the use of efficient tools leads to new municipal applications that optimize the processes within the city of tomorrow, sustainably and dynamically.

6.5.2 Urban ICT Infrastructure: Sensor Networks and M2M

The city of tomorrow will be shaped by M2M communication. M2M communication serves above all the monitoring, control, and localization of

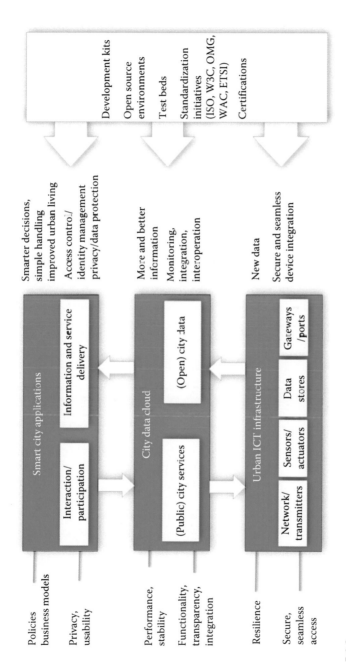

FIGURE 6.2
Urban ICT architecture.

units that were equipped with M2M technology. It allows for innovative applications for a variety of fields, for example, for building surveillance and control, electronic health care, and smart metering.

Modern cities, therefore, require a flexible M2M communication infrastructure [53] based on convergent networks that lay the foundation for new applications and their flexible combination. This leads to new possibilities for the use of previously isolated applications and data that are now the basis for new business models. For example, environmental data can be combined with up-to-date values attained from within the building to ensure the more efficient and automatic control of heating and ventilation systems. A uniform M2M communication infrastructure can make those and many other applications possible in both the private and industrial sectors. Sustainable information and telecommunication innovations are supported by the flexible availability of data and applications in all areas of urban life, such as in local electronic marketplaces.

Current research aims for the development of a generic M2M platform that is intended to act as a facilitator between different service platforms and the core network to enable the seamless communication management of different terminals, sensors, and actuators. It will thus be a priority to meet the special requirements of the M2M communication and the ability to flexibly adapt the platform for special M2M service domains such as the automotive sector.

6.5.3 City Data Cloud: Urban Data and Information

In a smart city, urban data, services, and applications are bundled on platforms that are required in both a technical and an organizational sense. On the one hand, a technical infrastructure for the integration and provision of the heterogeneous and distributed data sources is needed. On the other hand, an organizational framework must be established that controls the provision, processing, transfer, and use of the data for all actors involved—technologically, legally, and economically.

The objectives of a city data cloud are to provide enterprises, organizations, and citizens trustworthy access to urban and public data for the joint design of urban processes and operations. Commercial and public, current, and context-sensitive data together with accumulated information about urban infrastructures and resources in the city are made available, making current situations in the city transparent, facilitating decisions, and making them comprehensible. Basic requirements for a city data cloud include the following:

- Timeliness, accuracy, and quality of the data
- Data availability
- Possibilities for data aggregation and data analysis

- Uniform and device-independent access to data
- Data safety and legally compliant use and transfer of data

These demands can be addressed with a multilayered architecture whose core consists of the infrastructure layer that enables the provision and processing of comprehensive city-relevant data. Basic components of the infrastructure layer include discrete and continuous data sources, data memory, data descriptions and specifications, filters, transformers, and aggregators as well as combinable access services.

Such a city data cloud has already been created in Berlin [27] with companies following to support the idea of open data such as Vattenfall [41].

6.5.4 Smart City Applications and Services

City applications provide the citizens with access to up-to-date information and administrative services. Registration matters, house hunting, bulky waste pickup, study opportunities, traffic information, and lack of hygiene in restaurants are issues that all cities, communities, municipal companies, and citizens must deal with. The time that the appropriate agencies and municipal companies need to address these issues often seems endless, time-consuming, and too complicated for the citizen. City applications, no matter if they are provided by public authorities, companies, organizations, or citizens, remedy the situation: these applications must combine internal and external resources such as data, functionalities, and services and provide users with the opportunity to quickly contact the correct person in the local administration.

The establishment of platform- and device-independent city applications for administrative services and municipal service providers supports the (online) involvement of the citizen on the communal level and increases the transparency of administrative processes. With the help of city applications, ICTs can be set up in governmental bodies on a long-term basis. This results in the development of versatile access to administrative services and in providing citizens with up-to-date information about infrastructures and services.

For example, the citizens of a smart city must feel secure when moving around in the city and when using public transportation. Video surveillance is a key solution for enhancing citizen safety. However, this technology is challenged especially in the case of public and railway transportation. In video surveillance, it is assumed that the cars will be equipped with a set of appropriate cameras for surveillance and the corresponding video streams will be transmitted to a remote control server for investigation purposes. As video streaming is bandwidth demanding, even the use of broadband wireless technologies such as wireless local area network (WLAN), universal mobile telecommunications system (UMTS), and long term evolution (LTE) cannot completely solve the related issues, and support from the core network to control the video flows and prioritize them in case of congestion is paramount. In the context of smart cities, we will use our expertise in building a wireless-based video surveillance solution for railway transportation to enhance the safety of citizens using public transportation.

6.5.5 Seamless Access

Flexible access to offers, information, and services is provided through so-called open service platforms in a smart city. These open service platforms are based on convergent, heterogeneous network technologies, for example, wireless and wired telecommunication networks, next-generation networks (NGNs), and the Internet [42].

Smart city applications (SCAs) must be available everywhere in the city, implying a strong demand for mobile access (e.g., smartphones). They also must be scalable as the number of concurrent users can reach the millions. Also, in order to enable rapid SCA development and deployment, elementary services and properties such as authentication, authorization, accounting and QoS management must be part of a corresponding communication platform. The 3GPP evolved packet core (EPC) addresses these features in release eight and higher. The EPC follows the evolution of wired core networks and the occurrence of overlay control networks such as the IP multimedia subsystem (IMS). Its realization can become an operating system for SCAs.

Moreover, mobile cloud computing can be seen as an infrastructure allowing data storage and data processing outside the mobile device. Therefore, this new computing paradigm will enable the development of rich applications and services, especially in the context of e-Government, e-Health, and urban resource management. Such services cannot be deployed correctly without the support of the network in terms of connectivity, for instance

- Connection reliability: always best connected
- Security of the communication with the cloud
- Mobility and resource management

Another aspect of smart cities is the ability to use limited resources efficiently. Widely deployed sensor networks can be used to achieve a better understanding of floating requirements in the city (e.g., vehicular traffic control). Enabling sensor access and control in the EPC is still a research challenge with respect to communication volume, timing, real-time requirements, and security.

At the edge of any communication platform for various SCAs and services, there is an obvious need for *capillary networks* mainly based on sensor/actuator technologies enhanced by communication capabilities. These capillary edge networks in the field are responsible for all data gathering and provision functions required, obtaining close to real-time data in order to support all sorts of environmental and utility-related service and business process optimization.

Technologies for an on-demand establishment and configuration of communication infrastructure provide solutions for emergency communication and temporal peak loads (festivals, demonstrations, sports events) or in providing a backup infrastructure for disaster situations. The goal is the efficient utilization and cooperation of different networks and devices (fixed, mobile, ad hoc) that are deployed in the smart city. Resources provided by users, network operators, and authorities should be combined to provide coordinated emergency communication, temporal and backup communication networks, as well as security and prioritization policies across heterogeneous environments und administrative domains.

6.5.6 Engineering-Reliable and Secure ICT Solutions for Smart Cities

High-quality, reliable, and secure technical systems must constitute the ICT solutions for smart cities. The technical systems of cities include the infrastructure and administrative systems of a city. Municipal task responsibilities such as mobility, resource supply, waste deposal, and safety that were in the past most likely separate are assessed for common interfaces and interactions and linked in a logical way to benefit the city. This leads to increased demands on the information and communication systems that permeate through the various technical systems of a city. Integrated infrastructure and administrative systems raise the city of the future to a new quality level. They allow for analyses, safeguarding, and optimization beyond subject, organizational, and technological boundaries, whereby the interoperability, effectiveness, and the functionality and IT safety of these solutions must be designed and realized early on.

It is required to adjust and extend software engineering methods and tools for ascertaining the requirements of the ICT-based solutions as well as their design, specifications, implementation, and quality management. Particularly in the area of model-based development of software-intensive systems and the supportive development tool chains, quality management and certification processes are necessary.

For example, techniques for data analysis, privacy, anonymization, and others are essential to provide secure reliable communication infrastructures in a heterogeneous environment. In addition, data analysis techniques from the area of network traffic might also be useful in other contexts (e.g., other infrastructures).

Likewise, security and reliability are crucial to support cloud computing or communication networks for critical infrastructures. Solutions are needed to protect communication environments (robustness, availability, security). Network supervision for communication networks helps identify and defeat new attacks.

6.6 Open Data: A Driver toward Smartness

Open data are one of the main urban management trends that have emerged in recent years. It is based on the vision of open exchange of data under free licenses (e.g., CC-BY [37]), thereby providing users, companies, developer communities, and other stakeholders with a basis for creating novel services and applications. Thus, open data exchange would also provide opportunities for start-up companies since it fosters novel business models that rely on the availability of data. Correspondingly, open data have the potential to facilitate economic growth and contribute to prosperity. Furthermore, developing countries hope to improve their quality of life based on open data, given that it would allow for monitoring the environment and enable decision making with respect to vital aspects of everyday life, such as water quality.

Open data can originate from different sources. For instance, the data may be provided by a sensor measuring water quality at a particular location. Other data may be collected using certain crowdsourcing principles, or they could be provided by administration, NGOs, or science organizations as part of their daily data collection routines. One of the biggest challenges with open data is trustworthiness: No society can rely on wrong, manipulated, or compromised data. One possible remedy for this trust issue is to rely on data that solely come from the public sector, for instance, governmental institutions and different types of authorities. These institutions normally keep large amounts of data whose acquisition is financed by the tax payers. Thus, the data belong to the society since they have paid for its collection. Such data can also be considered trustworthy since they have gone through a lengthy process of collection and verification by experts in the institutions these data belong to. In addition, the institutions, being on the governmental side, have all rights to access different types of raw data that give them an advantage over data collection enthusiasts from the community (e.g., in case of crowdsourcing).

Institutions from the public sector but also companies and NGOs are intended to publish their data via so-called open data platforms. Such platforms are used to catalog the data with the help of appropriate metadata. A number of open data portals across the world are already operational—Berlin Open Data Platform [25], Kenya Open Data Platform [28], U.S. Open Data Portal [29], and the Open Data Portal of the United Kingdom [30], to mention some.

Within the EU Project *Open Cities* [39], Fraunhofer Institute FOKUS has developed a platform for publishing open data [33–35]. The primary benefit of the platform is a *one-stop-shop* experience, which means largely reduced efforts of developers and consumers when searching, accessing, downloading, and sharing the data. The platform is an integrated solution that utilizes open-source software: a data catalog based on Comprehensive Knowledge Archive Network (CKAN [40]) software, a data portal implemented in Liferay, an optional data store based on the Virtuoso database management system (DBMS), and an integration layer. The integration layer facilitates the interplay between the data portal, the data catalog, and the DBMS server toward an integrated open data platform solution. Technological specifics of the platform are (1) Java wrapper for accessing the API of the CKAN data catalog, (2) a metadata scheme for open data based on survey conducted in five major European cities (Amsterdam, Barcelona, Berlin, Helsinki, and Paris), (3) multi-language support, and (4) support for linked data (Figure 6.3).

However, the provisioning of open data still remains challenging for the stakeholders. Even though years ago the European Commission has issued regulations in the European Public Sector Information (PSI) directive [36], the amount of available open data across Europe remains far behind the identified potentials. This relates not only to the number of data sets but also to the timeliness, completeness, and accuracy of available data. Hence, the previously mentioned benefits of open data can be leveraged to a limited extent only.

The reasons are manifold. Above all, the required changes in working with data are that substantial that they require a new mind-set and paradigm changes [26] that take time for their implementation in large:

- Public vs. confidential: The decision to publish raw data needs to change from "everything is confidential if not explicitly marked as public" to "everything is public if not explicitly marked as confidential."
- Publication vs. withhold: The timing of publishing raw data needs to change from "an individual decision by administration that often means access on request only" to "a proactive timely publication of all data that is not subject to security or privacy constraints."
- Free vs. limited use: The license conditions for published raw data need to change from "availability for private use only, while all other usages need to be arranged separately," to "free licenses according to which published data are available for any use, including reuse and further processing—ideally at no cost."

Moreover, the platforms used to provide open data require improvements: The publication processes, that is, the way open data are provisioned, versioned, and provided, address the needs of data providers to a limited extent only. Data providers are required to go through a process of continuously updating the metadata as well as the data sets of static data or the APIs that provide access to dynamic real-time data.

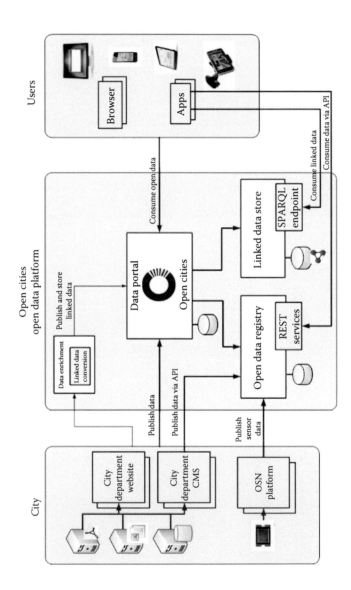

FIGURE 6.3

The open city platform architecture for open data.

This results in situations where after initial data publication the data and their metadata are often not continuously updated since it requires additional efforts from the data provider's side. To maximize the economic and social value of open data, tools for an easy provisioning, continuous maintenance, quality assurance, and data provenance are necessary, and a subject to current research.

6.7 Indicators for Smart City's ICT

Because of the growing amount of today's smart city initiatives [49–51] and since *smart city* is becoming a fashionable label, it is justified to raise the bar a little bit higher and to look for methods that will enable a more objective benchmarking. There are various studies and ICT indicator reports, provided, for example, by the United Nations or the ITU [44, 46–48]. However, these studies are basically related to countries, but not to cities. Moreover, while these studies count the number of mobile phone devices or the frequency and speed of Internet access in households (for e.g., [43]), they do not at all reflect on the various urban ICT technologies presented in the past chapters. Therefore, we would like to propose a series of new indicators that will lead to an advanced measurement of the significance of ICT for a smart city. In order to identify such indicators, we have stated the following three basic hypotheses:

- *Hypothesis 1*: A smart city is based on a highly dynamic policy- and business-driven community of ICT providers and ICT users. Continuous municipal and private investments but also ICT-related revenues for the city are the ground floor for flourishing urban ICT developments.

- *Hypothesis 2*: In smart cities, many aspects of urban systems and life-style (traffic, public transport system, public security, health care, e-Government, etc.) are supported by ICT. Citizens, businesses, public and private organizations, and visitors rely on secure and attractive services that are used in day-to-day routines and available wherever needed.

- *Hypothesis 3*: Smart cities are characterized by continuous growth of connectivity and integration. People, objects, vehicles, and assets are connecting to the diverse subsystems of a city ICT infrastructure for private, business, or public purposes. In parallel, there is a continuous trend of integrating systems and data in order to receive a consistent stream of data, an acceptable system performance, and finally, added value through information and services.

We used these three hypotheses as a basic structure to identify and group related indicators. As a result of our work, we would like to propose the following:

Indicator group 1: ICT policy and business. Urban ICT innovation is policy and business driven with a broad variety of objectives. A sustainable city ICT policy will encompass continuous long-term developments but also short-term projects heading for concrete solutions and addressing the specific key performance indicators (KPIs) of a city. As a specific part of urban planning, ICT innovation must include numerous stakeholders, may they be politicians or ICT professionals from the city municipality, local provider businesses and organizations, or local users (citizens, businesses, tourists) that influence the processes of ICT innovation in a city. A vital local ICT user and provider community helps to implement and quickly adopt the latest technology within the city but also to refinance municipal investments in technology innovation and thus contribute to the economic growth of a city.

INDICATORS: ICT POLICY AND BUSINESS

ICT-1.1: Existence of a road map for long-term ICT evolution in the city

ICT-1.2: Percentage of municipal investments in innovative local ICT projects as compared to the total investments (in the city infrastructure)

ICT-1.3: Municipal investment in ICT development relating to explicit objectives from city development plans as compared to the total amount of municipal ICT investments

ICT-1.4: Percentage of private investments in innovative local ICT projects as compared to the total investments in such projects

ICT-1.5: Share of the ICT sector in the economy of a city (% of GDP)

ICT-1.6: Share of workplaces in the ICT sector as compared to the total number of jobs in a city

Indicator group 2: ICT implementation rate and services adoption. In a smart city, ICT is implemented to realize a broad variety of purposes that relate to the city and its inhabitants. First, ICT can be used to optimize and automate typical urban processes. One example is a smart traffic system that helps avoid traffic jams and at the same time fosters the use of the public transport system. Furthermore, ICT is used to support innovation through the creation of new products, services, and markets (e.g., trading energy in neighborhoods). Other technologies aim at raising the empowerment of citizens, businesses, and organizations by providing information transparency and participation options. For example, open data portals allow for the development of services in widespread ecosystems including large ICT providers and small- and medium-sized enterprises and in a highly dynamic way. Last but not least, ICT plays an important role in governing a city by collecting and providing relevant data necessary for long-term urban planning and decision making and for control of sustainability.

Smart city services provide added value and a positive user experience to businesses, organizations, authorities, citizens, tourists, and other users. Services that are perceived as safe, simple, and user friendly will lead to a higher overall acceptance and adoption rate of ICT infrastructures and ICT services. With the growing use of smartphones and other mobile network devices, the demand for ad hoc access and service continuity is rising and the provision of these technologies is increasingly anticipated as normal.

INDICATORS: ICT IMPLEMENTATION RATE AND SERVICES ADOPTION

ICT-2.1: Percentage of individuals using local ICT services/information services via mobile broadband Internet every day or almost every day (private and work)

ICT-2.2: Percentage of individuals trading (selling or buying) goods and services in local malls or on service platforms

ICT-2.3: Percentage of local enterprises routinely exchanging data with public authorities via the Internet

ICT-2.4: Percentage of municipal procurements managed via the Internet

ICT-2.5: Number and usage rate of local health online services fostering access to health care, improved quality of care (e.g., health monitoring), illness prevention, and health-care efficiency

ICT-2.6: Number and usage rate of online services relating to the local public transportation system (e.g., real-time timetables, commute planning assistants, car-sharing services)

ICT-2.7: Number and usage rate of online services relating to the local traffic systems (e.g., fostering traffic jam prevention, street safety, environmental protection, parking lot finders)

ICT-2.8: Number and usage rate of transactional* online services fostering data exchange between local governments, citizens, businesses, and other organizations

---

* Transactional e-Government services are e-Government services on a higher maturity level. They are directly integrated in processes of the municipal government and thus go far beyond a pure information provision or the download of documents. Transactional services, such as the German tax return software ELSTER, provide direct interaction between the client of a citizen/enterprise and a relating server of the municipal government (http://bura.brunel.ac.uk/bitstream/2438/4486/1/250740082c.pdf).

ICT-2.9: Number of e-Government services based on open databases from the public sector

ICT-2.10: Share of households equipped with a smart metering solution

ICT-2.11: Share of households connected to a local smart grid

Indicator group 3: ICT connectivity and integration. Urban ICT usage is based on a high-level connectivity. Connectivity relates to the municipal infrastructure itself by mounting resilient, secure, and data protecting ICT systems to the physical infrastructures (buildings, streets, electricity, public transportation vehicles, etc.). The term connectivity also relates to citizens or tourists that ubiquitously connect through their mobile or static devices. It finally relates to vehicles, objects, and assets owned by citizens and businesses as well as by the municipality. Since connectivity always aims at the provision and the exchange of data and information, a concrete notion of data protection and the management of sensitive data are necessary.

The growing amount of enabling devices and platforms as well as the huge volume of data emerging from the physical world calls for the integration of data and information from various subsystems. For example, monitoring noise, pollution, structural health, and waste management demands data integration from different subsystems of an urban ICT infrastructure. These subsystems have grown into silos in the past. Data integration not only is a question of data protection and controlled data access but also puts numerous challenges on the stability and the performance of the system as well as on the data management (consistency, integrity, redundancy).

INDICATORS: CONNECTIVITY AND INTEGRATION

ICT-3.1: Percentage of individuals accessing the Internet through mobile devices every day or almost every day

ICT-3.2: Density of free Wi-Fi hotspots provided by the city

ICT-3.3: Number of sensors and cameras that are integrated into the physical infrastructure of the city*, per inhabitant

ICT-3.4: Amount of data streaming into municipal systems by time unit (MB/s or TB/s)

ICT-3.5: Share of municipal applications that aim at integrating data from various municipal subsystems/total amount of municipal applications

ICT-3.6: Percentage of local authorities and public service organizations that exchange data or access shared and integrated data pools

---

* This encompasses static elements (buildings, streets, street lamps) as well as mobile elements of the city infrastructure (city buses, metro trains).

The indicators will be used in an analysis of selected cities and their ICT strategies as part of the Fraunhofer Morgenstadt initiative [52].

6.8 Conclusion

Smart cities are to be addressed by cross-sectorial, multidisciplinary approaches and are to be understood as a continuous process of improvements in urban management.

This chapter discussed the central role of ICT for smart cities. While ICT itself does not make a city smart, there is no smart city without a strategic positioning of ICT and corresponding substantial investment.

A generic architecture for ICT of a smart city was presented, consisting of the city network infrastructure that provides ubiquitous access to environmental, urban management, supply and disposal, etc., data in raw, and machine-processable forms. The city network infrastructure reports into the city data cloud in which data are filtered, stored, aggregated, and processed into information and value-added services. These services are visualized and presented by dedicated urban applications to end users in business, administration, science, education, etc. In particular, value-added services for specific user groups become possible by the use of real-time information that is adjusted to the preferences and requirements of the users. The actions in response to the various needs can then be realized by interconnecting the different actors by means of ICT.

Subsequently, open data as a driving force for smartness in cities were discussed. Recent work in Berlin, in other European cities such as Helsinki [31], and broadly across Germany was presented. In September 2011, the first German Open Data Portal was launched for Berlin. A study for Berlin [25] about the status and potential of open data for the city in early 2012 was followed by a similar study for Germany [38] in summer 2012. The German study is more exhaustive in terms of quantitatively analyzing open data candidates, discussing legal issues along the data life cycle, and providing options for the operation of a German platform [32]. This study is the foundation for the ongoing pilot realization of that German platform. The German platform will offer data sets from German ministries, states, and municipalities as well as from domain-specific offers on statistical, environmental, and geo-data. It will result in a joint cross-level, cross-disciplinary platform that provides a centralized access to various open data offers in Germany. In a first phase, existing data offers from, for instance, Bremen or Berlin and from PortalU (environmental data sets) or GDI-DE (geo-data sets) are brought together into this German platform.

Finally, 23 ICT-related indicators for smart cities were presented. They were classified into indicators for ICT policy and business, for ICT implementation rate and services adoption, and for ICT connectivity and integration. These indicators can be used for the determination of ICT readiness of selected cities worldwide in order to measure the adoption of smart city concepts, and they will be evaluated as part of the Fraunhofer Morgenstadt initiative [52].

The concepts and approaches described in this chapter already resulted in new deployed solutions (e.g., in Berlin [27]) and will be further elaborated as part of the Fraunhofer FOKUS strategy toward ICT for smart cities [24].

Acknowledgments

This work was in part sponsored by Fraunhofer, under its Berlin 2030 and its Morgenstadt program. Any opinions, findings, and conclusions or recommendations are those of the authors and do not necessarily reflect the views of Fraunhofer FOKUS.

References

1. Roger W. et al.: Walshok transforming regions through information technology: Developing smart counties in California. http://www.smartcommunities.org/cal/articles.htm (as of December 23, 2012).
2. Koo J. et al.: Design status analysis for development of U-city education and training course, *JDCTA*, 3(1) (2009), 40–45.
3. Komninos N.: *Intelligent Cities: Innovation, Knowledge Systems and Digital Spaces.* Spon Press, London, U.K., 2002.
4. Ergazakis K. et al.: Towards knowledge cities: Conceptual analysis and success stories, *Journal of Knowledge Management*, 8(5) (2004), 5–15.
5. Gabbe C.: Bridging the digital divide in public participation: The roles of infra structure, hardware, software and social networks in Helsinki's Arabianranta and Maunula. Master thesis of urban planning, University of Washington, Seattle, WA, 2006.
6. Jones A.: *Ideopolis: Knowledge City-Regions.* The Work Foundation, London, U.K., 2006.
7. Gustafson P.: Digital disruptions: Technology innovations powering 21st century business. *Environmental Information Symposium 2008 "Transforming Information Into Solutions,"* December 10–12, Phoenix, AZ, 2008.
8. IBM Smarter Cities Conference (2009): Conference web site, Berlin, Germany, June 23–34, 2009, http://www.ibm.com/ibm/ideasfromibm/us/smartplanet/cities/index2.shtml (as of December 23, 2012).
9. Siemens City of the Future: http://www.telegraph.co.uk/sponsored/technology/siemens/6255585/The-city-of-the-future.html (as of December 23, 2012).

10. Siemens Annual Report 2009 How can we ensure sustainability while generating profitable growth?, 2009, http://www.siemens.com/annual/09/en/our_world/urbanization.htm (as of December 23, 2012).
11. Deutsche Telekom T-City Friedrichshafen: http://www.friedrichshafen.de/unsere-stadt/t-city/ (as of December 23, 2012).
12. SmartCity: http://www.smart-city.eu (as of December 23, 2012).
13. Living Labs: http://www.openlivinglabs.eu (as of December 23, 2012).
14. EuroCities initiative: http://www.eurocities.eu (as of December 23, 2012).
15. EIT ICT Labs: http://www.eitictlabs.eu (as of December 23, 2012).
16. Ina Schieferdecker (Eds.), Acatech, Germany: *Smart Cities Recommendation for Action*, January 2011, http://www.acatech.de/de/publikationen/stellungnahmen/kooperationen/detail/artikel/smart-cities-deutsche-hochtechnologie-fuer-die-stadt-der-zukunft.html (as of December 23, 2012).
17. ITU's Digital Cities initiative: www.itu.int/net/itunews/issues/2010/04 (as of December 23, 2012).
18. Massachusetts Institute of Technology (MIT), Cambridge, MA: Initiative on smart cities http://cities.media.mit.edu/ (as of December 23, 2012).
19. Intelligent Community Forum: http://www.intelligentcommunity.org/ (as of December 23, 2012).
20. Post-fossil living, Japan: http://www.deseretnews.com/article/700071721/Japan-looking-to-sell-smart-cities-to-the-world.html?s_cid=rss-5 (as of December 23, 2012).
21. Engaging citizens, Seoul, Korea: http://www.itu.int/net/itunews/issues/2010/04/40.aspx (as of December 23, 2012).
22. Sensing China: http://www.iot-week.eu/Iot-week-2011/presentations/presentations-iot-week-2011/2011-06-06/Jian_Sensing_China_Center.pdf (as of December 23, 2012).
23. IBM: Smarter planet http://www.ibm.com/smarterplanet/us/en/sustainable_cities/ideas/ (as of December 23, 2012).
24. Fraunhofer FOKUS ICT: For smart cities research portal. http://www.ict-smart-cities-center.com/ (as of December 23, 2012).
25. W. Both and I. Schieferdecker: Berlin open data strategy: Organisational, lawful and technical aspects of open data in Berlin—Concept, pilot system and recommendations. Fraunhofer Verlag, Berlin, Germany, 2012, online available in German on: http://www.berlin.de/projektzukunft/fileadmin/user_upload/pdf/sonstiges/Berliner_Open_Data-Strategie_2012.pdf (as of December 23, 2012).
26. Internet & Gesellschaft Co:llaboratory: Offene Staatskunst—Bessere Politik durch Open Government ? Abschlussbericht (in German), 1. Auflage, Berlin, Germany, 2010.
27. Berlin Open Data Platform: http://daten.berlin.de (as of December 23, 2012).
28. Kenya Open Data Platform: https://opendata.go.ke/ (as of December 23, 2012).
29. US Open Data Portal: http://www.data.gov/ (as of December 23, 2012).
30. Open Data Portal of the United Kingdom: http://www.data.gov.uk/ (as of December 23, 2012).
31. Helsinki Open Data Portal: http://www.hri.fi/fi/ (as of December 23, 2012).
32. German Open Government Portal: http://daten-deutschland.de/ (as of December 23, 2012).

33. Evanela L. et al.: Identification and utilization of components for a linked open data platform, in *Proceedings of the 1st IEEE International Workshop on Methods for Establishing Trust with Open Data (METHOD 2012) IEEE 36th Annual Computer Software and Application Conference Workshops (COMPSACW)*, 2012 (to appear).
34. Open Data Platform of Fraunhofer FOKUS: https://github.com/fraunhoferfokus/opendata-platform (as of December 23, 2012).
35. Tcholtchev N. et al.: On the interplay of open data, cloud services and network providers towards electric mobility in smart cities, in *Proceedings of the International Workshop on GlObal Trends in Smart Cities 2012*, October 22–25, 2012, Clearwater, FL, co-located with *IEEE LCN 2012* (to appear).
36. Directive 2003/98/EC of the European Parliament and of the Council of 17 November 2003 on the re-use of public sector information, Official Journal of the European Union 31.12.2003.
37. Creative Commons, CC-baseline rights: http://wiki.creativecommons.org/Baseline_Rights (as of November 9, 2012).
38. Klessmann J. et al.: Open government data Germany: Short version of the study on open government in Germany, Federal Ministry of the Interior, Berlin, Germany, online available: http://www.bmi.bund.de/SharedDocs/Downloads/DE/Themen/OED_Verwaltung/ModerneVerwaltung/opengovernment_kurzfassung_en.pdf (as of December 23, 2012).
39. Klessmann J. and Both, W.: Open cities—EU project on open data and crowd-sourcing. *eGov Präsenz*. (1) (2012), S76–S77.
40. Comprehensive Knowledge Archive Network: http://ckan.org/ (as of December 23, 2012).
41. Vattenfall/Fraunhofer FOKUS: Energy network data. http://netzdaten-berlin.de/ (as of December 23, 2012).
42. Schaffers H. et al.: Smart cities and the future Internet: Towards cooperation frameworks for open innovation. http://www.springerlink.com/content/h6v7x10n5w7hkj23/fulltext.pdf?MUD=MP (as of December 23, 2012).
43. i2010 High Level Group: Benchmarking digital Europe 2011–2015. A conceptual framework. http://ec.europa.eu/information_society/eeurope/i2010/docs/benchmarking/benchmarking_digital_europe_2011-2015.pdf (as of December 23, 2012).
44. Caragliu A. et al.: Smart cities in Europe, in *3rd Central European Conference in Regional Science—CERS*. 2009. http://www.cers.tuke.sk/cers2009/PDF/01_03_Nijkamp.pdf (as of December 23, 2012).
45. Microsoft: The smart city. Using IT to make cities more livable. http://www.microsoft.com/global/sv-se/offentlig-sektor/PublishingImages/The%20Smart%20City_Using%20IT%20to%20Make%20Cities%20More%20Livable.pdf (as of December 23, 2012).
46. International Telecommunication Union (ITU): Key ICT indicators for developed and developing countries and the world. http://www.itu.int/ITU-D/ict/statistics/at_glance/keytelecom.html (as of December 23, 2012).
47. UNESCO: Partnership on measuring ICT for development. Core ICT indicators 2010. http://www.uis.unesco.org/Communication/Documents/Core_ICT_Indicators_2010.pdf (as of December 23, 2012).
48. United Nations: Core ICT indicators. http://www.itu.int/ITU-D/ict/partnership/material/CoreICTIndicators.pdf (as of December 23, 2012).

49. Centre of Regional Science Vienna: Smart cities—Ranking of European medium-sized cities. http://www.anci.it/Contenuti/Allegati/Ranking%20EU%20citt%C3%96%20smarts.pdf (as of December 23, 2012).
50. Angoso J.: Rivas-Vaciamadrid—An example of energy efficiency in broadband smart cities. Smart cities—The ICT infrastructure for Ecoefficient cities, in *High Level OECD Conference ICTs, Environment, Climate Change*. Denmark, 27–28th May 2009. http://cdn.netamia.com/itst/ict2009/jose_luis_angoso_smart_buildings_rivas-vaciamadrid_municipality_a_spanish_case_for_energy_management_and_efficiency-202-1244189658.pdf (as of December 23, 2012).
51. Schulte M.A.: IDC smart cities benchmark. http://www.computacenter.de/services_solutions/branchen/IDC_Smart_Cities_Benchmark.pdf (as of December 23, 2012).
52. Fraunhofer Morgenstadt Initiative: Project web site, http://www.morgenstadt.de (as of December 23, 2012).
53. Fraunhofer FOKUS: OpenMTC platform. A generic M2M communication platform. White paper, November 2012. http://www.open-mtc.org/_files/WhitePaper-OpenMTC.pdf (as of December 23, 2012).
54. SAP: Urban matters initiative. http://www.sap.com/asset/index.epx?id=98222eb4-139b-4086-8b41-4d6c27a0ffab (June 2013).

Part III

Computational Discovery
and Ingenuity

7

Domain Science &* Engineering: A Foundation for Computation for Humanity

Dines Bjørner

CONTENTS

---

* We use the connection & instead of the more conventional *and*. The reason can be explained as follows: *A and B*, to us, signals that two topics, *A* and *B* are covered. whereas *A & B* signals that there is one topic named by the composite *A & B*.

7.1 Introduction

This section will discuss what might be meant by *computation for humanity*, what we mean by a domain and a domain description, and how domain science & engineering otherwise fits into the narrower context of software engineering.

7.1.1 What Can We Mean by *Computation for Humanity*?

There does not seem to be a generally accepted characterization of what might be meant by *computation for humanity*. We shall therefore examine reasonably broadly accepted characterizations of the term *humanity*, while focusing on those aspects that might be supported positively or negatively by computing, in particular on [3–5].

7.1.1.1 Humanity

Reasonably established uses of "humanity": We shall enumerate the characteristics of the *virtue* facet of *humanity*, one of six core virtues in [5]: *wisdom and knowledge* (strengths that involve the acquisition and use of knowledge)— such as creativity, curiosity, open-mindedness, love of learning, perspective and wisdom; *courage*—such as bravery, persistence, integrity and vitality; *humanity*—such as love, kindness and social intelligence; *justice*—such as active citizenship, social responsibility, loyalty, teamwork, fairness, and leadership; *temperance*—such as forgiveness and mercy, humility and modesty, prudence and self-regulation and self control; and *transcendence*—such as appreciation of beauty, appreciation of excellence gratitude, hope, humor and playfulness, and spirituality.

Maslow [3,4] proposed a hierarchy of human needs; from the *physiological,* basic needs of breathing, food, water, sex, homeostasis, and excretion; via the *safety and security* needs of body, employment, resources, family, health, and property; *love/belong* needs of friendship, family, and sexual intimacy; and *esteem* needs of self-esteem, confidence, achievement, respect of others, and respect by others; to *self-actualization* needs of morality, creativity, spontaneity, problem solving, lack of prejudice, and acceptance of facts.

Hijacked uses of "humanity": The earlier enumerations are the result of *humanities* studies. That is, of primarily analytical, critical, or speculative investigations. There are other speculative studies that purport to hinge on and represent issues of *humanity*. For example, human rights,*

---

* http://en.wikipedia.org/wiki/Human.rights, http://www.hrw.org/

humanity in action,* humanism,† etc. We somewhat controversially label these uses as being hijacked: there are the concepts listed in the previous paragraphs, and we consider these *universal*, and then there are the *movements* that, in one way or another, capitalize on these concepts, interpret them almost politically, religiously (or anti-such), etc. In our quest for *computation for humanity*, we shall try avoiding *hijacking* while striving for *universality*.

7.1.1.2 Computation: From Sciences to E-Government

In the early-to-mid-1940s, computations were focused on verifying or predicting physical behaviors‡: at the core of such computations were models of physics (including chemistry) and the computations either had as purpose to verify proposed models and, once these models could be trusted, then to predict physical (including chemical) behaviors.

The computational physics models had to satisfy exact mathematical models, and these models were not proprietary.

But the market for computation lay neither in government laboratories nor in the weapons industry but in commercial enterprises: initially banks and insurance companies, then in production (manufacturing) and in monitoring & control (e.g., the gas and oil industry), then in transport, logistics, etc., and in e-commerce, the Internet, the web, etc., public administration computation grew and e-government computation has arrived.

The computational models should satisfy some exact models since computations lure one to believe in their precision, but usually many such models, increasingly in public administration and in e-government in general, are implicit: have never been precisely formulated.

7.1.1.3 Computation for Humanity

We shall now put forward a reasonably objective and non-controversial interpretation of humanity. That interpretation is not complete; it does not claim to cover *as many*, let alone *all* aspects of humanity, that is simply not possible.

We shall now summarize Section 7.1.1.1's enumerations of issues that characterize a concept of *humanity*. We select a subset of the facets of humanity that may be supported by computation.

---

* http://www.humanityinaction.org/
† http://en.wikipedia.org/wiki/InternationalHumanist_and_EthicalUnion, an umbrella organization embracing humanist, atheist, rationalist, secular, skeptic, free-thought, and ethical culture organizations worldwide.
‡ The *Manhattan* program mandated such computations and John von Neumann thence conceived of the *von Neumann* computer architecture.

From the Peterson and Seligman's "Character Strengths & Virtues" [5] enumeration: *wisdom & knowledge*—creativity, curiosity, love of learning, perspective, wisdom; *humanity*—social intelligence; *justice*—active citizenship/ social responsibility/loyalty/teamwork, leadership; *temperance*—modesty, prudence; *transcendence*—appreciation of beauty, appreciation of excellence.

And from Maslow's "Hierarchy of Human Needs" [3,4]: *physiological*— food, water; *safety & security*—resources, health; *esteem*—confidence, achieve-ment, respect of others; *self-actualization*—creativity, problem solving, and acceptance of facts.

By computation for humanity, we shall mean any such computation that supports one or another or a combination of the concepts appearing in the two lists of this section.*

How domain science & engineering may support the concepts of human-ity of this section will be indicated as we now turn to the issue of what a domain description is.

7.1.2 What Is a Domain Description?

We first give a brief list of domain names. In doing so, we rely on the reader's familiarity with what these names stand for. Then we give a *summary* defini-tion of what we mean by *domain*. And finally we survey important ingredi-ents of a domain description.

7.1.2.1 Example Domains

Already in the abstract we listed names of some domains. We repeat and augment this list as follows: air traffic, banking, bus transport, consumer commerce, container lines, food processing, freight transport, health care, manufacturing, pipelines (gas, oil), railway systems, rail transport, resource management, the web, water supply & processing, or components thereof. Common to all of the earlier areas is that they involve humans in interaction with other humans and within man-made systems constructed, operated, and maintained by humans.

We shall, in Section 7.2, present rough sketch outlines of descriptions of some of these domains. In that connection, we shall then relate domain descriptions and hence software derived from domain descriptions to some of the issues listed in Section 7.1.1.3.

7.1.2.2 Domains

By a domain, we mean the observable phenomena of human-based—part, action, event, and behavior—entities that *make up* the domain, and where human behaviors interact with other human behaviors and with *other processes*.

---

* Or in lists *of that kind* and otherwise generally accepted.

7.1.2.3 Domain Descriptions

By a domain description, we mean an orderly set of statements that detail the observable or abstracted properties of parts, actions, events, and behaviors of the domain, both informally, in narrative form, and formally—thereby enabling verifiable theories about the domain.*

Domain descriptions tend to abstract: instead of describing specific instances (values) of entities, one describes classes (types) of these. Thus domain descriptions abstract concrete entities into concepts.

We shall soon give examples of domain descriptions.

Domain descriptions can, from both a pragmatic and a theoretical point, be composed of descriptions of at least six kinds of phenomena and concepts: intrinsics, support (technologies), rules & regulations, scripts, management & organization, and human behavior.

Intrinsics: By intrinsics, we mean those phenomena and concepts without which nothing can be described, that is, those phenomena and concepts that are basic to everything in the domain. *Example* intrinsics of a road transport system are thus: (parts) the road net, street segments, street intersections, vehicles, drivers, and passengers, etc.; (actions) inserting and removing vehicles, respectively, into and from a road net and starting, accelerating, decelerating, and stopping vehicles along street segments and around street intersections, etc.; (events) vehicle crashes, etc.; (behaviors) driving a vehicle from point A to point B, etc.

Support technologies: By support technologies we mean those, primarily, technologies that support parts, actions, events, and behaviors of the domain. *Example* support technologies for a road transport system are street signals, level railway crossing gates, and parking meters.

Rules & regulations: By *rules & regulations*, we mean a set of pairs of rules and regulations pertaining to a well-delineated set of domain parts. For a domain, there can be many such sets. *Rule*: By a rule, we mean the prescription of a predicate over pairs of domain states (i.e., designated domain parts), a *before* and an *after* state, which must hold for any given action, event, or behavior; if not, the rule is said to have been violated. *Regulation*: By a regulation, we mean the prescription of a set of one or more actions that should be applied if a corresponding rule has been violated—with the effect of creating a new pair of *before/after* states in which the rule now holds.

Example rules & regulations for road traffic are thus: speed—rules for speed limits per road type and fine regulations if caught speeding; travel direction—stop for red traffic signal and a fine regulation if caught crossing while the red signal is lighted.

---

* There are interesting relations between *domain descriptions* and *ontologies*. The latter, in a sense, is contained within the former, but the former goes well beyond accumulating properties (of mostly endurants).

Scripts: By a script we mean a (structured) set of rules and regulations to be respectively applied and (conditionally) enacted according to the set structure. An *example* script could be the set of questions to be asked of and the actions to be carried out with respect to a potential hospital patient for the anamnese, analysis, diagnosis, and treatment plan when this patient is first admitted to a hospital.

Management & organization: We analyze the conjoined (hence the &) concept of *management & organization* into its three components: *management*, *organization*, and *management & organization*. *Organization*: By organization we mean an iterative partitioning of resources, and, where applicable, their related behaviors, into some (hierarchical, matrix or other) structures. An organization *example* is the divisioning of an automobile company into personal car, truck, motor, transmission, parts, divisions, etc., and some of these again into continent/country divisions. *Management*: By management, we mean the monitoring and control of resources (i.e., parts): human staff, time, monies, and physical plant and materials—where monitoring & control implies that staff and/or machines, as behaviors, are charged with the performance of actions while obeying rules & regulations and scripts. A management example could be the overall management of the main divisions of an automobile company with its directives and lines of monitoring and control (including change) of the organizational structure and strategies, tactics, that is, reactions to or preparations for market conditions, and operational handling of, for example, the day-to-day business. *Management & organization*: By management & organization, we mean that management primarily follows the organizational structure but that some rules & regulations and/or scripts mandate deviation from the organizational structure.

Human behavior: By human behavior, we mean the manner in which a person conducts herself with respect to performing (or not performing) mandated or expected actions and to respond to expected or unexpected events. The behavioral description usually includes descriptions that reflect a spectrum from diligent via sloppy and delinquent to outright criminal behavior.

7.1.3 Role of Domains in Software Development

Three phases of software engineering: Computations are usually based on computer software. Before software can be designed, one must understand its *requirements*: what is expected from the software, not how it operates. Before *requirements* can be expressed, one must understand the underlying *domain*: not how it operates, but its properties.

From this *S, R, D* triplet, we conclude that we must first do some *domain science* & engineering, then *derive* the *requirements* before we finally design the *software*.

Yes, there are rather precise methods for *discovering*, based on a domain description and in collaboration with stakeholders of the domain, a significant part of the desired requirements.

"Deriving" requirements from domain descriptions: We review some of the (almost) algebraic operations that the domain cum requirements engineers perform in collaboration with both domain and requirements stakeholders. We do so in order to better understand how both domain and requirements engineering can contribute to *computing for humanity*. We shall overview some of the (almost) algebraic operations, namely, projection, instantiation, determination, and extension.

Projection: By the projection of a domain, we mean the removal of, from the domain description, those part, action, event, and behavior descriptions that are deemed irrelevant for expressing the requirements. The result of the projection is a requirements prescription. An *example* projection arises for a road net building application in which we are not concerned with vehicle traffic and hence also not with drivers and passengers. So we remove vehicles, drivers, and passengers. Another *example* projection arises for a road pricing application in which we are not concerned with drivers and passengers. So we remove drivers and passengers.

Instantiation: By instantiation, we mean the refinement* of parts, actions, events, and behaviors: where before an entity specification allowed many, for example, part compositions, an instantiated specification usually is *less flexible*, suggesting a more specific entity. An *example* instantiation arises for a toll way application: has road nets as orderly, finite sequences of pairs of one-way toll road segments between adjacent toll road intersections which later one can also access or leave through two-way plaza road segments from or to toll plaza *intersections*.

Determination: By determination, we mean the refinement of non-deterministic choices into less non-deterministic, including only deterministic choices as concerned parts, actions, events, and behaviors. An *example* determination arises for a toll way application in which the state of toll roads always and only allow traffic only and always in one direction, in which the state of toll plaza roads always allow traffic both directions, and in which the state of toll road intersections always allow traffic from any toll [plaza] road incident upon the intersection to any toll [plaza] road eminent from the intersection.

Extension: By an extension, we mean the inclusion of a further domain description of phenomena and concepts that were not feasible without computation. An *example* extension is a road pricing arrangement whereby car

---

* Refinement is understood as a concretization of the specification, that is, from a more abstract specification is obtained a more concrete one, one that is *closer* in some sense to how one might specify the software.

movements are tracked and records kept on car routes from which payment can then be calculated.

Discussion: We have included the earlier aspects of requirements engineering for the following reasons. The listed requirements operations are all described with no reference to specific computation concepts and hence invite *technology-free* creativity, innovation, and problem solving all of whom contribute toward satisfying some of the *humanity* criteria.

7.2 Informal Description of Some Domains

We shall hint at some domains.

7.2.1 Banking System

We describe a generic class of main street banks, that is, banks that service ordinary citizens, hold demand/deposit accounts with credit limits, and offer mortgages (and loans), hence with mortgage cum loan accounts, repayments, etc. We abstract away from branch offices, that is, a bank with all its branch offices appears as one unit, yet there may be several banks serving several communities.

7.2.1.1 Intrinsics

Parts and their attributes: We exemplify a narrative description. We itemize this description. Had we also shown the omitted formal description, then we would have enumerated the narrative so as to allow easy cross-reference to formulas.

1. A banking system is here delineated by
   a. A set of (uniquely identified) clients
   b. A set of (uniquely identified) banks
   c. A banking *watchdog*: an authority that monitors banking practices, etc.*
   d. Client and bank identifications are further undefined

We shall focus, in this chapter, just on the client aspects of banks.

---

* The watchdog authority is: in the United States, FEC—Federal Exchange and Securities Commission; in the United Kingdom, FSA—Financial Services Authority; and in Germany, BaFin—Bundesanstalt fr Finanzdienstleistungsaufsich.

2. A client is here considered an atomic part with the following attributes:

   a. Name

   b. Addresses (physical and electronic)

   c. National identifier

   d. A set of zero, one, or more demand/deposit bank account numbers (of accounts held)

   e. A set of zero, one or more mortgage account numbers (of outstanding mortgages), such that each mortgage account is outstanding against exactly one client, etc.

One observes that a demand/deposit bank account may be shared between two or more clients. It is assumed that all accounts listed do indeed exist.

3. A bank

   a. Is uniquely identified (an attribute), and holds, as separate parts

   b. A cash register

   c. Zero or more clients, each of which is an atomic part

   d. Zero or more client demand/deposit accounts, each of which is an atomic part

   e. Zero, one, or more mortgage accounts, each of which is an atomic part

4. The cash register holds coins and bank notes.

5. A client demand/deposit account (which is an atomic part) has the following attributes:

   a. A unique identifier

   b. A balance that is a money designation above a credit limit

   c. An interest rate to be paid quarterly to the bank on averaged client demand/deposit account balances between a possibly negative balance credit limit and zero money units

   d. A yield rate to be paid quarterly by the bank into the account when averaged client demand/deposit account balances are above zero money units

   e. A list of transaction designators

   f. A date when established, etc.

We observe that each individual client demand/deposit account has its own interest and yield rates. Should a particular bank not offer differentiated client demand/deposit account interest and yield rates then that is modeled by all the individual client demand/deposit accounts having identical interest and yield rates.

6. A client mortgage (cum loan) account (which is an atomic part) has the following attributes:

   a. A unique identifier

   b. A repayment balance

   c. An original mortgage (loan) amount

   d. An interest rate

   e. A repayment schedule

   f. A repayment fee

   g. A deed (referring to some property held by the client as security against the mortgage [loan])

Actions: The following actions are performed on demand/deposit accounts:

7. Open demand/deposit account, where

   a. A client states the bank name, the client name, addresses and national personal identifier, and a cash deposit amount with which to open the account.

   b. That bank provides that client with an account number, interest, and yield rates and establishes the account balance and a credit limit while recording this time-stamped transaction as the first of that new account's transaction list.

8. Deposit money where

   a. An identified client states the bank name, an appropriate account, and a cash amount to be deposited.

   b. The bank accepts the amount, that is, increments of the identified account's balance accordingly while recording this time-stamped transaction as the most recent of that account's transaction list.

The aforementioned descriptions can be made more precise.

We omit description of further demand/deposit transactions, for example,

9. Withdraw cash

10. Transfer monies

11. Open share count

12. Close share account

13. Request transaction statement

14. Close demand/deposit account

The following actions are performed on mortgage/loan accounts for which we also omit more suggestive descriptions:

15. Negotiate mortgage/loan
16. Open mortgage/loan account
17. Mortgage/loan repayment
18. Default mortgage/loan repayment
19. Close on mortgage/loan account

Events

20. Overdraft on demand/deposit account where
    a. A cash withdrawal (Item 9) exceeds the balance and credit limit on the account.
21. Bankruptcy where
    a. The bank's outstanding debt exceeds its cash and credits.

Behaviors

22. Demand/deposit account behavior as a series of zero or more cash deposits, cash withdrawals, transfers, open and close share accounts, and transaction statement requests, in any order, prefixed and suffixed by, respectively, open and close account transactions—all of these with respect to a specific account and interwoven with possible events.
23. Mortgage account behaviors as a series of zero or more mortgage repayment, and default mortgage repayment prefixed by a pair of negotiate mortgage and open mortgage account, and suffixed by a close on mortgage/loan account transaction—all of these with respect to a specific account and interwoven with possible events.
24. Client behavior as a series of zero or more *interwoven* demand/deposit account behavior and zero one or more mortgage account behaviors—with respect to the same client.
25. Bank behavior as a series of *interwoven* client behaviors—over all of its clients—together with internal bank transactions—so far not mentioned: calculate demand/deposit interests, calculate demand/deposit yields, calculate mortgage interests, and interwoven with possible events.

7.2.1.2 Support Technologies

We omit careful descriptions of *standard* support technologies such as

26. Credit cards
27. Automated teller machines
28. E-banking
29. Etc.

7.2.1.3 Rules & Regulations

We omit careful descriptions of rules & regulations such as

30. Related to overdraft
31. Default on loans
32. Bankruptcy
33. Etc.

7.2.1.4 Scripts

We omit careful descriptions of scripts such as

34. Calculation of interest or yield
35. Calculation of loan repayments
36. Estimation of loan risks
37. Etc.

Note that the borderline between the concepts of *rules & regulations* and of *scripts* can be fuzzy.

7.2.1.5 Management & Organization

We omit careful descriptions of management & organization matters such as

38. Main versus branch offices
39. Bank teller versus *back office* staff
40. Signatory rights (say on loans)
41. etc.

7.2.1.6 Human Behavior

We omit careful descriptions of human behaviors such as

42. Interest or yield calculation
43. Loan repayment calculation
44. Loan risk assessment
45. Etc.

7.2.1.7 Discussion

We have just barely rough-sketched a partial narrative bank description. A proper bank description would focus on each of the numbered items and formulate these with far more precision than shown here. And a proper bank description would be *paired* with a formal description, numbered item by numbered item. The formalization would follow the narrative wording very closely. The formalization would then be the basis for stating and proving theorems, that is, laws of banking, such as seen by the entire banking plus client domain, by any individual bank, and by any individual client.

7.2.2 Health Care System

Another domain is *health care*.

Health-care systems can be domain-described from a variety of rather distinct views. One could be singled out: the monitoring and control of flow of patients, medical staff, medical supplies, patient information, health-care management information, visitors, etc., within and between private physicians, hospitals, policlinics, pharmacies, health insurance companies, health industry monitors, etc.

Other, perhaps *orthogonal*, view-points are possible. But just with the one given, we can see the overall complexity. It seems a good idea to domain-describe the health-care sector. To give an example of a component of a *flow-oriented* description, we refer to one aspect of the patient/medical staff/ medical supplies/patient information flow, one that can be captured by the flow/state diagrams of Figure 7.1. We let the labels of the patient handling action rectangles and the medical staff diamond-shaped decision *box speak* for the interpretation of the diagrams.

So, instead of describing as a domain only this aspect of the health service as a domain, we indicate, to the reader, that a domain description could very well follow the lines of the banking domain description.

7.2.3 Other Domains

Over the years, many domain descriptions have been worked out. Some can be mentioned: airports, air traffic, banking, container lines, the consumer market, logistics, manufacturing (water, oil, gas, etc.), pipelines, railways, transport systems, stock exchanges, etc.

7.3 Humanities and Domain Science & Engineering

We now review the *products* of domain science & engineering in the light of our characterization of *humanity* in Section 7.1.1.1.

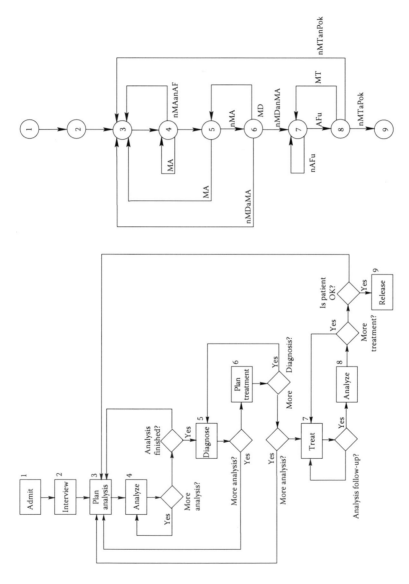

FIGURE 7.1
A hospitalization plan: flowchart and finite state machine.

7.3.1 Banking for Humanity

Let us review the special font terms in the first of the two enumerations of Section 7.1.1.3.

From the "character strengths & virtues" enumeration:

Wisdom & knowledge: studying (i.e., *learning*) a domain model for banking (creativity) should enable a bank employee and/or a software engineer to better create new banking products, and/or new computing supports for banking, respectively; (curiosity) provoke more in-depth studies of banking; (love of learning) if the domain description is of a reasonable high standard, its study should entice the reader towards further studies; (perspective) while giving those new in *banking* who understand the domain model a perspective they cannot have had before.

Humanity: understanding and possessing a domain model for banking (social intelligence) should significantly facilitate the human capacity to effectively navigate and negotiate complex banking relationships and environments.

Justice: understanding and possessing a domain model for banking (active citizenship/social responsibility/loyalty/teamwork) should allow this domain *expert* to better deploy her insight of these listed areas; and (leadership) thus show leadership.

Temperance: understanding and possessing a domain model for banking (modesty) (with all its complexities, even when expressed in an as simple manner) ought to instill some modesty with, for example, that domain expert's opinion of his own fallibility; and (prudence) enable that *expert* to govern and discipline by the use of reason.

Transcendence: (appreciation of beauty) a domain description, that is, a domain model, must be beautiful [2]—hence it must instill appreciation for beauty also in matters of banking; (appreciation of excellence) grasping a domain model should be easier than learning it *by osmosis*. The domain model should be an excellent piece of intellectual work.

7.3.2 Other Domains for Humanity

The observations of Section 7.3.1 apply, with suitable term changes, to basically any other domain. Understanding a well-described domain—in general—enhances, we claim, creativity, curiosity, love of learning, perspective, and wisdom; social intelligence, active citizenship/social responsibility/loyalty/teamwork, and leadership; modesty and prudence; and appreciation of beauty and excellence.

We now relate Maslow's "hierarchy of human needs" to domains in general.

Physiological: (Food) domain descriptions of eventually all aspects of the food chain and subsequent IT support for relevant facets should enable improved production, distribution, and sale of foodstuffs. Today there are no such comprehensive sets of related domain models for *all* sides of the food chain. There are many IT applications, but only where these can be commercially marketed. To develop domain models, for example, for the food chain, is a task, not of commerce or industry, but, it seems, of public, international research. We claim that once such part models emerge, new IT supports will be commercially feasible, that is, computations that *interface* smoothly with other such IT supports. (Water) The above paragraph (for food) can be rephrased, inter alia, for water.

Safety & security: Resource domain descriptions of eventually all aspects of the resources on which we, as humans, rely will emerge. We may distinguish between the following kinds of resources: economic (commodity, service, or other asset used to produce goods and services that meet human needs and wants); biological (a resource is defined as a substance or object required by a living organism for normal growth, maintenance, and reproduction); IT equipment, natural resources (land, water, air, minerals); labor, capital, infrastructure (basic physical and organizational structures needed for the operation of a society or enterprise); and intangible resources (corporate images, brands and patents, and other intellectual property—these exist in abstraction). Researching, teaching, and learning such domain models should empower the individual, already from youth, to better understand the role of resources in all their variety and to objectively know what factual problems of resources are: shortage, price, availability, and accessibility. Domain models of resources may be *implemented* in terms of computation* [1]. Computational models may, for example, demonstrate (*demo*) *non-renewable resource chains*: from their origin, via extraction, transport, and processing, to their eventual consumption. Computational models may simulate *what-if* scenarios. And computational models may calculate risks and benefits. We claim that *demo,* simulator, and calculator computations based on domain descriptions (with which the users of these computations are familiar) represent a significant advantage, that is, *can be far more human,* compared to computations not based on serious domain descriptions. Health: On the basis of domain descriptions, such as very sketchily hinted at in Section 7.2.2, we can claim the following. Studying and learning domain models of increasingly wider areas of the health service sector empowers ordinary citizen— and their elected politicians—to better understand, organize, monitor, and control that sector, thus enabling computations that are better founded, hence more human. The health sector and its current computerization are especially problematic. Most electronic patient journal systems thus fail in communicating with one another, and most patient hospitalization systems likewise.

---

* There are certainly many IT applications with respect to resources. But none or few are based on publicly available domain descriptions of the kind we are advocating, descriptions which, for example, can be the basis for secondary school education.

Esteem: The key attributes here are confidence, achievement, and respect of others. Researching, developing, studying, and learning domains through well-structured, scientifically objective domain descriptions should empower citizens, from researchers and developers via domain *owners* to domain users, to be far more confident in the understanding and use of that domain than when no such properly researched and documented understandings are available. At the same time, these citizens should achieve far better and technology-independent use of such domains and thereby be able to have a well-founded esteem for all stakeholders of the domain.

Self-actualization: The key attributes here are creativity, problem solving, and acceptance of facts. Researching, developing, studying, and learning domains through well-structured, scientifically objective domain descriptions give, we claim, those scientists and engineers who have researched and developed these domain descriptions hitherto unrivaled abilities to (business process) reengineer (i.e., create) several facets of these domains.* In this way, they can solve existing usually human relations domain problems. These researchers cum engineers can bring to those other domain stakeholders (who have studied and understood descriptions of domains [which would have been near-impossible to fully grasp without such domain descriptions]) a *peace-of-mind* acceptance of domain facts, while empowering them to better voice their possible dissatisfaction with existing domain practices.

7.3.3 Natural Sciences versus Domain Sciences

It is commonly accepted that we teach and that we are expected to learn and be reasonably capable in reckoning (i.e., calculating) and mathematics, and natural and life sciences: physics, chemistry, botanic, zoology, geology, biology, etc. To study logic and learn to reason logically is, however, not considered so necessary! We shall now plead that we must expand what is taught from primary school onward to also include studies of man-made domains.

The study of the natural and life sciences is motivated with such statements as "we must understand the world that surrounds us," "engineering is based on these sciences and, to stay competitive, we must study them," "health care is based on these sciences and, to survive, we must thus study them," etc.

The study of the domain sciences, that is, the study of domains such as the financial service industry,† transportation,‡ manufacturing (in its widest

---

* For example, new support technologies; new rules, regulations, and scripts; and new management & organization structures.
† Banking, insurance, portfolio management, the trading of securities instruments (stocks, bonds and other commodities (oil, gas, minerals, grain, meat, etc.)), etc.
‡ Road nets, bus, rail, ship and air transport, shipping, logistics, pipelines, etc.

sense*), the web, the Internet, etc., can be likewise motivated: "many aspects of the processes of man-made domains are hidden from the human eye, and thus they all too easily attain an undesirable 'air' of mystification,"† "many processes of man-made domains are prone to be 'process engineered' in such a manner as to reflect misuse of power of those who manage these processes over those whom these processes deal with," etc.

Therefore, in a free (read: human) society, it seems reasonable to also expect that we properly educate and train coming generations—from school—in the seeming complexities of their society.

Why should the natural and life sciences be taught and learned and not the domain sciences & engineering? To make full use of domain knowledge, we must also strengthen logical reasoning, that is, we must teach basic logic in school.

7.4 Conclusion

What have we achieved? We have introduced the notion of domain. First we have characterized its descriptional ingredients and then we have given an albeit very rough sketch example of a fragment of a banking domain and an even sketchier example of a fragment of a hospital domain. Before these examples, we have enumerated a number of concepts that we claim characterize one set of facets of *humanity*. There, surely, may be other such sets of facets. But these are the ones that we shall *bring to bear* on *domain sciences and engineering*. After the examples, we reviewed the selection of *humanity* concepts of Section 7.1.1.1 in light of the examples—and in general. This review is not a clear-cut *scientific* review. That is, it does not argue by way of (formal) logical reasoning. Instead it reasons by way of more-or-less persuasive statements. Two subject areas have been brought together, contrasted with one another: domain sciences & engineering, a reasonably *strict* discipline whose expressions (i.e., the domain descriptions) are reasonably precise, and *humanities*, a far more *fluent* discipline, whose enumerations, in Section 7.1.1.1, are not based on mathematics. If you think that *humanities* can (also) be characterized by the enumerations of Section 7.1.1.1 and can be supported by computation as sketched in Section 7.3, then domain science & engineering, as exemplified in Section 7.2, can contribute in significant ways to computations for humanity, otherwise our claims are just that: claims.

---

* Energy production and distribution, petrochemical industry, water supplies, iron ore to steel processing, metal working, electronics, etc.
† In *ye olde days* children could easily grasp the *industry* of their parents: father at the woodwork bench making tools for farming and furniture, mother at the weave, etc.

There is no one and only concept of *humanity*. The discipline area of formulating a concept of *humanity* is fraught with problems. It easily becomes a *battle ground* for political or other opinions and a *feel good position* to claim to be *human* or to speak for *humanity*. In this chapter, we have tried to view *computation for humanity* in a cool, detached manner, relying on reasonably noncontroversial (the enumerated) concepts. Whether we have succeeded is, partly, up to you to decide. At least we hope that your critical sense, when you are next confronted with the term *humanity*, has been alerted and primed differently from that before you read this chapter!

References

1. D. Bjørner. Domains: Their simulation, monitoring and control—A divertimento of ideas and suggestions. In *Festschrift for Hermann Maurer: Rainbow of Computer Science Lecture Notes in Computer Science*, Vol. 6570 (eds. C. Calude, G. Rozenberg, and A. Saloma). Springer, Heidelberg, Germany, January, pp 167–183, 2011.
2. W. Feijen, A. van Gasteren, D. Gries, and J. Misra, eds. *Beauty Is Our Business: A Birthday Salute to Edsger W. Dijkstra*. Texts and Monographs in Computer Science, Springer, New York, 1990, http://c2.com/cgi/wiki?BeautyIsOurBusiness.
3. A. H. Maslow. Theory of human motivation. *Psychological Review* 50(1):370–396, 1943.
4. A. H. Maslow. *Motivation and Personality*, 3rd edn. Harper and Row Publishers, New York, 1954.
5. C. Peterson and M. E. P. Seligman. *Character Strengths and Virtues: A Handbook and Classification*. Oxford University Press, Oxford, U.K., 2004.

8

Human–Robot Interaction

Alexandra Kirsch

CONTENTS

Industrial robots have been utilized for decades to perform *dirty, dull, and dangerous* tasks. But maybe such machines could do more. In Hollywood movies, we have seen robots that do the shopping, act as secretaries, or help people in dangerous situations. These applications, in contrast to today's industrial robots, all have in common that a robot must move and act in an environment that is tailored to human necessities and where people live and work. In addition, it must take commands and advice from the user and be able to ask for clarification if necessary. Moreover, when humans interact, there is always a social component, which cannot be ignored when developing interactive robots.

This chapter gives an overview of the field of human–robot interaction. It starts with a classification of interactive robot applications, showing the wide variety of interaction modes and goals of the interaction. Then a summary of the technical challenges follows, which cannot be exhaustive but should provide an intuition of the breadth of challenges that must be faced before we will have truly interactive robots. Besides the technological challenges, one must not forget that the development process for human–robot interaction is a challenge by itself, and subsequently a glimpse is provided of development and evaluation methods currently utilized or under development. The chapter concludes with an outlook on possible future development by presenting current projects and recent products.

8.1 Facets of Human–Robot Interaction

The term *human–robot interaction* is used in very different contexts. Sometimes interaction is specified as dialogue activity; in other cases, a robot and a human interact silently when trying to achieve a joint goal. The interaction can also take place on a very technical level, for example, when a robot helps a person out of a chair or on a more social level when the robot's task is to entertain the person. Here we classify the applications along two dimensions:

- The closeness of the interaction: The interaction of robots and humans can be very close, for example, when they carry things together, or it can be almost independent, when both have independent tasks, but have to take the presence of the other into consideration.

- The purpose of the robot and the interaction: Some applications are motivated by specific tasks that can be performed more efficiently when a human and a robot interact. In other cases, robots are intended for entertainment.

The chosen dimensions for classification are only two specific aspects and are supposed to give an idea of the wide spectrum of robotics applications and dimensions of interaction to be considered. A common theme for most applications is the social aspect of the interaction [30].

8.1.1 Closeness of Interaction

First, we can classify robots by the closeness of their interaction. In its most general notion, human–robot interaction takes place as soon as a robot is deployed in an environment in which humans work or live. As a minimal requirement, such a robot is required to sense the presence of humans in order to ensure safety. Besides, it would be desirable if the robot would act in a manner that is understandable for present humans. For example, a delivery robot in a factory should move along paths that are intended for object delivery instead of using any free space that may be available.

In a nearer sense, human–robot interaction starts when there is an interface for a person to give commands to the robot. This interface can be a keyboard, a touch screen, or a microphone that allows the robot to recognize spoken commands. As an example, consider a robot that delivers coffee in an office environment such as the Nesbot of BlueRobotics.* This robot takes user commands via a web interface or a pocket PC.

Complementary to giving instructions for a robot to execute, a person can remote-control the actuators of a robot, something which is referred to as *teleoperation*. This form of interaction can enhance human manipulation skills or allow access to otherwise unreachable (or badly accessible) environments. An example of enhancing human manipulation skills is using robots for surgery, allowing minimally invasive surgery by using small instruments and cameras that are attached to a robotic arm, which is controlled by a surgeon. Challenges for such systems are to control the instruments in a smooth manner, so as not to hurt the patient, and to provide a well-usable interface to the surgeon [114].

The closest form of interaction is when a team of a human and a robot (or more generally several humans and several robots) attempts to achieve a common goal. For example, consider a household robot that assists elderly people in their daily routine tasks. A person and a robot may have the joint goal to prepare a meal. Ideally, the robot would know which tasks are difficult for the person and perform them proactively, such as bringing objects that are stored in a wall cupboard and are not reachable for a small person. Joint tasks are the most challenging ones, not only because the robot needs particularly sophisticated sensing and action capabilities, but also because the interaction should feel natural to the user. A robot taking commands and executing them autonomously should respect social constraints, but it still has a high degree of freedom on how to execute the action. In contrast, a teammate for a human must have a very thorough understanding of the situation and the timing of actions, so that a person can rely on its help.

When robots and humans act as a team, it is possible that both perform independent actions to achieve the common goal or that they interact physically. The latter can include the handing over of objects and the joint manipulation of objects (e.g., a robot coworker holding a part so that a person can work on it). In some applications, the robot directly touches the person, for example, robots helping to lift patients in hospitals from their beds [86].

---

* http://www.bluebotics.com

8.1.2 Purpose of Interaction

Another dimension by which interactive robots can be classified is their purpose. Some applications arise from the unavailability or scarcity of human work power. For example, nursing faces difficult problems as more and more people need care (in particular, older people in our aging society) and less people are willing to provide this care. In addition, costs play a crucial role and not everyone can afford a full-day private nurse. Robots have already been utilized in hospitals for delivering medicines, lifting patients from beds, and for providing social interaction with relatives through telepresence systems. The purpose of these robots is not to replace nursing staff, but to take over routine tasks so that nurses have more time to spend on their actual jobs to care for patients.

In other applications, the complementary skills of humans and robots are exploited. A good example is rescue robots, in which robots are often teleoperated [48]: the robots can endure dangerous situations, operate without oxygen, and do not tire. On the other hand, human operators contribute their cognitive capabilities to control the robot in a manner that achieves a desired goal. Another example is a factory assistant robot that can bring components, steady objects, and provide necessary tools. Together with a skilled human worker, it is possible to produce diverse, high-quality products, which would not be possible either by a completely automated process or by manual work alone.

Similar to complementing skills, robots can support humans to make tasks more efficient. For example, sales robots can help customers to find a product faster in a store. It can directly show the available products to customers and guide them to the desired product.

There is another line of development, in which robots do not fulfill any practical task, but act purely as a social entity. One class of such robots are entertainment robots such as the robot dog Aibo,* which used to be produced by Sony, or the Pleo dinosaur,† which can be petted and that plays. Social robots are also developed for medical purposes, for example, to entertain elderly people suffering from dementia [119] or to help autistic children to acquire social skills [21,64].

8.2 Technological Challenges

Basically, human–robot interaction faces all of the challenges of autonomous robots for the implementation of interesting and useful applications. But there are some additional aspects that must be taken into account when

---

* http://aiboplus.sourceforge.net/
† http://www.pleoworld.com

interacting with humans, such as social constraints and specific natural interfaces such as speech or gestures. However, as described in the preceding section, social robots can have applications without any matter-of-fact practical use by serving as a social companion. In these cases, the technical challenges are less demanding, but it is still important to ensure a stable, safe behavior of the robot. We have also seen in the preceding section that the applications of human–robot interaction are very diverse. Not every technological aspect is relevant for every robot, and depending on the intended application, the challenges can be more on the technological side or on the integration into a social context.

8.2.1 Robot Hardware

In general, a robot is designed for a specific purpose: a vacuum cleaning robot must be small and have wheels to move around, a manufacturing robot must have an arm, etc. The same principles hold for interactive robots. A social toy robot must appear cute and have the potential to show social behavior (e.g., by facial expressions or by wagging a tail), or a household assistant robot should be able to move around in the house, have at least one arm to manipulate objects, and must have a suitable size to perform the necessary tasks (such as opening doors) and to fit into a normal household.

Some design properties may not be motivated by the primary purpose of the robot, but can contribute to the acceptance of the overall system. For instance, in the Nursebot project, the robot whose main purpose was to supply cognitive support to elderly users was equipped with sturdy handlebars to help people to rise from a chair [94].

But users do not only judge a device by its functionality. The robot design, considered from a subjective, emotional point of view, has been subject to research. One prerequisite for an acceptable robot is that its design tells the user immediately what the robot can or cannot do and what its intended purpose is [34,41]. For example, when a robot has an arm, people assume that it can grasp or hand over objects. In many cases, very simple adjustments of the design can help to make the purpose of a robot clear. If a robot is intended as a butler, it can be enrobed as a typical butler with a black suit (or possibly a black surface) and a bow tie.

Being utilized in environments that have been designed for humans, it could make sense to provide robots with a human form and human capabilities such as walking, climbing stairs, and opening doors. The field of *humanoid robotics* develops the hardware and technical capabilities for robots resembling humans, that is, robots that have a head, two arms, and two legs. Challenges include walking capabilities of robots, in particular under changing weight conditions (i.e., enabling a robot to carry objects while walking) [39] and the use of two arms together [127].

A human form can also foster social interaction. In fact, the more human-like a robot appears, the more it is accepted for social interaction—up to a certain point! There is a borderline, referred to as *uncanny valley*, at which the effect becomes reversed [51,84,101]. A robot that appears almost humanlike but is obviously not human is perceived as a sort of zombie and thus evaluated as creepy rather than as a social partner. However, when the form is identical to a human form, the acceptance rises again.

On the *far end* of the uncanny valley, some researchers attempt to imitate humans very accurately. Such robots are referred to as *androids*. Hiroshi Ishiguro has made headlines by constructing a robotic copy of himself [44]. However, such human replicas do not possess the entire functionality of a human body and are still unable to walk. This problem is even more difficult for androids than for other humanoids, because the form is determined by the human model and cannot be adapted to meet physical constraints.

In addition to the overall robot form, researchers work on functional, economic, and noiseless robot components that are necessary for everyday tasks. This includes work on robot platforms (with legs or wheels), arms [91], as well as grippers and hands [12,77]. Also the sensory input of robots still lags behind human capabilities, which restricts the possibilities of intelligent behavior and interaction. Robot skins [81,104,122] are developed to sense even small pressure on the robot's body while artificial ears help estimate the direction of a sound [98].

To facilitate social interaction, different models for robot heads have been developed. Kismet, which was designed at MIT [9], is one of the first attempts to design a robot face. It had two eyes with eyebrows, a mouth, and ears that can be utilized not only for perception, but also for showing facial expressions such as smiling or looking surprised. The robot head EDDIE, developed at TU München [69], shows a similar design as Kismet, but additionally includes animallike features such as the crown of a cockatoo, which enriches the possibilities of showing different types of emotions such as alertness or sadness. Both Kismet and EDDIE appear explicitly robot-like. The robot-head Flobi of Bielefeld University [78], for instance, appears more like a cartoon character, also providing mechanisms for showing emotions. These are only a few examples of different constructions for robotic heads, which cover the entire spectrum from anthropomorphic through comic-like and animallike [117] all the way to robot-like.*

8.2.2 Perception and Communication

Dedicated hardware is often necessary for the interfaces to communicate with the user. In addition, sophisticated algorithms from artificial intelligence are necessary to interpret the input and react appropriately in a given situation. First we will concentrate on specific interaction interfaces such as speech, gestures, and facial expressions to convey information from one partner to another. The subsequent section concentrates on the more abstract

---

* For an overview of different designs, see http://mindtrans.narod.ru/heads/heads.htm

interpretation of sensory inputs, in particular those related to the user and how this information is processed for decision making.

8.2.2.1 Interfaces

The most natural approach to issue commands and exchange information for humans is language, where spoken language is usually the most flexible option since it leaves the hands free and does not require physical contact between a person and an input device [2,11]. For human–robot interaction, both speech recognition and generation are important. In most cases, a robot mastering only one of those skills will not be perceived as an intelligent partner. When a robot speaks, users expect the robot also to understand the language. But language understanding is not the only concern: to achieve results in a team situation, the dialog forms part of the overall problem-solving strategy.

People also communicate via body language, or more specifically by using gestures. Understanding gestures of the user can be important for a robot, for example, an assistant robot that is supposed to bring an object should understand pointing gestures, thus relieving the user from the necessity to name every object explicitly [26,96,120]. Vice versa, a robot can emphasize or disambiguate a spoken sentence or indicate its intentions by gestures [52,126].

Since both speech and gesture recognition are still not sufficiently developed to be employed in general situations, one can use simpler alternatives to communicate commands to a robot. For example, scientists from the Healthcare Robotics Laboratory at the Georgia Institute of Technology have employed laser pointers for persons with severe motor impairments to indicate to a robot which object in the room it should fetch [89]. This proved to be a pragmatic solution for this particular application: on the one hand, the users of this specific user group would not be able to use arm gestures to indicate their intentions, while on the other hand, identifying a laser pointer is a lot more stable for a robot than interpreting hand gestures.

Another option to address the limitations of available interaction modalities is to combine them. *Multimodal interaction* is a wide research field, in which different interfaces are combined to ensure both a reliable and intuitive communication [45,109].

Interaction with the user is not only about clarifying goals and issuing commands, but emotional features play a role as well. Human emotions can be recognized via voice [105,118], facial expressions [16], body posture [19], gait [46,76], or a combination of features [3,22]. With this information, a robot can adapt its behavior to the user's mood, which is particularly important for social robots. Also for robots, it can make sense to show some type of emotions to indicate that a situation is unclear or that the robot is not able to fulfill its goal. Options for showing robot emotions include facial expressions, where again the hardware design plays a significant role (see Section 8.2.1), or simple indicators such as a green and a red light.

Some robots do not have a physical face or arms, but interact socially via a virtual character that is displayed on a screen on the robot [18,125]. Such characters can either be displayed as a face on top of the robot, allowing for facial expressions and lip movements when speaking, or as complete characters that also have the possibility to use gestures and other forms of body language.

In general, interfaces from human–computer interaction can also be used for human–robot interaction, not just to display information, but also to receive information. Traditional input devices such as keyboard or mouse have the advantage that users are familiar with them and they operate very reliably. A drawback is the physical proximity that is necessary to control a robot in such a case. An intermediate solution is the use of known input modalities such as a mobile phone to issue commands or receive feedback from a robot. Additionally, user interface controls that were originally designed for computer games have been explored for their benefit to control robot actions [35].

8.2.2.2 Interpretation

The success of team work depends to a large degree on a common perception of a situation, in human teams referred to as *shared mental models* [111]. Therefore, when interacting with a person autonomously, a robot must establish common ground with the person regarding the current situation [56]. Interaction mechanisms as described before make an important contribution to this goal. But if a robot has to ask every detail about the current situation from the user, the interaction will be very cumbersome. A better solution is a robot that can perceive many aspects of the world by itself and uses methods of user interaction to clarify situations and to add social components (such as looking at a person, greeting the user, etc.).

Typical sensors for autonomous robots are cameras, laser range finders, sonar sensors, and accelerometers. But also microphones and touch sensors are increasingly utilized for social interaction. From the readings provided by the sensors, a robot must determine a wide range of information about the world such as its own position, the position of its arms (if present), the position of objects in its environment, and the position and movement of people.

Different sensors contribute different information about the situation. For example, laser range sensors very reliably detect the presence of all types of objects, while a camera can ease the classification of the objects (whether a detected leg belongs to a table, a chair, or a person). Research on *sensor fusion* strives to combine the readings from different sensors to obtain a more complete and more reliable interpretation [50,83,100]. This is, however, a challenging task. The sensors must be synchronized in time, and objects detected by one sensor must be matched with objects detected by other sensors. Since all the readings are defective to some degree, the fusion of those readings can be ambiguous. For example, the laser range finder may report a pot at coordinate (3.6, 4.7, 2.1) while the camera detects a bowl at position (3.1, 4.4, 2.2). Are the detected pot and bowl the same object, which is interpreted in a slightly different manner by the two sensors (and is located somewhere between the two position estimates) or are they two different objects that are possibly further apart than estimated by the sensors?

The pure positions of objects and humans are not sufficient to interpret a situation correctly. For example, a robot may detect a chair *floating* in the air and a human closely behind this chair, both moving in the same direction with the same speed. An intelligent robot should find out that the person is carrying the chair. In general, not only the positions and movements of a person are of interest, but also their actions (e.g., running, sitting, sleeping, reading, watching TV, preparing coffee). Research on semantic interpretation of human actions is still in its infancy. Algorithms can detect different modes of movements such as running, walking, and jumping [47,49]. Other approaches draw inferences from the movement of objects (if an object has moved from the cupboard to the table and the robot is not responsible for it, it is very likely that a person has moved the object) [6]. Inferring more abstract actions or even the intentions of humans from such basic actions is subject to ongoing research [37,70].

Another aspect of establishing common ground is to perceive the world from the perspective of the interaction partner. This task is referred to as *perspective taking* and includes questions such as which objects the person is seeing from her perspective to infer which object she is likely referring to [10,65,97] and disambiguating the use of *left* and *right* (which can be understood from the perspective of the speaker, the listener, or the object) [82,116].

In all, it is critical for an interactive robot to interpret the situation correctly, in particular when it is supposed to achieve a joint goal with a person, as the understanding of the overall situation is a prerequisite for making the correct decisions.

8.2.3 Decision Making

Ultimately, robots must take action in the world, either to contribute to a joint goal or to signal its readiness for social interaction. Decisions on which action to perform must be taken on different levels of abstraction and for different purposes [60], for example, basic navigation and manipulation actions (taking into account social constraints); deciding which actions to take next, possibly based on a preconceived plan (e.g., go to the fridge, open the fridge, take the butter, hand the butter to the person); or the initialization of a dialog.

8.2.3.1 Human-Aware Navigation

For most practical applications, it is necessary that a robot be able to move around in the world. In contrast to general robot navigation, in human–robot interaction, the robot must take into account social constraints and movement characteristics of humans. This comprises different aspects such as how to approach a person correctly, how to plan a path, how to move in human-structured environments, and how to move in crowds of people. Some of these issues are very much linked to research on how humans navigate and avoid collisions.

The navigation problem for human–robot interaction starts by deciding where to navigate to. This depends on whether a person wishes to interact with the robot, for example, give something to the robot or initialize a dialog, or whether the person is currently not interested in an interaction. Based on such an assessment of the situation, a robot must decide where to position itself relative to the person [38,99]. In particular, when a robot is required to hand over an object to a person, special care is necessary to stand at a position where the robot is visible to the person and the person can easily take the object [20].

When navigating in the presence of moving humans, it would be advantageous to know where the people will be going. However, this is extremely difficult as the robot does not know the person's more abstract goals. This is why in many cases it is implicitly or explicitly assumed that a moving person will keep on moving on a straight line with constant velocity [68]. Some researchers, though, explicitly take into account the expectation of future human movement. Bennewitz et al. [7] learn patterns of human motion as a probability distribution. A motion pattern is a typical way of behaving in a certain situation. By observing the initial part of a motion pattern, a robot then assigns a probability value to the person's intended motion pattern and uses this information for its own path planning. Foka and Trahanias [29] predict human motion in crowd navigation. Because estimates become very inaccurate, the more one attempts to project into the future, they use a combination of a short-term model, which predicts one step into the future, and long-term model, estimating the goal point of the person.

Without completely predicting the movements of people, it can be helpful to identify a flow of people in a certain direction, for example, in the passageway of an office building. A robot in such an environment should move with the flow of people, that is, move into the same direction as most of the people around it [88]. The movement direction is also important when following a single person. Gockley et al. [33] have observed that a movement in the same direction is perceived as more natural than following the path of a person.

The classical core problem of robot navigation is how to move to a given goal position. In the presence of humans, two aspects are of particular importance: safety for the users and social acceptability. The first aspect is considered by Tadokoro et al. [112], who use a probabilistic prediction function of human motion and evaluate the danger of a possible path, which is then employed to optimize the robot path for safety. The second requirement is addressed by the human-aware navigation planner HANP [107], which also considers human safety, but in addition avoids situations in which people feel uncomfortable, for example, when the robot is hidden from the view (e.g., by a large object) or moves behind the back of the person.

HANP has been designed for moving among standing or sitting humans and for approaching people. When people are moving constantly, for example, when tidying a room, the cost model must be changed. For instance, in this case, it is more appropriate to pass behind the back of the person than in front, because this would block the person's path [67,68].

The navigation behavior of humans changes when they move in dense crowds. Research on how to move in large groups often uses physics models to describe the movement, similar to the movement of billiard balls [110]. Alternatively, the individuals in a crowd can be modeled as cognitive entities with a simple heuristic navigation strategy [85].

8.2.3.2 Joint Manipulation

Many assistive tasks require special actions in which both the human and the robot must participate, for example, jointly carrying an object, lifting an elderly person from the bed, or handing over an object from one agent to another [55].

The task of a robot handing over an object to a person has been studied by several researchers. Apart from perception and motor control, the main challenges that have been examined are the determination of a handing-over position of the object [25], the trajectory to move the object to this position [42], and the velocity of the movement [23].

In most robots, navigation and manipulation are distinct modules, which operate sequentially. When investigating human behavior in hand-over tasks, Basili et al. [4] have found, not surprisingly, that the phases of approaching and arm movement blend smoothly into each other. The exact parameters of when to lift the arm and at which speed varied among participants, but the overall profile of the task execution was comparable. In a second experiment, the receiving human was replaced by a table. Even though the goal position for *handing over* was different in the table case, the parameters for preparing this action were similar. In another study, Koay et al. [61] have studied how a robot should approach a person to hand over an object. They have shown that users prefer a robot to start its arm movement about 1 m before it reaches the handing-over position.

Whereas in a hand-over task the agents jointly hold the object for a very short time, there are other forms of interaction, in which an object is manipulated jointly over a longer time period. In this type of interaction, haptic feedback, which is the measurements from force sensors, can be utilized for controlling the robot's movements. The direct coupling of the actions of humans and robots requires special attention regarding the stability of the coupled dynamic systems as well as human comfort and safety, and raises the demand for immediate compliant reaction under real-time constraints. The classical approach for joint manipulation tasks is that the robot is merely a passively following tool for the human. In this paradigm, the human leads the joint manipulation operation, while the robot serves as an additional load carrier, which reactively adapts its behavior based on force and position input from the human to share the overall task load [62,63,123]. Combining abilities of both agents can make it possible to achieve goals that are out of reach with the human/master—robot/slave concept. For one thing, it has been shown mathematically that for two agents to navigate an object jointly in narrow space, two initiative-taking partners are necessary to achieve the desired trajectory [71]. Also, the different perception and manipulation capabilities of both agents and limited visibility are arguments to combine the cognitive capabilities.

The interaction of robots with humans can be even closer. For example, the nursing robot RIBA [86] lifts hospital patients from their bed into a wheelchair. This application is of great practical importance, since this task puts high physical demands on hospital nurses and must be performed frequently. In this application, haptic feedback and force control are the technical basis as well. In addition, since the patient must rely on the robot because otherwise helpless, the safety and subjective feeling of safety and trust must receive special consideration.

8.2.3.3 Action Selection and Planning

From navigation and manipulation actions, robots can build more abstract plans to achieve useful goals such as preparing a meal. Joint action plans are necessary to coordinate the activities of a robot and a human to achieve a common goal. The planning problem in this collaborative setting comes with two additional requirements compared to single-robot planning: (1) there are special types of actions involving contributions from the human and the robot and (2) the primary objective measure to evaluate plans are legibility (the ability of a person to intuitively understand the robot's actions) and human comfort.

These requirements imply the necessity for new plan representation mechanisms. A plan can not only be composed of robot actions with pre- and postconditions, but must also include the actions of humans. This problem has received only little attention so far. Joint intention theory [17] provides a basis for modeling the commitment of the cooperation partners to a joint task. The Human-Aware Task Planner [1] combines this theory with a classical planning approach. It uses hierarchical task planning to plan joint courses of action and evaluates the generated plans with respect to social acceptance.

A problem with classical planning approaches, which generate a fully or partially ordered list of actions to be executed sequentially, is their limited flexibility in changing situations. This flexibility, however, is indispensable in collaboration scenarios. One possibility is the use of reactive planning [5,28,80], which means that plans are represented in a richer language than partially ordered action sequences and allows to specify contingencies, parallel task execution, and failure handling [54]. Reactive plans can take into account several courses of action of the human and thus are valid for a wider range of human behavior.

Reactive plans can be generated and modified by transformational planning [87]. To make a robot behavior natural, the robot should recognize opportunities to achieve its goals faster or more reliably than it was possible at the time the plan was made [66]. For example, for a household robot at planning time, all cupboards in the kitchen are assumed to be closed. However, in the course of a joint activity, a person may open a cupboard and leave it open for a while. A robot could make use of this change of situation to retrieve the objects it needs from this cupboard.

Opportunities may not necessarily arise from a change in the world by a person; it can also result from new knowledge or slightly changed timing in plan execution [103]. Recognizing and exploiting opportunities are very natural to humans, and a robot showing a similar behavior can act more efficiently and naturally than a preprogrammed one.

8.2.3.4 Dialog and Action

As stated before, a correct interpretation of the situation around the robot is necessary for taking good actions. But the challenge is not as simple as initiating or continuing a conversation with a person in, on the one hand, a form of perception (acquiring information about the situation) and on the other hand a form of action (generating speech output, looking at a person, or making gestures). Instead, dialog actions must be integrated into the conversation.

Dialog actions can be modeled into a robot plan similar to other actions [27]. This allows a robot to include communication actions in its plan, to adjust the timing, and to integrate them with other actions. A complication is that in many cases communication is necessary because of a change in the world, and in this case, a predefined plan would not help. Rather, the robot should use its communication skills to clarify the situation and adapt its current plan [15].

8.2.4 Learning

To be acceptable as a natural interaction partner, robots must adapt to human necessities in general and to the preferences of specific users in particular. Besides, implementing robot skills can be facilitated by watching and imitating human skills. All these issues fall under the broad topic of (social) robot learning.

A basic type of knowledge necessary for assistant robots is models about abilities and preferences of users. The most convenient manner to acquire this knowledge and adapt it over time is to observe the user during interaction. This requires a robot to have learning capabilities integrated into its normal control program to be constantly on the watch for useful information to learn from [57]. However, because of very noisy data and the fact that the robot should learn from only few examples, this research is still in its infancy [59,79].

In addition to the passive observation, a user can actively teach a robot [13,90]. The challenge is to learn from instructions that a person gives in a manner that he would instruct another person. In this context, it is important to emphasize the role of affect and emotions in learning, which is a basis for human learning and has to be taken into account when a robot is to be taught by a person [93].

A specific aspect of learning from humans is the imitation of gestures [14,73,102] and movements. The perception for imitation learning is often facilitated by using tracking systems, where markers are placed on the limbs of the person (or are integrated into a suit that is worn by the person). This can then be transformed into a model of the movement. But as robots are constructed differently than humans—they have different types of joints, which are controlled in a very different manner—the problem to match the human movement into robot movement remains. Since most of the time an action is connected to some type of goal (such as hitting a tennis ball with a racket), the robot must also understand the intent of the action [72].

8.2.5 Integration of Components

Finally, the different aspects of technology described so far do not make an intelligent robot by themselves. An important question is the integration of the different techniques. The robot hardware does not exist independently of the software to control the joints and make the robot move, and the movements of the robot depend on the task the robot must perform.

In particular, the integration of perception and action is a big challenge. Robot middleware such as the Robot Operating System [95] provides the technological basis to exchange data between different processes. However, this does not solve the problem of integrating (possibly contradictory) information from different sensors [36]. Furthermore, it does not solve the problem of using the information for action and making the complete process efficient. For example, the single reading of a laser scanner may be considered as sufficient evidence for an obstacle to make the robot stop for safety reasons. In this case, this information is sufficient to cause a reaction of the robot. However, when the robot decides how to circumvent the obstacle, a better understanding of the situation may be necessary, such as whether the obstacle is moving and in which direction, what the obstacle is (e.g., a human or a cat), and the task of the robot (where it needs to go). A big challenge here is still how to describe general situations and processes in space and time [92].

Also the role of dialogue is an interesting one: on the one hand, it serves as a perception technique to obtain input from the user, and, on the other hand, it is a form of action, which can in some cases serve as an alternative to a movement. Moreover, *pure* perception methods such as visual input can profit from a tight coupling with actions, because the purpose of an action can also be to gather information [8]. Also, humans actively look at surprising scenes consciously or move toward a point of interest. With this consideration, action selection becomes more complex, because different modules have different priorities for possible next actions.

Several architectures for integrating different components have been proposed. Some are structured along levels of abstraction [31], and others use concepts such as beliefs, desires, and intentions [32] to guide the robot behavior or use a central information structure, called a blackboard, for information exchange between otherwise independent processes [40,74].

The integration of the components is a specifically hard problem, because most algorithms in use have been developed independently of complete systems. For example, image understanding is an independent field of research that develops the basic algorithms, but does not take into account the complete architecture. So for a robot, it may not be necessary to analyze a complete image, but some aspects such as the position of an object that is to be grasped are of particular importance and must be identified quickly. The isolated research on specific problems is founded in the fact that all the individual aspects already present challenges that require experts, and in addition, setting up the complete system introduces an entire set of further challenges that require different expertise. Building complete robots is a huge effort that requires large teams of experts coming from different research disciplines, expensive robot hardware, and long-term financing to allow for an interactive design process that produces generalizable and reproducible results.

8.3 Development Approaches

The problem of component integration also has an implication for the testing of interactive robots. Ideally, researchers need complete integrated systems to find out if the system is acceptable for humans and provides the required functionality. However, such integrated systems are rare and exist only in laboratories, especially in case of systems designed to handle complex tasks. However, waiting for autonomous robot technology to further mature before starting to think of interaction with humans bears the risk that the developed functionalities are not acceptable and will have to be reconsidered.

This is why the field proceeds in two directions: (1) from the perspectives of humans to clarify the requirements for potential robot applications and

(2) from the perspective of robot technology to test it toward its useful-ness for everyday tasks in cooperation with humans. The first approach often makes assumptions about robot behavior that can currently not be implemented but give an impression of possible future applications. With the results of these tests, the technology can be developed in a more goal-directed way—there is no need to spend money on technology develop-ment for functionality that is not required or socially unacceptable. The second approach often develops single components or general means on how to integrate them. For future products, of course, both aspects play an important role.

8.3.1 Human Perspective

To find out which requirements humans have with respect to a robot—regarding its purpose, skills, and visual appearance—it is most desirable to have people interact with an actual robot. But how do we do so if the robot functionality is not yet available?

A popular approach is so-called Wizard-of-Oz studies, in which a person teleoperates a robot to interact with a study participant. Teleoperation is by itself still subject to research, but it is far easier to move a robot in a room with a joystick than having it navigate by itself. In particular, the cognitive capabilities of the teleoperator such as perception and language understand-ing can be utilized to their full extent. For instance, one Wizard-of-Oz study has identified spatial relationships for interaction, such as the distance a robot should keep to a person in different situations [43]. Another example is the research on a museum tour guide robot and how it can interact with several people while explaining exhibits [106].

Sometimes human expertise can be combined with autonomous behavior. For example, in the work of Kanda et al. [52], a robot helps customers to find their way in a shopping center. The robot autonomously moves toward and addresses people, but language understanding and generation are assumed by a human operator.

Another approach for testing different robot designs or behaviors is the use of movie sequences. In movies, the robot can be remote-controlled or autonomous, but the interaction does not have to be as smooth and stable as in a live test, because the movie can be recorded several times and be composed from several trials. This allows researchers to record more real-istic scenes than that would be possible with Wizard-of-Oz experiments, and it is a relatively efficient approach for obtaining the opinion of a large number of users [121,124]. For instance, in the Italian RoboCare project [18], researchers attempted to find out which functionalities, behavior, and configuration of a robot are desirable for elderly users. They recorded dif-ferent scenes of a robot assisting a senior person. For example, in one sce-nario in which the person put a kettle on a stove, turned on the stove, and then moved into the living room. After some time, the robot reminded the person that the kettle was still on the stove. The video scenes showed different behavior and configurations of a robot, one with a social character displayed on a screen and one without such social capabilities. The study examined which assistive tasks were considered as most useful and which robot design would be preferred by elderly users. Using videos made it possible to conduct the experiment in Sweden and Italy and thus to allow a cross-cultural evaluation.

8.3.2 Technology Perspective

To advance technology, it is important to have efficient facilities to work on algorithms without the necessity to conduct user studies continually. Steinfeld et al. [108] suggest the Oz-of-Wizard approach, with reference to the previously mentioned Wizard-of-Oz experiments. They argue that when technology is in focus, it is desirable to simulate the human counterpart to some extent. Such simulation can have different forms: it can be a set of heuristics of how humans behave or prerecorded data from human behavior.

Instead of abstracting only the human or only the robot behavior, one can simulate both and have both interact directly [58]. For example, the MORSE simulator [24] is a physical 3D simulation of the world. It makes it possible to model robots at different levels of detail: with or without physics, using simulated perception or ground truth data, and controlling the robot at different levels of abstraction (e.g., giving motor commands vs. specifying navigation goals). In addition, this simulation includes a human avatar that can be controlled with the keyboard and mouse as in 3D computer games. In this manner, a person (be it the developer or a study participant) and a robot can perform joint tasks. The person has the freedom to decide on the avatar's actions and their timing, while the robot must decide autonomously what to do next, where to move, and how to react to actions of the human.

Simulation-based studies are very useful to develop the underlying technology and to acquire some proof-of-concept validation. Eventually, however, all functionality must be tested with actual robots in user studies to verify the validity. In both cases, simulation and real-world user studies, one must define a set of metrics for evaluating the robot behavior. Such metrics can include standard robotics metrics such as success rate or time to complete a task, but one must also consider human-centric metrics so as to ensure that people appreciate interacting with the robot.

One important concept for the overall acceptability of a robot is the legibility (or readability) of robot behavior [75]. This means that a person must be able to understand the robot's actions and intentions without any specific instruction, just as one is able to anticipate the goals and actions of other people. Takayama et al. [113] present a simulation-based study to assess the effect of animation techniques on the readability of robot behavior. This included a preparation for an action (such as looking at a door to be opened) and showing a reaction when a task has failed.

8.4 Outlook

The trends of technology in human–robot interaction can best be illustrated by ongoing research projects and already available products. The research efforts are targeted along two axes: (1) the development of robot technology and the impact of the technology on society, and (2) specific applications for social robots.

The EU FP7 project *Cooperative Human Robot Interaction Systems** investigates service robots capable of safe cooperative physical interaction with humans under the assumption that safe interaction between human and robot can be engineered physically and cognitively for joint physical tasks requiring cooperative manipulation of objects.

Living with Robots and Interactive Companions† investigates the technology and the social, psychological, and cognitive foundations and consequences of technological artifacts entering our daily lives—at work, or in the home—for extended periods of time.

While robots in everyday life are often regarded as frightening or unnatural in Europe and America, in Japan, the expectation of a future with robot companions is regarded more positively [53]. The *Intelligent Systems Research Institute* as part of the National Institute of Advanced Industrial Science and Technology is one example of a major robotics research effort in Japan, including interaction technologies.

Many projects explore the use of robots for elderly people—as walking aid, communication platform, entertainment device, or assistant (e.g., EU FP7 projects *KSERA*,‡ *Florence*,§ EU Ambient Assisted Living project *Alias*¶). Often the robot is integrated into a sensor-equipped living environment easing the perception of humans and the situation together with a mobile robot as a social interaction partner (*CompanionAble***). In the United States, the *Quality of Life Technology Center*†† follows similar objectives.

Using interactive robots for rescue tasks is also a focus of the current research. The mission of the *Center for Robot-Assisted Search and Rescue*‡‡ at Texas A&M University is to transfer knowledge from science to applications to make rescue robotics easily available for future disaster scenarios. A similar goal is pursued by the *International Rescue System Institute*§§ in Japan, in which industry, government, and academia have their share.

---

* http://www.chrisfp7.eu
† http://www.lirec.org
‡ http://www.ksera-project.eu
§ http://www.florence-project.eu
¶ http://www.aal-alias.eu
** http://www.companionable.net
†† http://www.qolt.org
‡‡ http://www.crasar.org
§§ http://www.rescuesystem.org

Recently, robotic technology has been transferred from research to business. Autonomous vacuum cleaners are a good example of using robots in homes. Even though these robots currently do not interact significantly with people, the deployment of such devices is a first step to making robots part of everyday life.

Robots may also soon become part of our working environments. The Texai robot* is a remote-presence system that allows a person to be virtually present through a remote-controlled robot. This idea can be regarded as an extension of videoconference systems by providing the remote partner with mobility. Thus, the presence is not limited to pictures and voice, but a remote team member can move around and interact with colleagues.

Robots for therapy, education, and entertainment can be purchased today as well. The PARO robot,[†] which has the form of a seal and acts as a companion for elderly people, is available as a product for care institutions and individuals. The NAO robot of Aldebaran Robotics[‡] is a small humanoid robot that is used widely in education and research. Because of its size and smooth design, it is well-suited for interaction with humans and has been utilized in projects to treat autistic children and to assist and entertain elderly citizens.

Entertainment robots were one of the first robotic products. The robot dog Aibo,[§] which used to be produced by Sony, is one of the most well-known examples. The dinosaur robot Pleo[¶] is a more recent instance that can even be programmed by a user. Users can interact with robot pets by petting them and playing games.

One of the first applications in which robots were *let lose* in research projects were robotic tour guides, for example, in a museum or other crowded places [115]. About a decade later, such robots are available as commercial products: The BlueBotix RoboX** is a dedicated robot guide for trade fairs and exhibitions. A similar model, Gilberto,[††] additionally has a touch screen and can thus be utilized as a mobile information desk. Similar functionality is provided by the robot Scitos A5[‡‡] of MetraLabs, which is currently guiding customers to their desired products in a chain of home improvement stores in Germany.

Robots in care giving institutions such as hospitals include robots for patient transfer such as the Care Assist Robot of Toyota[§§] and transportation systems to distribute food and medicine such as the AMV-1[¶¶] of BlueBotix.

---

* http://www.suitabletech.com
† http://www.parorobots.com
‡ http://www.aldebaran-robotics.com/en/
§ http://aiboplus.sourceforge.net/
¶ http://www.pleoworld.com
** http://www.bluebotics.com/entertainment/RoboX/
†† http://www.bluebotics.com/entertainment/Gilberto/
‡‡ http://metralabs.com/index.php?option=com_content&view=article&id=67&Itemid=66
§§ http://www.toyota-global.com/innovation/partner_robot/family_2.html#h204
¶¶ http://www.bluebotics.com/automation/AMV-1/?domain=SERVICE%20ROBOTICS

All these examples demonstrate that robots will increasingly move into human everyday life. However, the interaction capabilities are still limited, and there is a long way to go till we have fully intelligent robots that do our laundry or help our children with their homework.

References

1. S. Alili, R. Alami, and V. Montreuil. A task planner for an autonomous social robot. In H. Asama, H. Kurokawa, J. Ota, K. Sekiyama, eds., *Distributed Autonomous Robotic Systems 8*. Springer Verlag, Berlin, Germany, 2009.
2. A. Atrash, R. Kaplow, J. Villemure, R. West, H. Yamani, and J. Pineau. Development and validation of a robust speech interface for improved human–robot interaction. *International Journal of Social Robotics*, 1(4):345–356, 2009.
3. T. Bänziger, D. Grandjean, and K. R. Scherer. Emotion recognition from expressions in face, voice, and body. The multimodal emotion recognition test (mert). *Emotion*, 9:691–704, 2009.
4. P. Basili, M. Huber, T. Brandt, S. Hirche, and S. Glasauer. Investigating human–human approach and hand-over. In H. Ritter, G. Sagerer, R. Oillmann, M. Buss, eds., *Human Centered Robot Systems: Cognition, Interaction, Technology*, pp. 151–160. Springer-Verlag, Berlin, Germany, 2009.
5. M. Beetz, L. Mösenlechner, and M. Tenorth. CRAM—A cognitive robot abstract machine for everyday manipulation in human environments. In *IEEE/RSJ International Conference on Intelligent RObots and Systems*, Taipei, Taiwan, 2010.
6. M. Beetz, M. Tenorth, D. Jain, and J. Bandouch. Towards automated models of activities of daily life. *Technology and Disability*, 22(1–2):27–40, 2010.
7. M. Bennewitz, W. Burgard, and S. Thrun. Adapting navigation strategies using motions patterns of people. In *Proceedings of the IEEE International Conference on Robotics and Automation (ICRA)*, Taipei, Taiwan, 2003.
8. A. Blake and A. Yuille, eds. Active vision. *Artificial Intelligence*. MIT Press, Cambridge, MA, 1992.
9. C. Breazeal. *Designing Sociable Robots*. The MIT Press, Cambridge, MA, 2002.
10. C. Breazeal, J. Gray, and M. Berlin. An embodied cognition approach to mind-reading skills for socially intelligent robots. *The International Journal of Robotics Research*, 28(5):656–680, 2009.
11. D. Burileanu. Spoken language interfaces for embedded applications. In D. Gardner-Bonneau, H. E. Blanchard, eds., *Human Factors and Voice Interactive Systems. Signals and Communication Technology*. Springer, Boston, MA, 2008, http://www.springer.com/engineering/signals/book/978-0-387-25482-1.
12. J. Butterfass, M. Grebenstein, H. Liu, and G. Hirzinger. Dlr-hand ii: Next generation of a dextrous robot hand. In *IEEE International Conference Robotics and Automation (ICRA)*, Seoul, Korea, 2001.
13. M. Cakmak, N. DePalma, A.L. Thomaz, and R. Arriaga. Social learning mechanisms for robots. *Autonomous Robots*, 31(3–4):309–329, 2010.

14. S. Calinon, F. D'halluin, E. L. Sauser, D. G. Caldwell, and A. G. Billard. Learning and reproduction of gestures by imitation: An approach based on hidden Markov model and Gaussian mixture regression. *IEEE Robotics and Automation Magazine*, 17(2):44–54, 2010.

15. A. Clodic, R. Alami, V. Montreuil, S. Li, B. Wrede, and A. Swadzba. A study of interaction between dialog and decision for human–robot collaborative task achievement. In *The 16th IEEE International Symposium on Robot and Human interactive Communication (RO-MAN)*, Jeju Island, Korea, 2007.

16. I. Cohen, A. Garg, and T. S. Huang. Emotion recognition from facial expressions using multilevel HMM. In *Neural Information Processing Systems*, 2000.

17. P. Cohen and H. Levesque. Teamwork. *Nous, Special Issue on Cognitive Science and AI*, 25(4):487–512, 1991.

18. G. Cortellessa, G. K. Svedberg, A. Loutfi, F. Pecora, M. Scopelliti, and L. Tiberio. A cross-cultural evaluation of domestic assistive robots. In *Proceedings of the AAAI Fall Symposium on AI and Eldercare*, Washington, DC, 2008.

19. M. Coulson. Attributing emotion to static body postures: Recognition accuracy, confusions, and viewpoint dependence. *Journal of Nonverbal Behavior*, 28(2):117–139, 2004.

20. K. Dautenhahn, M. Walters, S. Woods, K. L. Koay, C. L. Nehaniv, E. A. Sisbot, R. Alami, and T. Siméon. How may I serve you? A robot companion approaching a seated person in a helping context. In *ACM SIGCHI/SIGART International Conference on Human Robot Interaction (HRI)*, pp. 172–179, Salt Lake City, Utah, 2006.

21. K. Dautenhahna, C. L. Nehaniv, M. L. Walters, B. Robins, H. Kose-Bagci, N. A. Mirza, and M. Blow. Kaspar—A minimally expressive humanoid robot for human–robot interaction research. *Applied Bionics and Biomechanics*, 6(3–4):369–397, 2009.

22. B. de Gelder and J. Vroomen. The perception of emotions by ear and eye. *Cognition and Emotion*, 14(3):289–377, 2000.

23. F. Dehais, E. A. Sisbot, M. Causse, and R. Alami. Physiological and subjective evaluation of a human–robot object hand over task. *Applied Ergonomics*, 42(6):785–791, 2011.

24. G. Echeverria, N. Lassabe, A. Degroote, and S. Lemaignan. Modular openrobots simulation engine: Morse. In *Proceedings of the IEEE ICRA*, Shanghai, China, 2011.

25. A. Edsinger and C. C. Kemp. Human–robot interaction for cooperative manipulation: Handing objects to one another. In *The 16th IEEE International Symposium on Robot and Human Interactive Communication, 2007 (RO-MAN 2007)*, Jeju Island, Korea, 2007.

26. A. Erol, G. Bebis, M. Nicolescu, R. D. Boyle, and X. Twombly. Vision-based hand pose estimation: A review. *Computer Vision and Image Understanding, Special Issue on Vision for Human–Computer Interaction*, 108 (1, 2): 52–73, 2007.

27. G. Ferguson and J. Allen. A cognitive model for collaborative agents. In *Proceedings of the AAAI 2011 Fall Symposium on Advances in Cognitive Systems*, Arlington, VA, 2011.

28. J. Firby. An investigation into reactive planning in complex domains. In *Proceedings of the Sixth National Conference on Artificial Intelligence*, pp. 202–206, Seattle, WA, 1987.

29. A. Foka and P. Trahanias. Predictive autonomous robot navigation. In *IEEE/RSJ International Conference on Intelligent Robots and Systems*, Lausanne, Switzerland, 2002.

30. T. Fong, I. Nourbakhsh, and K. Dautenhahn. A survey of socially interactive robots. *Robotics and Autonomous Systems*, 42(3–4):143–166, 2003.

31. E. Gat. Three-layer architectures. *Artificial Intelligence and Mobile Robots*, pp. 195–210. MIT Press, Cambridge, MA, 1998.

32. M. Georgeff, B. Pell, M. Pollack, M. Tambe, and M. Wooldridge. The belief-desire-intention model of agency. In J. P. Muller, M. Singh, and A. Rao, eds., *Intelligent Agents V*, vol. 1365 of lecture notes in AI. Springer-Verlag, New York, 1999.

33. R. Gockley, J. Forlizzi, and R. Simmons. Natural person-following behavior for social robots. In *Proceedings of the ACM/IEEE International Conference on Human–Robot Interaction, HRI'07*, Arlington, VA, 2007.

34. J. Goetz, S. Kiesler, and A. Powers. Matching robot appearance and behavior to tasks to improve human–robot cooperation. In *12th IEEE Workshop Robot and Human Interactive Communication*, Millbrae, CA, 2003.

35. C. Guo and E. Sharlin. Exploring the use of tangible user interfaces for human–robot interaction: A comparative study. In *Proceedings of ACM CHI 2008 Conference on Human Factors in Computing Systems*, pp. 121–130, Florence, Italy, 2008.

36. D. L. Hall and J. Llinas. *Handbook of Multisensor Data Fusion*. Electrical engineering and applied signal processing series. CRC Press, Boca Raton, FL, 2001.

37. T. A. Han, L. M. Pereira, and F. C. Santos. The role of intention recognition in the evolution of cooperative behavior. In *Proceedings of the Twenty-Second International Joint Conference on Artificial Intelligence*, Barcelona, Spain, 2011.

38. S. T. Hansen, M. Svenstrup, H. J. Andersen, and T. Bak. Adaptive human aware navigation based on motion pattern analysis. In *Robot and Human Interactive Communication, 2009*, Toyama, Japan, September–October, 2009.

39. K. Harada, S. Kajita, H. Saito, M. Morisawa, F. Kanehiro, K. Fujiwara, K. Kaneko, and H. Hirukawa. A humanoid robot carrying a heavy object. In *Proceedings of the 2005 IEEE International Conference on Robotics and Automation (ICRA)*, Barcelona, Spain, 2005.

40. B. Hayes-Roth and F. Hayes-Roth. A cognitive model of planning. *Cognitive Science*, 3(4):275–310, 1979.

41. F. Hegel, M. Lohse, and B. Wrede. Effects of visual appearance on the attribution of applications in social robotics. In *18th IEEE International Symposium on Robot and Human Interactive Communication (RO-MAN'09)*, Toyama, Japan, 2009.

42. M. Huber, M. Rickert, A. Knoll, T. Brandt, and S. Glasauer. Human–robot interaction in handing-over tasks. In *The 17th IEEE International Symposium on Robot and Human Interactive Communication, 2008 (RO-MAN 2008)*, Munich, Germany, 2008.

43. H. Huettenrauch, K. S. Eklundh, A. Green, and E. A. Topp. Investigating spatial relationships in human–robot interaction. In *Proceedings of the 2006 IEEE/RSJ International Conference on Intelligent Robots and Systems*, Beijing, China, 2006.

44. H. Ishiguro. Studies on humanlike robots—Humanoid, android and geminoid. In *Proceedings of the 1st International Conference on Simulation, Modeling, and Programming for Autonomous Robots*, Venice, Italy, 2008.

45. A. Jaimesa and N. Sebe. Multimodal humancomputer interaction: A survey. *Computer Vision and Image Understanding, Special Issue on Vision for Human–Computer Interaction*, 108 (112): 116–134, 2007.

46. D. Janssen, W. Schöllhorn, J. Lubienetzki, K. Fölling, H. Kokenge, and K. Davids. Recognition of emotions in gait patterns by means of artificial neural nets. *Journal of Nonverbal Behavior*, 32(2):79–92, 2008.

47. O. C. Jenkins, G. González, and M. M. Loper. Recognizing human pose and actions for interactive robots. In *Human–Robot Interaction*, Chapter 6, pp. 119–138. InTech, New York, 2007.

48. K. S. Jones, B. R. Johnson, and E. A. Schmidlin. Teleoperation through apertures: Passability versus driveability. *Journal of Cognitive Engineering and Decision Making*, 5:10–28, 2011.

49. I. Junejo, E. Dexter, I. Laptev, and P. Pérez. View-independent action recognition from temporal self-similarities. *IEEE Transactions on Pattern Analysis and Machine Intelligence*, 33(1):172–185, 2010.

50. M. Kam, X. Zhu, and P. Kalata. Sensor fusion for mobile robot navigation. *Proceedings of the IEEE*, 85(1):108–119, 1997.

51. T. Kanda, T. Miyashita, T. Osada, Y. Haikawa, and H. Ishiguro. Analysis of humanoid appearances in humanrobot interaction. *IEEE Transactions on Robotics*, 24(3):725–735, 2008.

52. T. Kanda, M. Shiomi, Z. Miyashita, H. Ishiguro, and N. Hagita. An affective guide robot in a shopping mall. In *Proceedings of the 4th ACM/IEEE International Conference on Human Robot Interaction*, La Jolla, CA, 2009.

53. F. Kaplan. Who is afraid of the humanoid? Investigating cultural differences in the acceptance of robots. *International Journal of Humanoid Robotics*, 1(3):465–480, 2004.

54. M. Karg, M. Sachenbacher, and A. Kirsch. Towards expectation-based failure recognition for human robot interaction. In *22nd International Workshop on Principles of Diagnosis, Special Track on Open Problem Descriptions*, Murnau, Germany, 2011.

55. C. Kemp, A. Edsinger, and E. Torres-Jara. Challenges for robot manipulation in human environments. *IEEE Robotics and Automation Magazine*, 14(1):20–29, 2007.

56. S. Kiesler. Fostering common ground in human–robot interaction. In *Robot and Human Interactive Communication (ROMAN)*, Nashville, TN, 2005.

57. A. Kirsch. Robot learning language—Integrating programming and learning for cognitive systems. *Robotics and Autonomous Systems Journal*, 57(9):943–954, 2009.

58. A. Kirsch and Y. Chen. A testbed for adaptive human–robot collaboration. In *33rd Annual German Conference on Artificial Intelligence (KI 2010)*, Karlsruhe, Germany, 2010.

59. A. Kirsch and F. Cheng. Learning ability models for human–robot collaboration. In *Robotics: Science and Systems (RSS)—Workshop on Learning for Human–Robot Interaction Modeling*, Zaragoza, Spain, 2010.

60. A. Kirsch, T. Kruse, E. A. Sisbot, R. Alami, M. Lawitzky, D. Brščić, S. Hirche, P. Basili, and S. Glasauer. Plan-based control of joint human–robot activities. *Künstliche Intelligenz*, 24(3):223–231, 2010.

61. K. L. Koay, E. A. Sisbot, D. S. Syrdal, M. L. Walters, K. Dautenhahn, and R. Alami. Exploratory studies of a robot approaching a person in the context of handing over an object. In *Proceedings of the AAAI—Spring Symposium 2007: Multidisciplinary Collaboration for Socially Assistive Robotics*, Palo Alto, CA, 2007.

62. K. Kosuge and N. Kazamura. Control of a robot handling an object in cooperation with a human. In *6th IEEE International Workshop on Robot and Human Communication, RO-MAN'97*, Sendai, Japan, 1997.

63. K. Kosuge, H. Yoshida, and T. Fukuda. Dynamic control for robot–human collaboration. In *2nd IEEE International Workshop on Robot and Human Communication*, pp. 398–401, Tokyo, Japan, 1993.

64. H. Kozima, M. P. Michalowski, and C. Nakagawa. Keepon—A playful robot for research, therapy, and entertainment. *International Journal of Social Robotics*, 1(1):3–18, 2009.

65. G. J. Kruijff and M. Brenner. Modelling spatio-temporal comprehension in situated human–robot dialogue as reasoning about intentions and plans. In *Proceedings of the Symposium on Intentions in Intelligent Systems*. Stanford University, Palo Alto, CA, 2007. AAAI Spring Symposium Series 2007.

66. T. Kruse and A. Kirsch. Towards opportunistic action selection in human–robot cooperation. In *33rd Annual German Conference on Artificial Intelligence (KI 2010)*, Karlsruhe, Germany, 2010.

67. T. Kruse, A. Kirsch, E. A. Sisbot, and R. Alami. Dynamic generation and execution of human aware navigation plans. In *Proceedings of the Ninth International Conference on Autonomous Agents and Multiagent Systems (AAMAS)*, Toronto, Canada, 2010.

68. T. Kruse, A. Kirsch, E. A. Sisbot, and R. Alami. Exploiting human cooperation in human-centered robot navigation. In *IEEE International Symposium in Robot and Human Interactive Communication (RO-MAN)*, Viareggio, Italy, 2010.

69. K. Kühnlenz, S. Sosnowski, and M. Buss. The impact of animal-like features on emotion expression of robot head Eddie. *Advanced Robotics*, 24(8–9):1239–1255, 2010.

70. T. Lan, Y. Wang, W. Yang, S. Robinovitch, and G. Mori. Discriminative latent models for recognizing contextual group activities. In IEEE Trans Pattern Anal Mach Intell, 34(8): 1549–1562, 2012.

71. M. Lawitzky, A. Mörtl, and S. Hirche. Human–robot cooperative task sharing. In *19th IEEE International Symposium in Robot and Human Interactive Communication (RO-MAN)*, Viareggio, Italia, 2010.

72. D. Lee and Y. Nakamura. Mimesis model from partial observations for a humanoid robot. *The International Journal of Robotics Research*, 29(1):60–80, 2010.

73. D. Lee and C. Ott. Incremental kinesthetic teaching of motion primitives using the motion refinement tube. *Autonomous Robots*, 31(2):115–131, 2011.

74. V. R. Lesser and L. D. Erman. A retrospective view of the hearsay-II architecture. In *Proceedings of the Fifth International Joint Conference on Artificial Intelligence*, pp. 790–800, Cambridge, MA, 1977.

75. C. Lichtenthaeler, T. Lorenz, and A. Kirsch. Towards a legibility metric: How to measure the perceived value of a robot. In *International Conference on Social Robotics, ICSR 2011, Work-In-Progress-Track*, Amsterdam, The Netherlands, 2011.

76. F. Loula, S. Prasad, K. Harber, and M. Shiffrar. Recognizing people from their movement. *Journal of Experimental Psychology: Human Perception and Performance*, 31(1):210–220, 2005.

77. C. S. Lovchik and M. A. Diftler. The robonaut hand: A dexterous robot hand for space. In *IEEE International Conference Robotics and Automation (ICRA)*, Detroit, MI, 1999.

78. I. Lütkebohle, F. Hegel, S. Schulz, M. Hackel, B. Wrede, S. Wachsmuth, and G. Sagerer. The bielefeld anthropomorphic robot head "flobi." In *2010 IEEE International Conference on Robotics and Automation*, Anchorage, AK, 2010.

79. M. Mason and M. Lopes. Robot self-initiative and personalization by learning through repeated interactions. In *Robotics: Science and Systems (RSS)—Workshop on Learning for Human–Robot Interaction Modeling*, Zaragoza, Spain, 2010.

80. D. McDermott. A reactive plan language. Technical report, Yale University, Computer Science Department, New Haven, CT, 1993.

81. P. Mittendorfer and G. Cheng. Humanoid multi-modal tactile sensing modules. *IEEE Transactions on Robotics—Special Issue on Robotic Sense of Touch*, 27(3):401–410, 2011.

82. R. Moratz, T. Tenbrink, J. Bateman, and K. Fischer. Spatial knowledge representation for human–robot interaction. In C. Freksa et al., ed., *Spatial Cognition III*, Vol. 2685 of *LNAI*, pp. 263–286. Springer-Verlag, Berlin, Germany, 2003.

83. H. P. Moravec. Sensor fusion in certainty grids for mobile robots. *AI Magazine*, 9(2):61–74, 1988.

84. M. Mori. The uncanny valley. *Energy*, 7(4):33–35, 1970. Originally in Japanese, translated by K. F. MacDorman and T. Minato.

85. M. Moussaïd, D. Helbing, and G. Theraulaz. How simple rules determine pedestrian behavior and crowd disasters. In *Proceedings of the National Academy of Sciences of the United States of America*, 108(17):6884–6888, 2011.

86. T. Mukai, S. Hirano, H. Nakashima, Y. Sakaida, and S. Guo. Realization and safety measures of patient transfer by nursing-care assistant robot riba with tactile sensors. *Journal of Robotics and Mechatronics*, 23(3):360–369, 2011.

87. A. Müller, A. Kirsch, and M. Beetz. Transformational planning for everyday activity. In *Proceedings of the 17th International Conference on Automated Planning and Scheduling (ICAPS'07)*, pp. 248–255, Providence, RI, September 2007.

88. J. Müller, C. Stachniss, K. O. Arras, and W. Burgard. Socially inspired motion planning for mobile robots in populated environments. In *International Conference on Cognitive Systems (CogSys'08)*, Karlsruhe, Germany, 2008.

89. H. Nguyen, A. Jain, C. Anderson, and C. C. Kemp. A clickable world: Behavior selection through pointing and context for mobile manipulation. In *IEEE/RSJ International Conference on Intelligent Robots and Systems, 2008 (IROS 2008)*, pp. 787–793, Nice, France, 2008.

90. N. Otero, J. Saunders, K. Dautenhahn, and C. L. Nehaniv. Teaching robot companions: The role of scaffolding and event structuring. *Connection Science*, 20(2–3):111–134, 2008.

91. C. Ott, R. Mukherjee, and Y. Nakamura. Unified impedance and admittance control. In *IEEE International Conference on Robotics and Automation (ICRA)*, Anchorage, AK, 2010.

92. D. J. Peuquet. *Representations of Space and Time*. Guilford Press, New York, 2002.

93. R. W. Picard, S. Papert, W. Bender, B. Blumberg, C. Breazeal, D. Cavallo, T. Machover, M. Resnick, D. Roy, and C. Strohecker. Affective learning—A manifesto. *BT Technology Journal*, 22(4):253–269, 2004.

94. J. Pineau, M. Montemerlo, M. Pollack, N. Roy, and S. Thrun. Towards robotic assistants in nursing homes: Challenges and results. *Robotics and Autonomous Systems*, 42:271–281, 2003.

95. M. Quigley, K. Conley, B. Gerkey, J. Faust, T. Foote, J. Leibs, R. Wheeler, and A. Ng. Ros: An open-source robot operating system. In *IEEE International Conference on Robotics and Automation (ICRA 2009)*, Kobe, Japan, 2009.

96. M. Rehm, N. Bee, and E. André. Wave like an Egyptian: Accelerometer based gesture recognition for culture specific interactions. In *Proceedings of the 22nd British HCI Group Annual Conference on People and Computers: Culture, Creativity, Interaction—Volume 1*, Liverpool, UK, 2008.

97. R. Ros, E. A. Sisbot, R. Alami, J. Steinwender, K. Hamann, and F. Warneken. Solving ambiguities with perspective taking. In *Proceeding of the 5th ACM/IEEE International Conference on Human-Robot Interaction (HRI'10)*, Osaka, Japan, 2010.

98. M. Rothbucher, D. Kronmüller, H. Shen, and K. Diepold. Underdetermined binaural 3d sound localization algorithm for simultaneous active sources. In *Audio Engineering Society Convention*, London, U.K., 2010.

99. S. Satake, T. Kanda, D. F. Glas, M. Imai, H. Ishiguro, and N. Hagita. How to approach humans? Strategies for social robots to initiate interaction. In *ACM/IEEE International Conference on Human-Robot Interaction*, La Jolla, CA, 2009.

100. A. Sanfeliu, J. Andrade-Cetto, M. Barbosa, R. Bowden, J. Capitn, A. Corominas, A. Gilbert et al. Decentralized sensor fusion for ubiquitous networking robotics in urban areas. *Sensors*, 10(3):2274–2314, 2010.

101. A. P. Saygin, T. Chaminade, H. Ishiguro, J. Driver, and C. Frith. The thing that should not be: Predictive coding and the uncanny valley in perceiving human and humanoid robot actions. *Social Cognitive and Affective Neuroscience*, 7(4): 413–422, 2011.

102. S. Schaal, A.-J. Ijspeert, and A. Billard. Computational approaches to motor learning by imitation. *The Neuroscience of Social Interaction*, pp. 199–218. Oxford University Press, Oxford, U.K., 2004.

103. P. Schermerhorn, J. Benton, M. Scheutz, K. Talamadupula, and S. Kambhampati. Finding and exploiting goal opportunities in real-time during plan execution. *Robotics*, pp. 3912–3917, 2009.

104. A. Schmitz, P. Maiolino, M. Maggiali, L. Natale, G. Cannata, and G. Metta. Methods and technologies for the implementation of large-scale robot tactile sensors. *IEEE Transactions on Robotics—Special Issue on Robotic Sense of Touch*, 27(3):389–400, 2011.

105. B. Schuller, S. Reiter, R. Muller, M. Al-Hames, M. Lang, and G. Rigoll. Speaker independent speech emotion recognition by ensemble classification. In *IEEE International Conference on Multimedia and Expo, 2005 (ICME 2005)*, Amsterdam, the Netherlands, 2005.

106. M. Shiomi, T. Kanda, S. Koizumi, H. Ishiguro, and N. Hagita. Group attention control for communication robots with wizard of oz approach. In *Proceedings of the ACM/IEEE International Conference on Human–Robot Interaction*, Arlington, VA, 2007.

107. E. A. Sisbot, L. F. Marin-Urias, R. Alami, and T. Simeon. A human aware mobile robot motion planner. *IEEE Transactions on Robotics*, 23:874–883, 2007.

108. A. Steinfeld, O. C. Jenkins, and B. Scassellati. The oz of wizard: Simulating the human for interaction research. In *Proceedings of the 4th ACM/IEEE International Conference on Human Robot Interaction*, La Jolla, CA, 2009.

109. R. Stiefelhagen, H. K. Ekenel, C. Fugen, P. Gieselmann, H. Holzapfel, F. Kraft, K. Nickel, M. Voit, and A. Waibel. Enabling multimodal human–robot interaction for the karlsruhe humanoid robot. *Robotics, IEEE Transactions on*, 23(5):840–851, 2007.

110. G. K. Still. *Crowd Dynamics*. PhD thesis, Mathematics department, Warwick University, Coventry, U.K., 2000.

111. R. J. Stout, J. A. Cannon-Bowers, E. Salas, and D. M. Milanovich. Planning, shared mental models, and coordinated performance: An empirical link is established. *Human Factors*, 41(1):61–71, 1999.

112. S. Tadokoro, M. Hayashi, Y. Manabe, Y. Nakami, and T. Takamori. On motion planning of mobile robots which coexist and cooperate with human. *IROS, IEEE/RSJ*, 2:2518, 1995.

113. L. Takayama, D. Dooley, and W. Ju. Expressing thought: Improving robot readability with animation principles. In *Proceedings of the 6th International Conference on Human–Robot Interaction, HRI'11*, pp. 69–76, ACM, New York, 2011.

114. M. Tavakoli and R. D. Howe. Haptic effects of surgical teleoperator flexibility. *The International Journal of Robotics Research*, 28(10):1289–1302, 2009.

115. S. Thrun, M. Beetz, M. Bennewitz, A. Cremers, F. Dellaert, D. Fox, D. Hähnel et al. Probabilistic algorithms and the interactive museum tour-guide robot Minerva. *International Journal of Robotics Research*, 19(11):972–999, 2000.

116. J. G. Trafton, N. L. Cassimatis, M. D. Bugasjska, D. P. Brock, F. E. Mintz, and A. C. Schultz. Enabling effective human–robot interaction using perspective taking in robots. *IEEE Transactions on Systems, Man and Cybernetics*, 35(4):460–470, 2005.

117. A. J. N. van Breemen. ICAT: Experimenting with animabotics. In *AISB 2005 Creative Robotics Symposium*, Hatfield, England, 2005.

118. D. Ververidis and C. Kotropoulos. Emotional speech recognition: Resources, features, and methods. *Speech Communication*, 48(9):1162–1181, 2006.

119. K. Wada, T. Shibata, T. Asada, and T. Musha. Robot therapy for prevention of dementia at home—Results of preliminary experiment. *Journal of Robotics and Mechatronics*, 19(6):691–697, 2007.

120. S. Waldherr, S. Thrun, and R. Romero. A gesture-based interface for human–robot interaction. *Autonomous Robots*, 9(2):151–173, 2000.

121. M. L. Walters, M. Lohse, M. Hanheide, B. Wrede, K. L. Koay, D. S. Syrdal, A. Green, H. Hüttenrauch, K. Dautenhahn, G. Sagerer, and K. Severinson-Eklundh. Evaluating the robot personality and verbal behaviour of domestic robots using video-based studies. *Advanced Robotics*, 25:2233–2254, 2011.

122. M. Wiertlewski, J. Lozada, and V. Hayward. The spatial spectrum of tangential skin displacement can encode tactual texture. *IEEE Transactions on Robotics— Special Issue on Robotic Sense of Touch*, 27(3):461–472, 2011.

123. T. Wojtara, M. Uchihara, H. Murayama, S. Shimoda, S. Sakai, H. Fujimoto, and H. Kimura. Human–robot collaboration in precise positioning of a three-dimensional object. *Automatica*, 45(2):333–342, 2009.

124. S. N. Woods, M. L. Walters, K. L. Koay, and K. Dautenhahn. Comparing human robot interaction scenarios using live and video based methods: Towards a novel methodological approach. In *9th IEEE International Workshop on Advanced Motion Control*, Istanbul, Turkey, 2006.

125. B. Wrede, S. Kopp, K. Rohlfing, M. Lohse, and C. Muhl. Appropriate feedback in asymmetric interactions. *Journal of Pragmatics*, 42:2369–2384, 2010.

126. Y. Xu, M. Guillemot, and T. Nishida. An experiment study of gesture-based human–robot interface. In *IEEE/ICME International Conference on Complex Medical Engineering*, Harbin, China, 2007.

127. R. Zollner, T. Asfour, and R. Dillmann. Programming by demonstration: Dual-arm manipulation tasks for humanoid robots. In *Proceedings of the IEEE/RSJ International Conference on Intelligent Robots and Systems, 2004 (IROS 2004)*, pp. 479–484, Sendai, Japan, 2004.

9

Multimedia Retrieval and Adaptive Systems

Frank Hopfgartner

CONTENTS

9.1 Introduction

Recently, *The Guardian*, one of Britain's most popular daily newspapers, published an online article,* recognizing the fifth anniversary of the video sharing portal YouTube.† YouTube is at the forefront of a recent development that, in 2006, convinced the renowned *Time* magazine to dedicate their Person of the Year award to *You*. *You* represents the millions of people that started to voluntarily generate (user-generated) content, for example, in Wikipedia, Facebook, and, of course, YouTube. More and more people do not only actively consume content, but they have also started to create their own content. Thus, we are observing a paradigm change from the rather passive information consumption habit to a more active information search. Tim Berners-Lee, credited for inventing the World Wide Web (WWW), is convinced that eventually this development will completely change the way in which we engage information. During a discussion following his keynote speech "The Web at 20" at the BBC documentary "Digital Revolution," he envisioned that

> As a consumer, if I have an internet connection I should be able to get at, and pay for if necessary, anything that has ever been broadcast. So I'm looking forward to when the BBC, for example, [offers] a complete random access library so that I can follow a link, so that somebody can tweet about some really, really cool thing or some fun show, or some otherwise boring show, and I can follow a link directly to that. Whether it's pay or free, its per view, and I get it by following a link, one way or another. I won't be searching channels. I think the concept of a channel is going to be history very quickly on the internet. It's not relevant.

This chapter illustrates how users can be assisted in exploring such digital video libraries. First, a brief introduction is provided of basic concepts of video retrieval, a research area that has drawn more and more attention since the rise of YouTube. Arguing that personalization techniques can be employed that assist the users in identifying videos they are interested in, different approaches are then surveyed that are used to gather user information and introduce techniques to exploit this information to recommend video documents that match users' personal interests.

9.2 Basic Concepts of Video Retrieval

Video retrieval is a specialization of information retrieval (IR), a research domain that focuses on the effective storage and access of data. In a classical

---

* http://www.guardian.co.uk/technology/2010/apr/23/youtube-five-years-on, accessed on 7 September 2011
† http://www.youtube.com/, accessed on 7 September 2011.

IR scenario, a user aims to satisfy their *information need* by formulating a *search query*. This action triggers a retrieval process that results in a list of ranked documents, usually presented in decreasing order of relevance. The activity of performing a search is called the *information-seeking* process. A *document* can be any type of data accessible by a retrieval system. In the text retrieval domain, documents can be textual documents such as e-mails or websites. Image documents can be photos, graphics, or other types of visual illustrations. Video documents are any type of moving images. In Section 9.2.1, we introduce the structure of a typical video document. A repository of documents that is managed by an IR system is referred to as a *document collection*. In Section 9.2.2, we argue that video documents should be analyzed and processed first in order to ease access to their content. Preferably, retrieved documents are ranked in accordance to their relevance to the user's information need. In Section 9.2.3, we discuss this retrieval and ranking process further. Aiming to visualize retrieval results for the user, graphical user interfaces are required that allow the user to input their information need and to inspect the retrieved results, thus to access the document collection. Section 9.2.4 introduces challenges of graphical user interfaces in the video retrieval domain.

9.2.1 Structure of Video Documents

Computers serving multimedia and other devices are going to change the handling of videos completely. Films are consistently broadcast, recorded, and stored in *digital* form. A video is a sequence of still images, accompanied by an audio stream. Classical digital video standards are the MPEG-1 and MPEG-2 formats. They were released by the Motion Pictures Expert Group (MPEG). Videos in MPEG-1 format have a resolution of 352 × 240 pixels and a bit rate of up to 1.5 Mbps with a frame rate of 25 frames per second. MPEG-1 achieves a high compression rate by the use of motion estimation and its compensation between frames. Considering the resulting low quality of the encoded videos, MPEG-1 videos are often compared to old-fashioned video cassette recorder (VCR) recordings. The newer MPEG-2 video format is used to encode videos in digital versatile disk (DVD) quality. It is the standard used for digital video broadcasting television (DVB-T, DVB-S, DVB-C) [122].

Another ISO standard that has been defined by MPEG is MPEG-7. Its purpose is to provide a unified standard for the description of multimedia data using meta-information. Within this standard, various descriptors have been defined to describe visual content, including color descriptor, shape descriptor, motion descriptor, face descriptor, and textual descriptor [78]. An overview of MPEG-7 descriptors is given by Manjunath et al. [78].

Besides the video's own text and audiovisual data streams, video documents can be enriched with additional data, the so-called metadata. Blanken et al. [13] survey various types of metadata: (1) a description of the video document, (2) textual annotation, and (3) semantic annotations. All approaches aim to provide annotations in textual form that attempt to bridge the semantic gap. The remainder of this section provides a brief overview over these metadata types. For a more detailed description, the reader is referred to Blanken et al. [13].

Descriptive Data: Descriptive data provide valuable information about the video document. Examples are the creation date, director or editor, length of the video, and so on. A standard format for descriptive data is called Dublin Core [123]. It is a list of data elements designed to describe resources of any kind. Descriptive metadata can be very useful when documents within the video collection shall be filtered based on certain document facets. Think, for instance, of a user who wants to retrieve all video documents that have been created within the last month or all videos from one specific director.

Text Annotations: Text annotations are textual descriptions of the content of video documents. More recent state-of-the-art online systems, such as YouTube and Dailymotion,* rely on using annotations provided by users to provide descriptions of videos. However, quite often users can have very different perceptions about the same video and annotate that video differently. This can result in synonymy, polysemy, and homonymy, which makes it difficult for other users to retrieve the same video. It has also been found that users are reluctant to provide an abundance of annotations unless there is some benefit to the user [45]. Van Zwol et al. [117] approach this problem by transferring video annotation into an online gaming scenario.

Apart from manually created content, an important source for textual annotations is speech transcripts. Huang [56] argues that speech contains most of the semantic information that can be extracted from audio features. Further, Chang et al. [25] argue that text from speech data plays an important role in video analysis. In literature, the most common text sources are teletext (also called closed-caption), automatic speech recognition (ASR) transcripts, and optical character recognition (OCR) output. Considering that textual annotations can be a valuable source for IR systems aiming to retrieve the video documents, various approaches have been studied to automatically determine textual annotations. A survey of state-of-the-art approaches is given by Magalhães and Rüger [76]. More recent examples include [72,73,91,110,121].

Semantic Annotations: Another type of annotation is semantic annotations. The idea here is to identify concepts and define their relationship between each other and the video document. Concepts can hence set the content of video documents into a semantic context. This is especially useful for semantic retrieval approaches. We will survey such approaches in Section 9.2.3. The previously mentioned MPEG-7 standard allows for describing multimedia documents and their semantic descriptions. Promising extensions include Core Ontology for Multimedia (COMM), an ontology introduced by Arndt et al. [6]. Ontologies are *content-specific agreements* on vocabulary usage and

---

* http://www.dailymotion.com/, accessed on 7 September 2011.

sharing of knowledge [43]. Other metadata models include [11,39,115], which aim to enrich interactive television broadcast data with additional information by combining existing standards. All approaches therefore build upon similar ideas.

Semantic annotations can either be derived from textual annotations or from the videos' low-level features, that is, by identifying high-level concepts. Magalhães and Rüger [76] provide a survey on state-of-the-art methodologies to create semantic annotations for multimedia content. They distinguish between three semantic annotation types: (1) hierarchical models, (2) network models, and (3) knowledge-based models. *Hierarchical* models aim to identify hierarchical relations or interdependencies between elements in an image or key frame. Examples include [8,14]. *Network models* aim to infer concepts given the existence of other concepts. Surveyed approaches are [52,67]. The third approach, *knowledge-based* models, relies on prior knowledge to infer the existence of concepts. Bürger et al. [18], for example, enrich news video data with a thesaurus of geographic names. Therefore, they determine location names within news reports' transcripts and map these with their thesaurus. Further, they identify thematic categories by mapping terms in the transcript with a controlled vocabulary. A similar approach is introduced by Neo et al. [85], who use the WordNet lexical database [41] to semantically enrich news video transcripts. Even though their approaches allow linking of related news videos, the main problem of their approaches is text ambiguity. Other examples are available in related work [103,112].

9.2.2 Video Segmentation

As will be shown in Section 9.2.3, IR systems index documents and retrieve these documents based on their relevance to a search query. This approach is problematic, however, if a document contains short paragraphs that are highly relevant to the query, while the majority of the document is not relevant at all. Classical ranking algorithms, for example, based on the document's term frequency, will result in a low ranking of this document. A promising approach to tackle this problem in the text domain is to split documents into shorter *passages* (e.g., [97]). Various potential advantages arise when considering these passages as unit of retrieval results. First of all, individual passages will be ranked higher than documents that contain the corresponding passages. Consequently, retrieval performance increases. Second, ranking problems because of variable document lengths is minimized, assuming that passages have a similar length. Third, short passages are easier to assess for the user than long documents. Users can easily browse through short results to search for their information need. The same problems apply to videos in the news video domain. However, because of the different nature of news video documents, successful passage retrieval approaches cannot easily be adopted. This section introduces typical segmentation of news videos.

Shot Segmentation: The atomic unit of access to video content is often considered to be the video *shot*. Monaco [81] defines a shot as a part of the video that results from one continuous recording by a single camera. It hence represents a continuous action in time and space in the video. Especially in the context of professional video editing, this segmentation is very useful. Consider, for example, a journalist who has to find shots in a video archive that visualize the context of a news event. Shot segmentation infers shot boundary detection, since each shot is delimited by two consecutive shot boundaries. Hanjalic [48] provides a comprehensive overview on issues and problems involved in automatic shot boundary detection. A more recent survey is given by Smeaton et al. [104].

9.2.2.1 News Story Segmentation

A more consumer-oriented approach is to segment videos into semantically coherent sequences. For instance, sports fans want to watch specific highlights of a game rather than short shots depicting only parts of this highlight.

In the news video domain, such coherent sequences are news stories. News stories are commonly seen as segments of a news broadcast with a coherent news focus that contain at least two independent declarative clauses. News bulletins consist of various continuous news stories, such as reports about political meetings, natural disasters, or sports events. Chaisorn and Chua [24] argue that the internal structure of news stories depends on the producer's style. While some stories consist of anchor person shots only, often with a changing background image, other stories can consist of multiple different shots, for example, other anchor persons, graphics or animations, interview scenes, or shots of meetings.

News story segmentation is essentially finding the boundaries where one story ends and the other begins. Various text-based, audiovisual-based, and combinations of all features have been studied to segment news videos accordingly. Detailed surveys are given by Arlandis et al. [5] and Chua et al. [27].

9.2.3 Document Representation

Video retrieval systems aim to retrieve relevant video documents that match the users' information needs. Various conditions must be fulfilled to enable this process. Snoek et al. [107] sketched a common framework that applies for most state-of-the-art video retrieval engines. Their framework can be divided into an indexing engine and a retrieval engine.

The first component involves the indexing of the video data, so that documents can be retrieved that match the users' information needs. With respect to current systems, this indexing can be incorporated on a visual, textual, and semantic level. Most video retrieval engines store their indexed data collection in a database. Techniques are required to match both information need and the video collection. These tasks are fulfilled by the retrieval engine. In the remainder of this section, state-of-the-art approaches that address these issues are briefly reviewed.

9.2.3.1 Video Indexing

Video indexing is the backbone of all video retrieval engines. Indexing approaches aim to develop effective and efficient methodologies for storing, organizing, and accessing video contents. As discussed in Section 9.2.1, a video document consists of several modalities, for example, a video document is made up of audio tracks, visual streams, and different types of annotations. Thus, video indexing has to take numerous modality features into consideration. Moreover, these features are of various nature. Video indexing techniques can be split into three main categories: content-based indexing, text-based indexing, and semantic indexing. Note that this chapter does not focus on video indexing, since it is out of the scope of this work. Hopfgartner et al. [55] present two multimedia indexing approaches that aim to address the two main challenges in content-based indexing: (1) the high-dimensional feature space of multimedia data and (2) the variable character of feature dimensions, that is, Boolean and multivalue features. A survey on content-based video indexing is given by Baeza-Yates and Ribeiro-Neto [7], and semantic indexing is surveyed by Snoek and Worring [105].

9.2.3.2 Retrieval

As discussed before, video data consist of multimodal features, including text, audiovisual features, and semantic annotations. Consequently, there are a number of different ways in which a user can query a video retrieval system. As Snoek et al. [107] pointed out, three query formulation paradigms exist in the video retrieval domain: query-by-textual-keyword, query-by-visual-example, and query-by-concept.

9.2.3.3 Query-by-Textual-Keyword

Query-by-textual-keyword is one of the most popular methods of searching for video [49]. It is simple and users are familiar with this paradigm from text-based searches. Query-by-text relies on the availability of sufficient textual descriptions, including descriptive data, transcripts, and annotations.

> *Query-by-Visual-Example*: Query-by-visual-example has its roots in content-based image retrieval. It allows the users to provide sample images or video clips as examples to retrieve more results. This approach uses the low-level features that are available in images and videos, such as color, texture, and shape, to retrieve results. The basic

idea is that visual similarity can be used to identify relevant documents. Content-based image retrieval has been well studied, and a
survey is given by Aigrain et al. [2].

Query-by-Concept: In an attempt to bridge the semantic gap, a great deal
of interest in the multimedia search community has been invested
in query-by-concept, also referred to as concept-based, conceptual,
or semantic retrieval. A survey on concept-based retrieval is given
by Snoek and Worring [106]. Conceptual retrieval relies on semantic annotations, that is, high-level concepts that have been associated with the video data. A well-known set of high-level semantic
concepts has been explored by the Large-Scale Concept Ontology
for Multimedia (LSCOM) initiative [84], a subset of which is used
within TRECVid* to study concept-based retrieval. Considering
semantic concepts as additional textual annotation, documents can
be retrieved by triggering textual search queries. Hildebrand et al.
[54] analyzed state-of-the-art semantic retrieval systems, concluding that semantic concepts are often used to filter retrieval results.
Query-by-concept is an extension to both query-by-textual-keyword
and query-by-visual-example, narrowing down corresponding
results. Indeed, the most successful video retrieval systems that
have been evaluated within TRECVid (e.g., [50,108]) employ these
two approaches to improve their retrieval results.

9.2.4 Interface Designs

According to Spink et al. [109], users are often uncertain of their information
needs and hence have problems finding a starting point for their information-seeking task. And even if users know exactly what they are intending
to retrieve, formulating a *good* search query can be a challenging task. This
problem even exacerbates when dealing with multimedia data. Graphical
user interfaces serve here as a mediator between the available data corpus
and the user. It is the retrieval system's interface that will provide users
facilities to formulate search queries and/or to dig into the available data.
Hearst [53] outlines various conditions that dominate the design of state-
of-the-art search interfaces. First of all, the process of searching is a means
toward satisfying an information need. Interfaces should therefore avoid
being intrusive, since this could disturb the users in their seeking process.
Moreover, satisfying an information need is already a mentally intensive
task. Consequently, the interface should not distract the users but rather support them in their assessment of the search results. Especially in the WWW
domain, search interfaces are used not by high-expertise librarians only but

---

* TRECVid is an annual evaluation effort, targeting research in multimedia information
retrieval.

also by the general public. Therefore, user interfaces must be intuitive to use by a diverse group of potential users.

Schoeffmann et al. [100] survey representative video browsing and exploration interfaces. They distinguish between interfaces that support video browsing that rely on interaction similar to classical video players, video exploration interfaces, and unconventional video visualization interfaces. Another survey is given by Snoek and Worring [106], who focus on concept-based video retrieval interfaces. For further reading, the reader is referred to these two publications.

9.2.5 Summary

This section introduced basic principles of video retrieval. It first presented the general structure of video documents. As discussed, videos consist of audiovisual data streams and are often accompanied with metadata. Metadata are mainly of textual nature. Further, it was argued that most retrieval scenarios require videos to be split into smaller units of retrieval. Two different segmentation units and their applications were surveyed: video shots and (news) video stories. Moreover, document representation and matching techniques were introduced, including indexing, retrieving, and ranking of video documents. Finally, different graphical user interfaces were introduced that support users in their information-seeking task.

The techniques and methods introduced combine all required parts of a video IR system. The next section surveys how these approaches can be applied to adaptively support users' information needs.

9.3 Personalized Video Search and Recommendation

The previous section surveyed basic concepts of video retrieval. It was argued that when interacting with a video retrieval system, users express their information need in search queries. The underlying retrieval engine then retrieves relevant retrieval results to the given queries. A necessary requisite for this IR scenario is to correctly interpret the users' information needs. As Spink et al. [109] indicate though, users very often are not certain about their information need. One problem they face is that they are often unfamiliar with the data collection, and thus, they do not know what information they can expect from the corpus [98]. Further, Jansen et al. [59] have shown that video search queries are rather short, usually consisting of approximately three terms. Considering these observations, it is hence challenging to satisfy users' information needs, especially when dealing with ambiguous queries. Without further knowledge, it is a demanding task to understand the users' intentions. Interactive IR aims at improving the classic

IR model that was introduced in the previous section by studying how to further engage users in the retrieval process, in a way that the system can have a more complete understanding of their information need. Thus, aiming to minimize the users' efforts to fulfill their information-seeking task, there is a need to personalize search.

This section will introduce basic concepts of personalized IR. Section 9.3.1 first surveys different sources that are used to gather users' interests, an important requisite for any type of personalization. Section 9.3.2 introduces different application techniques. Section 9.3.3 introduces state-of-the-art personalization approaches with a main focus on the video domain. Section 9.3.4 summarizes this personalization section.

9.3.1 Gathering and Representing Interest

Most of the approaches that follow the interactive IR model are based on relevance feedback techniques [98]. Relevance feedback is one of the most important techniques within the IR community. An overview of the large amount of research focusing on exploiting relevance feedback is given by Ruthven and Lalmas [95]. The principle of relevance feedback is to identify the user's information need and then, exploiting this knowledge, adapting search results. Rocchio [92] defines relevance feedback as follows: The retrieval system displays search results, users provide feedback by specifying keywords or judging the relevance of retrieved documents, and the system updates the results by incorporating this feedback. The main benefit of this approach is that it simplifies the information-seeking process, for example, by releasing the user from manually reformulating the search query, which may be problematic especially when the user is not exactly certain what they are looking for or does not know how to formulate their information need. There are three types of relevance feedback in interactive IR that will be introduced in the remainder of this section: explicit, implicit, and pseudo-relevance feedback. Further, the Ostensive Model of Developing Information Need is introduced to address the problem of non-static interest and to discuss approaches to represent user interests in personal profiles.

9.3.1.1 Explicit Relevance Feedback

A simple approach to identify users' interests is to explicitly ask them about their opinion. In a retrieval context, they can express their opinion by providing *explicit relevance feedback*. Hence, the user is asked during their retrieval process to actively indicate which documents are relevant in the result set. This relevance judgment can be given on a binary or graded relevance scale. A binary feedback indicates that the rated document is either relevant or nonrelevant for the user's current information need. Considering that binary relevance requires a rather strong judgment, a relevance scale allows the user to define different grades of relevance such as *highly relevant, relevant, maybe relevant,* or *somewhat relevant.* As of December 2011, the commercial video sharing platform YouTube supports binary feedback by providing a *Thumbs up* button in their interface. The Dailymotion platform opted for the graded relevance scale scheme. Registered users of their service can express their interest on a five-*star* scale. Explicit relevance feedback is very reliable. Although the impact of explicit relevance feedback in earlier systems remains unclear, it has been shown in text retrieval that giving explicit relevance feedback is a cognitively demanding task and can affect the search process. Also, previous evaluations have found that users of explicit feedback systems often do not provide sufficient levels of feedback in order for adaptive retrieval techniques to work [10,47].

9.3.1.2 Implicit Relevance Feedback

Deviating from the method of *explicitly* asking the user to rate results, the use of *implicit* feedback techniques helps the system to learn user interest unobtrusively. The main advantage is that this approach relieves the user from providing explicit feedback. As a large quantity of implicit data can be gathered without disturbing the user workflow, the implicit approach is an attractive alternative. According to Nichols [86], however, information gathered using implicit techniques are less accurate than information based on explicit feedback. Agrichtein et al. [1] evaluated the effect of user feedback on web retrieval using over 3000 queries and 12 million user interactions. They show that implicit relevance feedback can improve retrieval performance by as much as 31% relative to systems that do not incorporate any feedback. Furthermore, both implicit and explicit measures can be combined to provide an accurate representation of the user interests. Kelly and Teevan [65] provide a literature overview of the research that was done in the field.

Not all implicit measures are useful to infer relevance. Thus, various researches have been performed to detect those features that promise to be valid indicators of interest. From the psychological point of view, a promising indicator of interest is viewing time. People look at objects or things they find interesting for a longer time than uninteresting things. For instance, Faw and Nunnally [40] showed a positive correlation between *pleasant ratings* and viewing time, and Day [34] found that most people look longer at images they liked than at images they disliked. According to Oostendorp and Berlyne [87], people look longer at objects evoking pleasurable emotions. Transferring these findings into an IR context, users are expected to spend more time in viewing relevant documents than nonrelevant documents. Claypool et al. [28] introduce a categorization of both explicit and implicit interest indicators in web retrieval. They conclude that time spent on a page, the amount of scrolling on a page, and the combination of these two features are valid implicit indicators for interest. Furthermore, they found that individual scrolling measures and the number of mouse clicks are ineffective indicators. Morita and Shinoda [82] evaluated if user behavior while reading newsgroup articles could be used as an implicit indicator for interest. They measured the copying, saving, or following up of an entry and the time spent for reading the entries. They reveal that the reading time for documents rated as interesting was longer than for uninteresting documents. A relation between interest and following up, saving, or copying was not found. White et al. [127] consider reading time as a technique to automatically re-rank sentence-based summaries. Their results, however, were inconclusive. Kelly [64] criticizes the study approaches that focus on display time as relevance indicator, as she assumes that information-seeking behavior is not influenced by contextual factors such as topic, task, and collection. Therefore, she performed a study to investigate the relationship between information-seeking task and the display time. Her results cast doubt on the straightforward interpretation of dwell time as an indicator of interest or relevance.

Another indicator of interest that has been analyzed is the users' browsing behavior. Seo and Zhang [101] introduce a method to learn user preferences from inobtrusively observing their web-browsing behavior. They conclude that their approach can improve retrieval performance. However, the adaptation of user interest over a longer period of time has not been taken into account as their search sessions were set up for a short period only. Maglio et al. [77] suggest to infer attention from observing the eye movements. In the human computer interaction (HCI) community, this has become a common technique.

9.3.1.3 Pseudo-Relevance Feedback

A third relevance feedback approach is called *pseudo-*, *blind*, or *ad hoc relevance feedback*. It was first introduced by Croft and Harper [30]. Differing from the previous two approaches, pseudo-relevance feedback does not require users providing relevance assessments; the top ranked retrieval results are considered being relevant and used to adapt the initial search query. Considering the lack of manual input, its usage as source for personalization techniques is questionable.

9.3.1.4 Evolving User Interest

In a retrieval context, profiles can be used to contextualize the user's search queries within his or her interests and to re-rank retrieval results. This approach is based on the assumption that the user's information interest is static, which is, however, not appropriate in a retrieval context. Campbell [20] argues that the user's information need can change within different retrieval sessions and sometimes even within the same session. He states that the user search direction is directly influenced by the documents retrieved. The following example illustrates this observation:

> Imagine a user who is interested in red cars and uses a video retrieval system to find videos depicting such cars. Their first search query

returns several video clips, including videos of red Ferraris. Watching these video clips, he or she wants to find more Ferraris and adapts the search query accordingly. The new result list now consists of video clips showing red and green Ferraris. Fascinated by the rare colour for this type of car, he/she again reformulates the search query to find more green Ferraris. Within one session, the user's information need evolved from red cars to green Ferraris.

Based on this observation, Campbell and van Rijsbergen [22] introduce the Ostensive Model of Developing Information Need that incorporates this change of interest by considering *when* a user provided relevance feedback. In the Ostensive Model, providing feedback on a document is seen as ostensive evidence that this document is relevant for the user's current interest. The combination of this feedback over several search iterations provides ostensive evidence about the changing interest of the user. The model considers the user's changing focus of interest by granting the most recent feedback a higher impact over the combined evidence.

9.3.1.5 User Profiling

Considering the large amount of personal data that can be captured, most personalization systems rely on user profiles to manage these data. User profiling is the process of learning user interests over an extended period of time. User profiles may contain demographic information and user feedback that they expressed either explicitly or implicitly. This section highlights basic principles of user profiling. A state-of-the-art survey is given by Gausch et al. [42], who distinguish between two types of user profiles: *short-term* and *long-term* profiles. Short-term user profiles are used for personalization within one session, that is, any feedback that the user provides during their current information-seeking task is used to adapt the results. Long-term user profiles, on the other hand, aim to keep track of long-term user interests. Personalization services based upon such profiles can hence adapt results considering user feedback that was given over multiple sessions.

In literature, three types of user profile representations exist: weighted keywords or concepts, semantic networks, and association rules. Association rules are mainly applied in the field of web log mining. By identifying relations between variables, it is possible to identify popular variable combinations. Association rules rely on a large amount of data, often provided by different users. Considering this requirement, we therefore focus on weighting-based profiles and semantic network-based profiles, neglecting association rules. A survey on association rules is given by Mobasher [80].

Weighted keywords or concepts: The most popular representation of user interests is the weighted keyword or concept approach. Interests are represented as a vector of weighted terms that have either been extracted from those documents that users showed interest in or been provided manually by the users. The weighting indicates the importance of the corresponding term in the user profile. The main disadvantage of this approach is the so-called polysemy problem, hence the multiple meanings that each word can have.

An early example includes the *Amalthaea* system [83] where keywords, extracted from websites, are assigned with a weighting based on TF.IDF [7]. The term-weighting combination is represented in the profile as a vector. Similar approaches are studied by Sakagami and Kamba [96], who introduce personalized online newspapers; Lieberman [70], introducing a browsing assistant; and Pazzani et al. [89], who propose a recommender system that exploits weighted keyword profiles.

Even though weighted keyword profiling has been well studied in the text domain, hardly any work has been done on studying similar approaches in the video domain. Few exceptions include Weiß et al. [124], who, however, generate user profiles exploiting video metadata rather than audiovisual features.

Semantic networks: In the semantic network approach, keywords are replaced with concepts. User interests are represented as weighted nodes of a graph where each node is a concept in which the user showed interest. A similar approach is referred to as concept profiles. Differing from semantic networks, however, concept profiles consider abstract topics rather than specific words to represent user interests.

The advantage of semantic networks and concept profiles is that concepts can be organized in a hierarchical structure. In a web search scenario, for example, a user might have shown interest in a website that has been categorized in the Open Directory Project (ODP)* as being a website about "Travel and Tourism in Scotland." Within the ODP, "Travel and Tourism in Scotland" is a subcategory of "Travel and Tourism in the United Kingdom." Bloedorn et al. [15] argue to exploit such hierarchies, since they allow generalization of concepts. Personalization services could hence exploit this relationship, for instance, by recommending other websites belonging to the same or more general categories. Various approaches have been studied to exploit such public hierarchies.

Daoud et al. [32,33] analyze the documents that users provided implicit relevance feedback on, map the concepts of these documents with the ODP ontology, and store them in a hierarchical, graph-based user profile at the end of each search session. Other personalization techniques based on the ODP include [23,26,102,113], who show that incorporating this taxonomy can significantly outperform unpersonalized search techniques.

Dudev et al. [38] propose the creation of user profiles by creating knowledge graphs that model the relationship between different concepts in the

---

* http://dmoz.org/, accessed on 7 September 2011. The ODP is a manually edited catalog of websites. It is organized as a tree, where websites are leaf nodes and categories are internal nodes.

Linked Open Data Cloud.* Different from the ODP ontology, important parts of the Linked Open Data Cloud have been created automatically, for instance, by converting Wikipedia pages into an ontological representation. Consequently, the available data are of immense size but rather nonuniform.

Gauch et al. [42] argue that when exploiting such public directories, various preprocessing steps must be performed, including transforming the content into a concept hierarchy or dealing with situations where some concepts may have multiple entries while other concepts are less important. Moreover, they argue that the more levels the hierarchy used, the more general the profile representation may become.

9.3.1.6 Discussion

This section surveyed issues regarding gathering user interests. Relevance feedback was introduced as the most common technique used to identify user interest. Relevance feedback techniques were separated into three main categories: explicit, implicit, and pseudo-relevance feedback. Rui et al. [93] propose interactive relevance feedback as a method to bridge the semantic gap, assuming that high-level concepts can be identified using low-level features. In their approach, users have to rate images according to their relevance for the information need. The features are weighted automatically to model high-level concepts based on user feedback. The results of their study promise a reduction of query formulation efforts, as the relevance feedback technique appears to gather user information need effectively. According to Huang and Zhou [57], relevance feedback appears to be more promising in the image field than in the classical text field, as it is easier and faster to rank images according to their relevance than ranking text documents. Various researches have been performed to optimize parameters and algorithms [36,57,90,132]. Karthik and Jawahar [63] introduce a framework to evaluate different relevance feedback algorithms. They conclude that statistical models are the most promising algorithms. These models try to cluster images into relevant and nonrelevant images. Further, the Ostensive Model of Developing Information Need was introduced to emphasize the time when users provided relevance feedback. Various forms of this model have been developed and applied in image retrieval [21,69,116] and web search scenarios [62]. Finally, different approaches to user profiling were introduced. User profiling is one of the key challenges in adaptive search and recommendation. As discussed, two types of user profiling exist: short-term and long-term profiling.

---

* http://linkeddata.org/. Last time accessed on 5 May 2010. The Linked Open Data collection of ontologies unites information about many different freely available concepts, ODP being one of them.

9.3.2 Personalization Techniques

After introducing approaches to gather user interests, this section introduces personalization techniques that exploit such feedback. Jameson [58] lists various adaptation paradigms, including ability-based user interfaces, learning personal assistants, recommender systems, adaptation to situational impairments, personalized search, and user interfaces that adapt to the current task. Note that within this work, we will focus on recommender systems and personalized search, neglecting the other paradigms. Both paradigms will be introduced in the remainder of this section.

9.3.2.1 Personalized Search

In 2008, Marissa Mayer, the vice president of Search Products and User Experience of Google Inc., predicted in an interview held at the LeWeb conference that "in the future personalised search will be one of the traits of leading search engines" [79]. This statement reflects the increasing attention that personalized search draws from both academia and industry sides. Teevan et al. [114] argue that there is a big performance difference between personalized and nonpersonalized search engines. They hence argue that there is a big potential for personalization.

As discussed before, one of the main problems in IR is that users express their information need using short queries only. Matching these short queries with the document collection will return only a relative small amount of relevant results. Providing more search terms could improve the retrieval performance, as then more documents can be retrieved. Dependent on how much influence the user shall have, the expansion terms can either be added by the system in *automatic query expansion* or by the user in *interactive query expansion*. According to Ruthven [94], query expansion terms that have been provided in an interactive process are less useful than an automatically identified term. We therefore neglect interactive query expansion and focus on automatic query expansion techniques.

Automatic query expansion based on relevance feedback is a common technique to refine search queries (e.g., [92,98]). The reason for its success can be found by the users: they tend *not* to provide relevance feedback or to formulate their search queries appropriately. The first query expansion approach was introduced by Rocchio [92]. In a retrieval experiment, Rocchio added term vectors for all retrieved relevant documents and subtracted the term vectors for all irrelevant documents to refine search queries. Hence, terms are aligned with a weighting that can increase and decrease during the process. Järvelin [60] has shown that concept-based query expansion (i.e., exploiting ontologies) is helpful to improve retrieval performance. Multiple other studies show the effectiveness of ontology-based expansion [12].

Video retrieval-based query expansion approaches include Volkmer and Natsev [119], who rely on textual annotation (video transcripts) to expand search queries. Within their experiment, they significantly outperform a baseline run without any query expansion, hence indicating the potentials of query modification in video search. Similar results are reported by Porkaew and Chakrabarti [90] and Zhai et al. [131], who both expand search queries using content-based visual features.

9.3.2.2 Document Recommendation

Another personalization technique is document recommendation. Chris Anderson, editor in chief of the Wired Magazine, claims in his book [4] that "we are leaving the Information Age and entering the Recommendation Age." Differing from personalized search, where search results are adapted based on user interests, recommender systems provide additional items (documents). The main idea of this technique is to provide users faster and more information they might be interested in. Further, in an e-commerce scenario, recommendations shall influence the users. For example, Amazon.com* provides recommendations on other products that their customers might be interested in. Recommender systems hence inform users about things they might not be aware of and have not been actively searching for. They can be distinguished into two main categories: *content-based recommender systems* and *collaborative filtering systems*. In the remainder of this section, we will briefly introduce both approaches and discuss user profiling issues.

Content-based recommender systems: Content-based recommender systems determine the relevance of an item (e.g., a video document, website, or a product) based on the user's interest in other similar items. Items in the data collection are evaluated in accordance to previous feedback by users and the most similar items are recommended to the user. A survey is given by Pazzani and Billsus [88]. User interest is usually stored in personal user profiles.

Collaborative filtering: Collaborative filtering systems aim to exploit the opinion of people with similar interests. Thus, items are recommended when other users of the recommender system showed interest in it. Differing from content-based recommendation, where the content of the documents must be analyzed, the challenge in collaborative filtering is in identifying users with similar interests. A survey is given by Schafer et al. [99].

9.3.3 State-of-the-Art Personalization Approaches

Implicit feedback techniques have been successfully applied to retrieval systems in the past. For instance, White [125] and Joachims [61] defined and evaluated several implicit feedback models on a text-based retrieval system. Their results indicated that their implicit models were able to obtain

---

* http://www.amazon.com/, accessed on 7 September 2011.

a comparable performance to that obtained with explicit feedback models. However, their techniques were based on textual information and applied individually at runtime during the user's search session. As stated previously, the lack of textual annotation on video digital libraries prevents the adoption of this approach in video retrieval systems. One solution to this problem is to collect the implicit information from a large number of past users, following a collaborative recommendation strategy.

The exploitation of usage information from a community of past users is a widely researched approach to improve IR systems. However, most of past and present studies focus on the text retrieval domain. The main hypothesis of such systems is that when a user enters a query, the system can exploit the behavior of past users that were performing a similar task. For instance, Bauer and Leake [9] build up a task representation based on the user's sequence of accessed documents. This task representation is used by an information agent that proactively suggests documents to the user.

A commonly exploited past usage information structure is clickthrough data. Clickthrough data are limited to the query that the user executed into the system, the returned documents, and the subsequent documents that the user opened to view. Sun et al. [111] and Dou et al. [35] mine query log clickthrough information to perform a collaborative personalization of search results, giving preference to documents that similar users had clicked previously for similar queries. Sun et al. [111] apply a dimensional reduction preprocessing step on the clickthrough data in order to find latent semantic links between users, queries, and documents. Dou et al. [35] complement these latent relationships with user–topic and document–topic similarity measures. Craswell and Szummer [29] use a bipartite graph to represent the clickthrough data of an image retrieval system, where queries and documents are the nodes and links are the ones directly captured in clickthrough data. A random walk is then applied in order to recommend images based on the user's last query.

Following this line of investigations, White et al. [127] introduced the concept of query and search session trails, where the interaction between the user and the retrieval system is seen as a trail that leads from the user query to the last accessed document of the query session or the search session, which is constituted by multiple query sessions. They argue that the user's need of information was most likely satisfied with the last document of these trails, that is, the last document that the user accessed in the query or search session.

Earlier approaches rely on text to base personalization techniques. As discussed, however, multimedia documents consist of multiple modalities, including text and audiovisual features. A comprehensive survey on relevance feedback in the image domain is given by Zhou et al. [133]. The main problem when incorporating content-based features is to find out which feature represents the image best and, further, which approach should be followed to build an adaptive retrieval model. This problem even increases when dealing with video content, since additional features can be considered.

Early work focusing on relevance feedback in the video domain is presented by Rui et al. [130]. They illustrate that the content-rich nature of multimedia documents requires a more precise feedback. When a user provides relevance feedback on a video document, it is not clear which feature of this document should be exploited to identify similar documents. It could be, for example, the color feature of the video or the context of the video that is relevant to the user. In their work, Rui et al. [130] propose a relevance feedback framework for multimedia databases where search queries are adapted based on user relevance feedback. Their framework, however, is focusing on explicit relevance feedback only, neglecting the possibility to exploit implicit indicators of relevance. Another study focusing on explicit relevance feedback in the video domain is provided by Hauptmann et al. [51], who ask users to manually provide labels for video shots. After retrieving similar video documents for the label, users are then asked to explicitly mark those videos that match the corresponding label. Doulamis et al. [37] require explicit relevance feedback, given during the information-seeking process, to update video retrieval results. Hence, in all three studies that have been mentioned earlier, explicit relevance feedback is used to personalize search queries.

Luan et al. [74] consider text, high-level features, and multiple other modalities to base their relevance feedback algorithms on. In their approach, however, they do not focus on personalizing techniques but rather on improving video annotation. An extension of their work is presented in [75], where they propose multiple feedback strategies that support interactive video retrieval. They argue that the more complex nature of video data, when compared with textual data, requires different types of feedback strategies. Therefore, they distinguish between three categories of feedback types: recall-driven, precision-driven, and temporal locality-driven feedback. The first type focuses on analyzing the correlation of video features that can be extracted from positive- and negative-rated video documents. It aims to result in higher recall values. The second type employs active learning techniques and aims to constantly re-rank retrieval results. Its purpose is to increase precision. The third type, temporal locality-driven feedback, exploits the temporal coherence among neighbored shots. In their study, they show that giving the user the choice to choose between these different feedback types can effectively improve video retrieval performance.

Yan et al. [129] study video recommendations based on multimodal fusion and relevance feedback. They exploit viewing time to identify positive recommendations in their video recommender system. They interpret a very short video viewing time as negative relevance feedback, arguing that the user is not interested in the video content. Further, they argue that a longer viewing time indicates a higher interest in the video. Another study focusing on negative relevance feedback is introduced by Yan et al. [128], who consider the *lowest* ranked documents of a search query to be not relevant. Hence, they suggest to reformulate the initial search query by using these documents as negative example. Their approach, however, does not require any specific user input; it is based on pseudo-relevance feedback only. Nevertheless, their findings indicate that pseudo-relevance feedback can successfully be employed to adapt retrieval results.

Vrochidis et al. [120] aim to improve interactive video retrieval by incorporating additional implicit indicators of relevance. They consider the following user actions as implicit relevance feedback: (1) text-based search queries (given a search query, they assume that the keywords that form the search query are relevant), (2) visual queries (if a user provides a key frame as visual query, they assume this key frame to be relevant), and (3) side-shot and video-shot queries (when a user performs an action on a shot, they assume this shot to be relevant). Arguing that each indicator can be interpreted to a different degree to be relevant, they suggest to assign each feedback type a different weighting.

A purely content-based personalization approach is introduced by Aksoy and Cavus [3] who extract low-level visual features from explicitly relevant rated key frames. They suggest to extract different feature vectors from those key frames and to assign a relevance weighting for each vector.

A different understanding of implicit relevance feedback is introduced by Villa et al. [118]. Within their study, they aimed to elaborate whether the awareness of other user search activities can help users in their information-seeking task. They introduce a search scenario where two users remotely search for the same topic at the same time. Using their interface, each user is able to observe the other user's search activities. They conclude that awareness can have a direct influence in the information-seeking process, since users can learn from other user's search results and/or search queries. In case users copy the other user's search activity, this action can be interpreted as implicit relevance feedback, that the action has been performed on (or resulted in) relevant documents. A similar study is performed by Halvey et al. [46], who, however, study the impact of awareness in an asynchronous scenario. Within their experiment, users can interact with other user's previous search sessions. Again, continuing other user's search sessions can be seen as an implicit indication that the retrieved search results are partially relevant.

9.3.4 Summary

This section surveyed personalized video search and recommendation. An important prequisite for any kind of personalization service is to identify users' personal interests. As discussed, the most popular technique to gather this interest is relevance feedback. After introducing different feedback techniques and challenges such as evolving interest and user profiling, different personalization services were discussed, namely, personalized search and recommender systems. Finally, state-of-the-art personalization and recommendation systems were surveyed.

9.4 Summary

In recent years, multimedia content available to users has increased exponentially. This phenomenon has been accompanied by (and to great extent is the consequence of) a rapid development of tools, devices, and social services that facilitate the creation, storage, and sharing of personal multimedia content. A new landscape for business and innovation opportunities in multimedia content and technologies has naturally emerged from this evolution, and at the same time new problems and challenges arise. In particular, the hype around social services dealing with visual content, such as YouTube or Dailymotion, has led to a rather uncoordinated publishing of video data by users worldwide [31]. Because of the sheer amount of large data collections, there is a growing need to develop new methods that support the users in searching and finding videos they are interested in. This chapter first provided an overview on multimedia retrieval systems. The typical structure of video documents was introduced and an argument for the segmentation and representation of these documents was made. Personalization techniques that are required to identify those video documents that match the personal interest of a user were illustrated next. Different techniques to gather user interest were surveyed next. Finally, different personalization techniques were presented and state-of-the-art personalization approaches discussed.

References

1. E. Agichtein, E. Brill, and S. T. Dumais. Improving web search ranking by incorporating user behavior information. In *SIGIR 2006*, pp. 19–26, Seattle, WA, 2006.
2. P. Aigrain, H. J. Zhang, and D. Petkovic. Content-based representation and retrieval of visual media: A state-of-the-art review. *Multimedia Tools and Applications*, 3(3):179–202, 1996.
3. S. Aksoy and Ö. Çavuş. A relevance feedback technique for multimodal retrieval of news videos. In L. Milic, editor, *EUROCON'05: Proceedings of the International Conference on Computer as a Tool*, Belgrade, Serbia & Montenegro, pp. 139–142. IEEE, 11, 2005.
4. C. Anderson. *The Long Tail: Why the Future of Business is Selling Less of More.* Hyperion, New York, 7, 2006.
5. J. Arlandis, P. Over, and W. Kraaij. Boundary error analysis and categorization in the trecvid news story segmentation task. In *CIVR'05: Proceedings of the 4th International Conference on Image and Video Retrieval*, Singapore, pp. 103–112, July 20–22, 2005.
6. R. Arndt, R. Troncy, S. Staab, L. Hardman, and M. Vacura. COMM: Designing a well-founded multimedia ontology for the web. In K. Aberer, K. -S. Choi, N. Fridman Noy, D. Allemang, K.-Il Lee, L. J. B. Nixon, J. Golbeck, P. Mika,

D. Maynard, R. Mizoguchi, G. Schreiber, and P. Cudré-Mauroux, editors, *ISWC'07: Proceedings of the 6th International Semantic Web Conference, 2nd Asian Semantic Web Conference*, Busan, Korea, Vol. 4825 of *Lecture Notes in Computer Science*, pp. 30–43. Springer, 11, 2007.

7. R. Baeza-Yates and B. Ribeiro-Neto. *Modern Information Retrieval*. Addison Wesley, Harlow, Essex, England, 1st edition, 1999.

8. K. Barnard and D. A. Forsyth. Learning the semantics of words and pictures. In B. Werner, editor, *ICCV'01: Proceedings of the Eighth IEEE International Conference on Computer Vision*, Vancouver, British Columbia, Canada, pp. 408–415. IEEE, 2001.

9. T. Bauer and D. B. Leake. Real time user context modeling for information retrieval agents. In H. Paques, L. Liu, D. Grossman, and C. Pu, editors, *CIKM'01: Proceedings of the ACM CIKM International Conference on Information and Knowledge Management*, Atlanta, Georgia, pp. 568–570. ACM, 2001.

10. N. J. Belkin, C. Cool, D. Kelly, S. J. Lin, S. Park, J. Perez Carballo, and C. Sikora. Iterative exploration, design and evaluation of support for query reformulation in interactive information retrieval. *Information Processing and Management*, 37(3):403–434, 2001.

11. M. Bertini, A. Del Bimbo, and C. Torniai. Multimedia enriched ontologies for video digital libraries. *International Journal of Parallel, Emergent and Distributed Systems*, 22(6):407–416, 2007.

12. J. Bhogal, A. Macfarlane, and P. Smith. A review of ontology based query expansion. *Information Processing & Management: An International Journal*, 43(4): 866–886, 2007.

13. H. M. Blanken, A. P. de Vries, H. Ernst Bok, and L. Feng. *Multimedia Retrieval*. Springer Verlag, Heidelberg, Germany, 1st edition, 2007.

14. D. M. Blei and M. I. Jordan. Modeling annotated data. In J. Callan et al. [19], pp. 127–134.

15. E. Bloedorn, I. Mani, and T. R. MacMillan. Machine learning of user profiles: representational issues. In *AAAI/IAAI*, Vol. 1, pp. 433–438. AAAI, Portland, Oregon, 1996.

16. M. Boughanem, C. Berrut, J. Mothe, and C. Soulé-Dupuy, editors. *ECIR'09: Proceedings of the 31th European Conference on IR Research*, Toulouse, France, Vol. 5478 of *Lecture Notes in Computer Science*. Springer Verlag, 4, 2009.

17. P. Brusilovsky, A. Kobsa, and W. Nejdl, editors. *The Adaptive Web*. Springer Verlag, Berlin, Germany, New York, 2007.

18. T. Bürger, E. Gams, and G. Güntner. Smart content factory: assisting search for digital objects by generic linking concepts to multimedia content. In S. Reich and M. Tzagarakis, editors, *HT'05: Proceedings of the 16th ACM Conference on Hypertext and Hypermedia*, Salzburg, Austria, pp. 286–287. ACM, 2005.

19. J. Callan, G. Cormack, C. Clarke, D. Hawking, and A. F. Smeaton, editors. *SIGIR'03: Proceedings of the 26th Annual International ACM SIGIR Conference on Research and Development in Information Retrieval*, Toronto, Ontario, Canada. ACM, 2003.

20. I. Campbell. Supporting information needs by ostensive definition in an adaptive information space. In I. Ruthven, editor, *MIRO'95: Proceedings of the Final Workshop on Multimedia Information Retrieval*, Glasgow, Scotland, U.K., Workshops in Computing. BCS, 9, 1995.

21. I. Campbell. Interactive evaluation of the ostensive model using a new test collection of images with multiple relevance assessments. *Information Retrieval*, 2(1):85–112, 2000.

22. I. Campbell and C. J. van Rijsbergen. The ostensive model of developing information needs. In P. Ingwersen and N. Ole Pors, editors, *CoLIS'06: Proceedings of the 6th International Conference on Conceptions of Library and Information Sciences*, Copenhagen, Denmark, pp. 251–268. The Royal School of Librarianship, 10, 1996.

23. J. Chaffee and S. Gauch. Personal ontologies for web navigation. In A. Agah, J. Callan, E. Rundensteiner, and S. Gauch, editors, *CIKM'00: Proceedings of the 2000 ACM CIKM International Conference on Information and Knowledge Management*, McLean, VA, pp. 227–234. ACM, 11, 2000.

24. L. Chaisorn and T.-S. Chua. The segmentation and classification of story boundaries in news video. In X. Zhou and P. Pu, editors, *VDB'02: Proceedings of the IFIP TC2/WG2.6 Sixth Working Conference on Visual Database Systems*, Brisbane, Australia, Vol. 216 of *IFIP Conference Proceedings*, pp. 95–109. Kluwer, 2002.

25. S.-F. Chang, R. Manmatha, and T.-S. Chua. Combining text and audio-visual features in video indexing. In B. Mercer, editor, *ICASSP'05: Proceedings of IEEE International Conference on Acoustics, Speech, and Signal Processing*, pp. 1005–1008. IEEE, 3, Philadelphia, PN, 2005.

26. P.-A. Chirita, W. Nejdl, R. Paiu, and C. Kohlschütter. Using odp metadata to personalize search. In *SIGIR 2005*, pp. 178–185, Salvador, Bahia, Brazil, 2005.

27. T.-S. Chua, S.-F. Chang, L. Chaisorn, and W. H. Hsu. Story boundary detection in large broadcast news video archives: Techniques, experience and trends. In H. Schulzrinne, N. Dimitrova, M. Angela Sasse, S. B. Moon, and R. Lienhart, editors, *ACM MM'04: Proceedings of the 12th ACM International Conference on Multimedia*, New York, pp. 656–659. ACM, October 10–16, 2004.

28. M. Claypool, P. Le, M. Wased, and D. Brown. Implicit interest indicators. In C. Sidner and J. Moore, editors, *IUI'01: Proceedings of the International Conference on Intelligent User Interfaces*, Santa Fe, NM, pp. 33–40. ACM, 1 2001.

29. N. Craswell and M. Szummer. Random walks on the click graph. In W. Kraaij et al. [66], pp. 239–246.

30. W. B. Croft and D. J. Harper. Using probabilistic models of document retrieval without relevance information. *Readings in Information Retrieval*, 339–344, 1997.

31. S. J. Cunningham and D. M. Nichols. How people find videos. In R. Larsen et al. [68], pp. 201–210.

32. M. Daoud, L. Tamine-Lechani, and M. Boughanem. Using a graph-based ontological user profile for personalizing search. In J. G. Shanahan, S. Amer-Yahia, I. Manolescu, Y. Zhang, D. A. Evans, A. Kolcz, K.-S. Choi, and A. Chowdhury, editors, *CIKM'08: Proceedings of the 17th ACM Conference on Information and Knowledge Management*, Napa Valley, CA, pp. 1495–1496. ACM, 2008.

33. M. Daoud, L. Tamine-Lechani, M. Boughanem, and B. Chebaro. A session based personalized search using an ontological user profile. In S. Y. Shin and S. Ossowski, editors, *SAC'09: Proceedings of the 2009 ACM Symposium on Applied Computing*, Honolulu, HI, pp. 1732–1736. ACM, 2009.

34. H. Day. Looking time as a function of stimulus variables and individual differences. *Perceptual & Motor Skills*, 22(2):423–428, 1966.

35. Z. Dou, R. Song, and J.-R. Wen. A large-scale evaluation and analysis of personalized search strategies. In C. L. Williamson, M. E. Zurko, P. F. Patel-Schneider, and P. J. Shenoy, editors, *WWW'07: Proceedings of the 16th International Conference on World Wide Web*, Banff, Alberta, Canada, pp. 581–590. ACM, 2007.

36. A. Doulamis and N. Doulamis. Performance evaluation of Euclidean/correlation-based relevance feedback algorithms in content-based image retrieval systems. In L. Torres, editor, *ICIP'03: Proceedings of the International Conference on Image Processing*, Barcelona, Spain, Vol. 4, pp. 737–740. IEEE, 09, 2003.

37. A. D. Doulamis, Y. S. Avrithis, N. D. Doulamis, and S. D. Kollias. Interactive content-based retrieval in video databases using fuzzy classification and relevance feedback. In *ICMCS: IEEE International Conference on Multimedia Computing and Systems*, Florence, Italy, Vol. 2, pp. 954–958, 1999.

38. M. Dudev, S. Elbassuoni, J. Luxenburger, M. Ramanath, and G. Weikum. Personalizing the search for knowledge. In V. Christophides and G. Koutrika, editors, *PersDB'08: Proceedings of the Second International Workshop on Personalized Access, Profile Management, and Context Awareness: Databases*, pp. 1–8, Auckland, New Zealand, 2008.

39. G. Durand, G. Kazai, M. Lalmas, U. Rauschenbach, and P. Wolf. A metadata model supporting scalable interactive TV services. In Y.-P. Phoebe Chen, editor, *MMM'05: Proceedings of the 11th International Conference on Multi Media Modeling*, Melbourne, Victoria, Australia, pp. 386–391. IEEE Computer Society, 1, 2005.

40. T. Faw and J. Nunnally. The effects on eye movements of complexity, novelty, and affective tone. *Perception & Psychophysics*, 2(7):263–267, 1967.

41. C. Fellbaum, editor. *WordNet: An Electronic Lexical Database (Language, Speech, and Communication)*. The MIT Press, Cambridge, U.K., May 1998.

42. S. Gauch, M. Speretta, A. Chandramouli, and A. Micarelli. *User Profiles for Personalized Information Access*, chapter 2, pp. 54–89. Vol. 1 of P. Brusilovsky et al. [17], 2007.

43. T. R. Gruber. Toward principles for the design of ontologies used for knowledge sharing. *International Journal of Human-Computer Studies*, 43(5–6):907–928, 1995.

44. L. Guan and H.-J. Zhang, editors. *ICME'06: Proceedings of the 2006 IEEE International Conference on Multimedia and Expo*, Toronto, Ontario, Canada. IEEE, 2006.

45. M. Halvey and M. T. Keane. Analysis of online video search and sharing. In S. Harper, H. Ashman, M. Bernstein, A. I. Cristea, H. C. Davis, P. De Bra, V. L. Hanson, and D. E. Millard, editors, *HT'07: Proceedings of the 18th ACM Conference on Hypertext and Hypermedia*, Manchester, U.K., pp. 217–226. ACM, 9, 2007.

46. M. Halvey, D. Vallet, D. Hannah, and J. M. Jose. Vigor: a grouping oriented interface for search and retrieval in video libraries. In F. Heath, M. L. Rice-Lively, and R. Furuta, editors, *JCDL'09: Proceedings of the 2009 Joint International Conference on Digital Libraries*, Austin, TX, pp. 87–96. ACM, 2009.

47. M. Hancock-Beaulieu and S. Walker. An evaluation of automatic query expansion in an online library catalogue. *Journal of Documentation*, 48(4):406–421, 1992.

48. A. Hanjalic. Shot-boundary detection: Unraveled and resolved? *IEEE Transactions on Circuits and Systems for Video Technology*, 12(2):90–105, 2, 2002.

49. A. G. Hauptmann. Lessons for the future from a decade of informedia video analysis research. In *CIVR 2005*, Springer Verlag, pp. 1–10, Singapore, 2005.

50. A. Hauptmann, M. Christel, R. Concescu, J. Gao, Q. Jin, J. Y. Pan, S. M. Stevens, R. Yan, J. Yang, and Y. Zhang. CMU Informedia's TRECVID 2005 Skirmishes. In *TRECVid 2005*, Gaithersburg, MD, 2005.

51. A. G. Hauptmann, W.-H. Lin, R. Yan, J. Yang 0003, and M. yu Chen. Extreme video retrieval: joint maximization of human and computer performance. In K. Nahrstedt, M. Turk, Y. Rui, W. Klas, and K. Mayer-Patel, editors, *MM'06: Proceedings of the 14th ACM International Conference on Multimedia*, Santa Barbara, CA, pp. 385–394. ACM, 2006.

52. X. He, R. S. Zemel, and M. Á. Carreira-Perpiñán. Multiscale conditional random fields for image labeling. In F. Titsworth, editor, *CVPR'04: Proceedings of the IEEE Computer Society Conference on Computer Vision and Pattern Recognition*, Vol. 2, pp. 695–702. IEEE, Washington DC, 2004.

53. M. Hearst. *Search User Interfaces*. Cambridge University Press, Cambridge, NY, 2009.

54. M. Hildebrand, J. van Ossenbruggen, and L. Hardman. An analysis of search-based user interaction on the semantic web. Technical Report INS-E0706, Centrum voor Wiskunde en Informatica, Amsterdam, the Netherlands, 5, 2007.

55. F. Hopfgartner, R. Ren, T. Urruty, and J. M. Jose. *Information Organisation Issues in Multimedia Retrieval using Low-Level Features*, chapter 15. Multimedia Semantics: Metadata, Analysis and Interaction. Wiley, Hoboken, NJ, 1st edition, 2011.

56. C.-W. Huang. Automatic closed caption alignment based on speech recognition transcripts. Technical report, University of Columbia, New York, 2003. ADVENT Technical Report.

57. T. S. Huang and X. S. Zhou. Image retrieval with relevance feedback: From heuristic weight adjustment to optimal learning methods. In I. Pitas, A. N. Venetsanopoulos, and T. Pappas, editors, *ICIP'01: Proceedings of International Conference on Image Processing*, Thessaloniki, Greece, Vol. 3, pp. 2–5, IEEE, 2001

58. A. Jameson. Adaptive interfaces and agents. In *The Human-Computer Interaction Handbook: Evolving Technologies and Emerging Applications*, CRC Press, Boca Raton, FL, pp. 433–458, 2008.

59. B. J. Jansen, A. Goodrum, and A. Spink. Searching for multimedia: analysis of audio, video and image web queries. *World Wide Web*, 3(4):249–254, 2000.

60. K. Järvelin, J. Kekäläinen, and T. Niemi. ExpansionTool: Concept-based query expansion and construction. *Information Retrieval*, 4(3–4):231–255, 2001.

61. T. Joachims, L. A. Granka, B. Pan, H. Hembrooke, and G. Gay. Accurately interpreting clickthrough data as implicit feedback. In *SIGIR 2005*, pp. 154–161, Salvador, Bahia, Barzil, 2005.

62. H. Joho, D. Birbeck, and J. M. Jose. An ostensive browsing and searching on the web. In B.-L. Doan, M. Melucci, and J. M. Jose, editors, *CIR'07: Proceedings of the Second International Workshop on Context-Based Information Retrieval*, pp. 81–92, Roskilde University, Copenhagen, Denmark, 2007.

63. P. S. Karthik and C. V. Jawahar. Analysis of relevance feedback in content based image retrieval. *ICARCV'06: Proceedings of the 9th International Conference on Control, Automation, Robotics and Vision*, Singapore, 2006.

64. D. Kelly. Understanding implicit feedback and document preference: A naturalistic user study. PhD thesis, Rutgers University, Newark, NJ, 2004.

65. D. Kelly and J. Teevan. Implicit feedback for inferring user preference: A bibliography. *SIGIR Forum*, 37(2):18–28, 2003.

66. W. Kraaij, A. P. de Vries, C. L. A. Clarke, N. Fuhr, and N. Kando, editors. *SIGIR'07: Proceedings of the 30th Annual International ACM SIGIR Conference on Research and Development in Information Retrieval*, Amsterdam, the Netherlands. ACM, 2007.

67. S. Kumar and M. Hebert. Discriminative random fields: A discriminative framework for contextual interaction in classification. In K. Ikeuchi, O. Faugeras, and J. Malik, editors, *ICCV'03: 9th IEEE International Conference on Computer Vision*, Nice, France, pp. 1150–1159. IEEE Computer Society, 2003.

68. R. L. Larsen, A. Paepcke, J. L. Borbinha, and M. Naaman, editors. *JCDL'08: Proceedings of the ACM/IEEE Joint Conference on Digital Libraries*, Pittsburgh, PA. ACM, 2008.

69. T. Leelanupab, F. Hopfgartner, and J. M. Jose. User centred evaluation of a recommendation based image browsing system. In B. Prasad, P. Lingras, and A. Ram, editors, *IICAI'09: Proceedings of the 4th Indian International Conference on Artificial Intelligence*, Tumkur, Karnataka, India, pp. 558–573. IICAI, 2009.

70. H. Lieberman. Letizia: an agent that assists web browsing. In *IJCAI'95: Proceedings of the 14th International Joint Conference on Artificial Intelligence*, Montreal, Quebec, Canada, pp. 924–929. Morgan Kaufmann, 8, 1995.

71. R. Lienhart, A. R. Prasad, A. Hanjalic, S. Choi, B. P. Bailey, and N. Sebe, editors. *MM'07: Proceedings of the 15th International Conference on Multimedia*, Augsburg, Germany. ACM, 2007.

72. A. Llorente, S. E. Overell, H. Liu 0002, R. Hu, A. Rae, J. Zhu, D. Song, and S. M. Rüger. Exploiting term co-occurrence for enhancing automated image annotation. In C. Peters, T. Deselaers, N. Ferro, J. Gonzalo, G. J. F. Jones, M. Kurimo, T. Mandl, A. Peñas, and V. Petras, editors, *CLEF'08: Proceedings of the 9th Workshop of the Cross-Language Evaluation Forum*, Aarhus, Denmark, Revised Selected Papers, Vol. 5706 of *Lecture Notes in Computer Science*, pp. 632–639. Springer, 2008.

73. A. Llorente and S. M. Rüger. Using second order statistics to enhance automated image annotation. In M. Boughanem et al. [16], pp. 570–577.

74. H.-B. Luan, S.-Y. Neo, H.-K. Goh, Y.-D. Zhang, S. Lin, and T.-S. Chua. Segregated feedback with performance-based adaptive sampling for interactive news video retrieval. In R. Lienhart et al. [71], pp. 293–296.

75. H.-B. Luan, Y. Zheng, S.-Y. Neo, Y. Zhang, S. Lin, and T.-S. Chua. Adaptive multiple feedback strategies for interactive video search. In J. Luo, L. Guan, A. Hanjalic, M. S. Kankanhalli, and I. Lee, editors, *CIVR'08: Proceedings of the 7th ACM International Conference on Image and Video Retrieval*, Niagara Falls, Ontario, Canada, pp. 457–464. ACM, 2008.

76. J. Magalhães and S. M. Rüger. *Semantic Multimedia Information Analysis for Retrieval Applications*, pp. 334–354. IDEA Group Publishing, Hershey, PA, 2006.

77. P. P. Maglio, R. Barrett, C. S. Campbell, and T. Selker. Suitor: an attentive information system. In D. Riecken, D. Benyon, and H. Liebermann, editors, *IUI'00: Proceedings of the Fifth International Conference on Intelligent User Interfaces*, Santa Fe, NM, pp. 169–176, New York, ACM Press, 1 2000.

78. B. S. Manjunath, P. Salembier, and T. Sikora, editors. *Introduction to MPEG 7: Multimedia Content Description Language*. Wiley, Chichester, U.K., 6, 2002.

79. M. Mayer. *Interview at the "LeWeb 2008" Conference in Paris*, France, 2008.

80. B. Mobasher. *Data Mining for Web Personalization*, chapter 3, pp. 90–135. Vol. 1 of P. Brusilovsky et al. [17], 2007.

81. J. Monaco. *How to Read a Film*. Oxford Press, London, U.K., 4th edition, 2009.

82. M. Morita and Y. Shinoda. Information filtering based on user behavior analysis and best match text retrieval. In W. B. Croft and C. J. van Rijsbergen, editors, *SIGIR'94: Proceedings of the 17th Annual International ACM-SIGIR Conference on Research and Development in Information Retrieval*. Dublin, Ireland, (Special Issue of the SIGIR Forum), pp. 272–281. ACM/Springer, 1994.

83. A. Moukas and P. Maes. Amalthaea: an evolving multi-agent information filtering and discovery system for the WWW. *Autonomous Agents and Multi-Agent Systems*, 1(1):59–88, 1998.

84. M. R. Naphade, J. R. Smith, J. Tesic, S.-F. Chang, W. H. Hsu, L. S. Kennedy, A. G. Hauptmann, and J. Curtis. Large-scale concept ontology for multimedia. *IEEE MultiMedia*, 13(3):86–91, 2006.

85. S.-Y. Neo, J. Zhao, M.-Y. Kan, and T.-S. Chua. Video retrieval using high level features: exploiting query matching and confidence-based weighting. In H. Sundaram, M. R. Naphade, J. R. Smith, and Y. Rui, editors, *CIVR'06: Proceedings of the 5th International Conference on Image and Video Retrieval*, Tempe, AZ, Vol. 4071 of *Lecture Notes in Computer Science*, pp. 143–152. Springer, 2006.

86. D. M. Nichols. Implicit rating and filtering. In L. Kovacs, editor, *Proceedings of 5th DELOS Workshop on Filtering and Collaborative Filtering*, Budapest, Hungary, pp. 31–36. ERCIM, 11, 1998.

87. A. Oostendorp and D. E. Berlyne. Dimensions in the perception of architecture II: measures of exploratory behavior. *Scandinavian Journal of Psychology*, 19(1):83–89, 1978.

88. M. J. Pazzani and D. Billsus. *Content-Based Recommender Systems*, chapter 10, pp. 325–341. Vol. 1 of P. Brusilovsky et al. [17], 2007.

89. M. J. Pazzani, J. Muramatsu, and D. Billsus. Syskill & webert: identifying interesting web sites. In D. Weld and B. Clancey, editors, *AAAI/IAAI'96: Proceedings of the 13th National Conference on Artificial Intelligence*, Portland, OR, pp. 54–61. AAAI, 8 1996.

90. K. Porkaew and K. Chakrabarti. Query refinement for multimedia similarity retrieval in MARS. In J. Buford and S. Stevens, editors, *MM'99: Proceedings of the 7th ACM International Conference on Multimedia*, Orlando, FL, pp. 235–238. ACM, 10, 1999.

91. G.-J. Qi, X.-S. Hua, Y. Rui, J. Tang, T. Mei, and H.-J. Zhang. Correlative multi-label video annotation. In R. Lienhart et al. [71], pp. 17–26.

92. J. J. Rocchio. Relevance feedback in information retrieval. In Gerard Salton, editor, *The SMART Retrieval System: Experiments in Automatic Document Processing*, Englewood Cliffs, NJ, pp. 313–323, Prentice-Hall, 1971.

93. Y. Rui, T. S. Huang, M. Ortega, and S. Mehrotra. Relevance feedback: a power tool for interactive content-based image retrieval. *IEEE Transactions on Circuits and Systems for Video Technology*, 8(5):644–655, 1998.

94. I. Ruthven. Re-examining the potential effectiveness of interactive query expansion. In J. Callan et al. [19], pp. 213–220.

95. I. Ruthven and M. Lalmas. A survey on the use of relevance feedback for information access systems. *The Knowledge Engineering Review*, 18(2):95–145, 2003.

96. H. Sakagami and T. Kamba. Learning personal preferences on online newspaper articles from user behaviors. *Computer Networks*, 29(8–13):1447–1455, 1997.

97. G. Salton, J. Allan, and C. Buckley. Approaches to passage retrieval in full text information systems. In R. Korfhage, E. M. Rasmussen, and P. Willett, editors, *SIGIR'93: Proceedings of the 16th Annual International ACM Conference on Research and Development in Information Retrieval*, Pittsburgh, PA, pp. 49–58, ACM, June 27–July 1, 1993.

98. G. Salton and C. Buckley. Improving retrieval performance by relevance feedback. In *Readings in Information Retrieval*, Morgan Kaufmann Publishers, pp. 355–364, Cornell University, Ithaca, NY, 1997.

99. J. B. Schafer, D. Frankowski, J. Herlocker, and S. Sen. *Collaborative Filtering Recommender Systems*, chapter 9, pp. 291–324. Vol. 1 of P. Brusilovsky et al. [17], 2007.

100. K. Schoeffmann, F. Hopfgartner, O. Marques, L. Boeszoermenyi, and J. M. Jose. Video browsing interfaces and applications: a review. *SPIE Reviews*, 1(1):018004-1–018004-35, 2010.

101. Y. -W. Seo and B. -T. Zhang. Learning user's preferences by analyzing web-browsing behaviors. In *AGENTS'00: Proceedings of the Fourth International Conference on Autonomous Agents*, Barcelona, Catalonia, Spain, pp. 381–387, New York, ACM Press, 2000.

102. A. Sieg, B. Mobasher, and R. D. Burke. Web search personalization with ontological user profiles. In M. J. Silva, A. H. F. Laender, R. A. Baeza-Yates, D. L. McGuinness, B. Olstad, Ø. H. Olsen, and A. O. Falcão, editors, *CIKM'07: Proceedings of the Sixteenth ACM Conference on Information and Knowledge Management*, Lisbon, Portugal, pp. 525–534. ACM, 2007.

103. N. Simou, C. Saathoff, S. Dasiopoulou, V. Spyrou, N. Voisine, V. Tzouvaras, Y. Kompatsiaris, Y. Avrithis, and S. Staab. An ontology infrastructure for multimedia reasoning. In L. Atzori, D. D. Giusto, R. Leonardi, and F. Pereira, editors, *VLBV'05: Proceedings of the 9th International Workshop on Visual Content Processing and Representation*, Sardinia, Italy, Vol. 3893 of *Lecture Notes in Computer Science*. Springer, 9, 2005.

104. A. F. Smeaton, P. Over, and A. R. Doherty. Video shot boundary detection: seven years of TRECVid activity. *Computer Vision and Image Understanding*, 4(114):411–418, 2010.

105. C. G. M. Snoek and M. Worring. Multimodal video indexing: a review of the state-of-the-art. *Multimedia Tools & Applications*, 25(1):5–35, 2005.

106. C. G. Snoek and M. Worring. Concept-based video retrieval. *Foundations and Trends in Information Retrieval*, 2(4):215–322, 2009.

107. C. G. M. Snoek, M. Worring, D. C. Koelma, and A. W. M. Smeulders. A learned lexicon-driven paradigm for interactive video retrieval. *IEEE Transactions on Multimedia*, 9(2):280–292, 2, 2007.

108. C. G. M. Snoek, K. E. A. van de Sande, O. de Rooij, B. Huurnink, J. C. van Gemert, J. R. R. Uijlings, J. He et al. The mediamill trecvid 2008 semantic video search engine. In P. Over, W. Kraaij, and A. F. Smeaon, editors, *TRECVid'08: Notebook Papers and Results*, pp. 314–327, Gaithersburg, MD, National Institute of Standards and Technology, 11, 2008.

109. A. Spink, H. Greisdorf, and J. Bateman. From highly relevant to not relevant: examining different regions of relevance. *Information Processing & Management: An International Journal*, 34(5):599–621, 1998.

110. V. Stathopoulos and J. M. Jose. Bayesian mixture hierarchies for automatic image annotation. In M. Boughanem et al. [16], pp. 138–149.
111. J.-T. Sun, H.-J. Zeng, H. Liu, Y. Lu, and Z. Chen. CubeSVD: a novel approach to personalized web search. In A. Ellis and T. Hagino, editors, *WWW'05: Proceedings of the 14th international conference on World Wide Web*, Chiba, Japan, pp. 382–390. ACM, 2005.
112. H. Tansley. The multimedia thesaurus: adding a semantic layer to multimedia information. PhD thesis, University of Southampton, Southampton, U.K., 2000.
113. F. Tanudjaja and L. Mui. Persona: a contextualized and personalized web search. *HICSS'02: Proceedings of the 35th Hawaii International Conference on System Sciences*, Big Island, HA, 3:67, 2002.
114. J. Teevan, S. T. Dumais, and E. Horvitz. Potential for personalization. *ACM Transactions on Computer-Human Interaction*, 17(1):1–31, 2010.
115. C. Tsinaraki, P. Polydoros, F. G. Kazasis, and S. Christodoulakis. Ontology-based semantic indexing for MPEG-7 and TV-anytime audiovisual content. *Multimedia Tools and Applications*, 26(3):299–325, 2005.
116. J. Urban, J. M. Jose, and C. J. van Rijsbergen. An adaptive technique for content-based image retrieval. *Multimedia Tools & Applications*, 31(1):1–28, 2006.
117. R. van Zwol, L. Garcia Pueyo, G. Ramírez, B. Sigurbjörnsson, and M. Labad. Video tag game. In *WWW'08: Proceedings of the 17th International World Wide Web Conference (WWW developer track)*. ACM Press, Beijing, China, 2008.
118. R. Villa, N. Gildea, and J. M. Jose. A study of awareness in multimedia search. In R. L. Larsen et al. [68], pp. 221–230.
119. T. Volkmer and A. Natsev. Exploring automatic query refinement for text-based video retrieval. In L. Guan and H.-J. Zhang [44], pp. 765–768.
120. S. Vrochidis, I. Kompatsiaris, and I. Patras. Optimizing visual search with implicit user feedback in interactive video retrieval. In S. Li, editor, *CIVR'10: Proceedings of the 9th ACM International Conference on Image and Video Retrieval*, Xi'an, China. pp. 274–281, ACM, 2010.
121. M. Wang, X.-S. Hua, X. Yuan, Y. Song, and L.-R. Dai. Optimizing multi-graph learning: towards a unified video annotation scheme. In R. Lienhart et al. [71], pp. 862–871.
122. J. Watkinson. *The MPEG Handbook*. Focal Press, Burlington, MA, 2nd edition, 2001.
123. S. Weibel. The Dublin core: A simple content description model for electronic resources. *Bulletin of the American Society for Information Science and Technology*, 24(1):9–11, 2005.
124. D. Weiß, J. Scheuerer, M. Wenleder, A. Erk, M. Gülbahar, and C. Linnhoff-Popien. A user profile-based personalization system for digital multimedia content. In S. Tsekeridou, A. D. Cheok, K. Giannakis, and J. Karigiannis, editors, *DIMEA'08: Proceedings of the Third International Conference on Digital Interactive Media in Entertainment and Arts*, Athens, Greece, Vol. 349 of *ACM International Conference Proceeding Series*, pp. 281–288. ACM, 2008.
125. R. White. *Implicit Feedback for Interactive Information Retrieval*. PhD thesis, University of Glasgow, 2004.
126. R. W. White, M. Bilenko, and S. Cucerzan. Studying the use of popular destinations to enhance web search interaction. In W. Kraaij et al. [66], pp. 159–166.

127. R. White, I. Ruthven, and J. M. Jose. Finding relevant documents using top ranking sentences: an evaluation of two alternative schemes. In K. Järvelin, M. Beaulieu, R. Baeza-Yates, and S. H. Myaeng, editors, *SIGIR'02: Proceedings of the 25th Annual International ACM SIGIR Conference on Research and Development in Information Retrieval*, Tampere, Finland, pp. 57–64. ACM, 2002.

128. R. Yan, A. G. Hauptmann, and R. Jin. Multimedia search with pseudo-relevance feedback. In E. M. Bakker, T. S. Huang, M. S. Lew, N. Sebe, and X. S. Zhou, editors, *CIVR'03: Proceedings of the Second International Conference on Image and Video Retrieval*, Urbana-Champaign, IL, Vol. 2728 of *Lecture Notes in Computer Science*, pp. 238–247. Springer, 2003.

129. B. Yang, T. Mei, X.-S. Hua, L. Yang, S.-Q. Yang, and M. Li. Online video recommendation based on multimodal fusion and relevance feedback. In N. Sebe and M. Worring, editors, *CIVR'07: Proceedings of the 6th ACM International Conference on Image and Video Retrieval*, Amsterdam, the Netherlands, pp. 73–80. ACM, 2007.

130. R. Yong, T. S. Huang, S. Mehrotra, and M. Ortega. A relevance feedback architecture for content-based multimedia information systems. In P. Storms, editor, *Proceedings of IEEE Workshop on Content-Based Access of Image and Video Libraries*, San Juan, Puerto Rico, 1997.

131. Y. Zhai, J. Liu, and M. Shah. Automatic query expansion for news video retrieval. In L. Guan and H.-J. Zhang [44], pp. 965–968.

132. X. S. Zhou and T. S. Huang. Exploring the nature and variants of relevance feedback. In L. Palagi, editor, *CBAIVL '01: Proceedings of the IEEE Workshop on Content-based Access of Image and Video Libraries*, Fort Collins, CO, pp. 94–100, Washington, DC, IEEE Computer Society. 6, 2001.

133. X. S. Zhou and T. S. Huang. Relevance feedback in image retrieval: a comprehensive review. *Multimedia Systems*, 8(6):536–544, 2003.

10

Accelerated Acquisition, Visualization, and Analysis of Zoo-Anatomical Data

Alexander Ziegler and Björn H. Menze

CONTENTS

10.1 Introduction

The internal and external anatomical features of animals have been the focus of scientific study for thousands of years, starting with the early text-based descriptions of Aristotle and Plinius. Because complex 3D structures are in general difficult to convey using the written word, it was the invention of reproducible woodcut and block book techniques in the sixteenth century (Figure 10.1A) that led to a significant boost in the communication of anatomical findings (see, e.g., these seminal works: Belon (1553), Rondelet (1554), Gesner (1558), and Aldrovandi (1606)). Since then, 2D, paper-based illustrations have formed the standard for the presentation of anatomical results in scientific publications. Alas, the world that we live in is 3D by nature, and a considerable amount of information is therefore omitted when objects are portrayed two-dimensionally or described in written words. The numerous limitations inherent to analog data communication also extend to how anatomical information has been gathered and analyzed in the course of the centuries, for example, through photography, dissection, and histology (Figure 10.1B).

The digital revolution, which has substantially transformed multiple scientific disciplines over the past three decades, is now also beginning to change the way by which anatomists and zoologists look at the samples they are studying (see Budd and Olsson (2007), Schmidt-Rhaesa (2009), and Giribet (2010) for discussion). In contrast to conventional techniques such as manual dissection or histology, noninvasive imaging techniques such as micro-computed tomography (μCT) or magnetic resonance imaging (MRI) permit gathering anatomical data at high speed from the very onset of a study. Although these techniques rely on different physical principles, both acquire anatomical data in purely digital form and directly from the intact specimen itself (reviews by Baker (2010) as well as Walter et al. (2010) provide an introduction to imaging techniques currently employed in biology and medicine). Some of these modalities also allow unmasking anatomical information embedded within matter surrounding the specimen itself, such as for example, amber or rock. In contrast to conventional, primarily analog data that had been collected previously in the form of drawings and sections, significantly more information could be retrieved once digital image data of entire specimens became available (Figure 10.1C).

A prominent example for novel insights into zoology through primarily digital analyses is the large-scale genetic studies performed in recent years. Although here relatively abstract gene sequences instead of discrete anatomical structures are studied, the key to success is the availability of large amounts of digital data. For instance, using sequence data derived from a set of 12 genes taken from tissue samples of 2,871 different species, the evolutionary relationships of frogs, salamanders, and toads were reconstructed on a scale that was previously simply not achievable using the classical analog

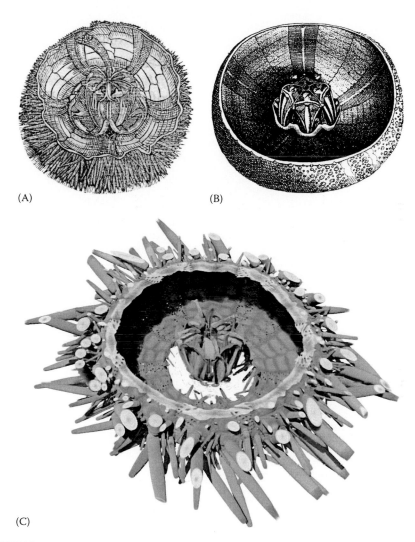

(A)

(B)

(C)

FIGURE 10.1
Selected anatomical representations illustrate the transition of data acquisition and data presentation from a traditional 2D and analog to a modern 3D and digital form. (A) Sketch of a dissected sea urchin. (Modified from Rondelet, G., *Libri de piscibus marinis, in quibus verae piscium effigies expressae sunt*, M. Bonhomme, Lyon, France, 1554.) The invention of reproducible print for text as well as drawings led to a significant increase in the scholarly communication of anatomical findings from the sixteenth century onward. (B) Drawing of a dissected sea urchin. (Modified from Tiedemann, F., *Anatomie der Röhren-Holothurie (Holothuria tubulosa), des Pomerantz-farbigen Seesterns (Astropecten aurantiacus) und des Stein-Seeigels (Echinus saxatilis)*, J. Thomann, Landshut, Germany, 1816.) For centuries, the standard for the presentation of anatomical results has been the paper-based publication, effectively restricting both authors and readers in the perception of complex 3D structures. (C) Volume rendering of a virtually dissected sea urchin. Noninvasive imaging techniques (here, μCT) allow gathering 3D digital anatomical datasets from zoological specimens noninvasively and at high speed. These image data can be interactively explored and manipulated using personal computers. Not to scale.

methods employed in systematics (Pyron and Wiens 2011). In a further genetic study (Hejnol et al. 2009), tissues were sampled from 94 species, and the final analyses included sequences from almost 1,500 genes, resulting in a significant improvement of the resolution of the tree of life of those animals that possess a bilateral body symmetry. Even larger amounts of data were processed in a study that employed DNA sequencing data derived from over 1,600 insect specimens in order to construct a database for future taxonomic reference and assessments of insect biodiversity in Madagascar (Monaghan et al. 2009). Such large-scale approaches have also found application in the elucidation of the genetic processes underlying anatomical variation in different dog races, resulting in data matrices encompassing hundreds of breeds and tens of thousands of genetic markers (Boyko et al. 2010, Vaysse et al. 2011). Obviously, such large-scale studies would not have been possible based solely on processing analog data.

Inherently digital imaging techniques such as MRI or μCT thus pave the road for similar progress in the study of animal anatomy. Not only is the application of these imaging modalities poised to result in a more transparent form of data presentation, but it will also enable large-scale zoo-anatomical studies, which simply would not have been undertaken before. As the previously mentioned high-throughput genetic studies suggest, the aim of future zoo-anatomical research should be to include similarly large numbers of species or specimens in order to come up with a picture as comprehensive as possible.

However, not only purely comparative anatomical projects would benefit from a consistent digitalization. Because of decreasing financial resources and an increasing lack of expert personnel in fundamental zoological disciplines such as taxonomy and biodiversity research, automatization and digitalization of the scientific workflow appear in fact to be the only way forward. For example, in order to classify and categorize the remaining species on Earth using conventional approaches, an estimated U.S. $ 263 billion and more than 360 years would be required at the current pace (Carbayo and Marques 2011)—although answers are needed much more rapidly. Faced with such a daunting task, taxonomists and biodiversity researchers are therefore urgently calling for a more interdisciplinary approach (see Wheeler et al. (2012) for a summary). In anatomical studies, this could primarily be achieved by incorporating digital imaging and digital analytical techniques, thus freeing scientists from routine investigations and allowing them to focus on their primary task, that is, the discovery and description of new species (MacLeod et al. 2010).

In the light of recent technical advances, this chapter aims to portray (i) how digital imaging techniques can be employed to generate 3D anatomical datasets of zoological specimens at high speed, (ii) how software can be utilized to effectively visualize the data generated in large-scale scanning studies, and, finally, (iii) how the resulting image data can be analyzed both qualitatively and quantitatively at a speed that was previously not attainable using nondigital techniques. We shall furthermore discuss the challenges that this new era of anatomical data acquisition, visualization, and analysis is currently facing and will outline how zoological studies may benefit from advances in other scientific disciplines, in particular medical imaging.

10.2 Accelerated Acquisition of Zoo-Anatomical Data

Various noninvasive imaging techniques allow gathering of 3D digital anatomical data of entire animals or parts thereof (Baker 2010, Walter et al. 2010). Some of the currently most successful 3D scanning modalities for zoological studies are MRI, computed tomography (CT), μCT, synchrotron radiation μCT (SRμCT), and optical projection tomography. These techniques rely on different physical principles, for example, MRI on nuclear magnetic resonance and CT as well as μCT on the application of x-rays. As a rule, the images produced by the respective scanners result in different types of datasets that may depict soft or hard tissue or a combination of both (see Boistel et al. (2011), Johnston (2011), and Laforsch et al. (2012) for further information about imaging techniques). Common to all these modalities is that they can be used to generate 3D datasets that are composed of many individual volume elements (voxels, basically 3D pixels). If the dimensions of the individual voxels contained within such datasets have been set to be cubic, the resulting volume file constitutes a true virtual replica of the scanned specimen at a given resolution. Luckily for zoologists, the development and improvement of noninvasive scanning technologies (in particular CT and MRI) are driven by their widespread application in human diagnostics. As opposed to the laborious, time-consuming, and often destructive techniques that have been dominating zoological studies until now, noninvasive imaging techniques enable gathering of anatomical data with considerable ease, at relatively high speed, and usually noninvasively.

10.2.1 Large-Scale Specimen Scanning Using MRI

The capacity of MRI to depict soft tissue has been employed, for example, to study internal and external anatomical features of more than 300 representative fish species (www.digitalfishlibrary.org). Because of the differently sized organisms employed in this study, scanning systems designed for human diagnostics and for studies on small animals such as mice or rats were used (Berquist et al. 2012). However, invertebrates of usually much smaller size have been the focus of a large-scale MRI scanning project as well, with almost 100 sea urchin species analyzed over the course of a few years (Ziegler et al. 2008, Ziegler 2012). The assembled data on sea urchins are representative for the soft tissue diversity observed in this group of marine animals (Figure 10.2). These two zoological MRI studies may serve as examples of what is yet to come, as many more animal groups have been identified as suitable candidates for an MRI-based, large-scale scanning approach (Ziegler et al. 2011a). Although technical as well as physical limitations restrict the resolution obtainable using MRI,

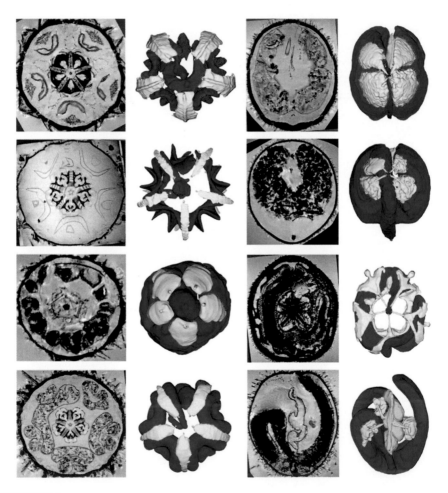

FIGURE 10.2

The large-scale application of MRI for the study of invertebrate anatomy reveals that internal anatomical features can be extracted from freshly fixed as well as century-old museum specimens. The images show virtual sections through 3D MRI datasets of different sea urchin species as well as corresponding surface renderings of selected internal soft tissue structures. Left column, from top to bottom: pencil urchin (*Eucidaris* sp.), diadem urchin (*Diadema* sp.), salenioid urchin (*Salenocidaris* sp.), and purple urchin (*Strongylocentrotus* sp.). Right column, from top to bottom: burrowing urchin (*Echinoneus* sp.), cake urchin (*Echinolampas* sp.), sand dollar (*Laganum* sp.), and heart urchin (*Abatus* sp.). Not to scale. Blue = digestive tract, cyan = gastric caecum, orange = buccal sacs, red = Stewart's organs, yellow = gonads.

the true strength of this imaging modality is its versatility, which allows analysis of live samples without the use of harmful ionizing radiation, small as well as large organisms, and use of a wide range of different imaging protocols (Lauridsen et al. 2011, Ziegler et al. 2011a). Furthermore, MRI can be carried out with or without the use of contrast agents (Gianolio et al. (2011) provide an overview). However, because of the previously mentioned limitations in resolution, zoo-anatomical studies using MRI will primarily have to focus on structures at the organ level. This field of zoological study has often been disregarded in the past because of the early availability of light and electron microscopy that enabled studies on the cellular and subcellular level. But, as recent studies on different internal organ systems in sea urchins demonstrate, MRI data can trigger new questions and can furthermore enable the integration of previously published anatomical data into a much broader evolutionary context (see, e.g., Ziegler et al. 2009, 2010a, 2012a).

10.2.2 Large-Scale Specimen Scanning Using CT

In contrast to MRI, x-ray-based imaging techniques such as CT, μCT, and SRμCT mainly enable obtaining hard tissue information from zoological specimens (Ritman (2011) provides an overview of recent μCT applications). Because of the application of harmful ionizing radiation in CT studies, experiments on live animals must be ethically justified—in contrast to studies where preserved material, for example, museum specimens, is utilized. The most notable strength of CT techniques is their capacity to generate high-resolution 3D datasets at much greater speed than would currently be possible with MRI. This important property has been employed in various comparative anatomical studies. For instance, CT has been used in a recent study to analyze physiological processes in multiple large animals such as crocodiles and snakes (Lauridsen et al. 2011). Similar to MRI, μCT techniques can be successfully used to obtain image data from different animal groups (Figure 10.3). By employing μCT, higher dataset resolutions are achievable, and this technique has been used, for example, to scan the heads of over 140 reptile species in order to infer bone anatomy noninvasively and from sometimes minute specimens (Rieppel et al. 2008). In a further study involving μCT, almost 200 sea urchin species have recently been scanned, resulting in a virtual specimen collection of several terabytes in size that is representative for most of the hard tissue diversity encountered in this group of marine animals (Ziegler 2012). However, the largest single collection of μCT and CT datasets, with well over 1,000 fossil and living species analyzed so far (Rowe and Frank 2011), has been created by paleontologists and zoologists for the Digital Morphology website project (www.digimorph. org). Further CT-based datasets, primarily of whales and dolphins, can be explored on the homepage of the Computerized Scanning and Imaging Facility (http://csi.whoi.edu).

FIGURE 10.3
The large-scale application of μCT for the study of invertebrate and vertebrate anatomy reveals that multiple animal groups can be successfully scanned using digital x-ray techniques. The volume renderings shown here are based on 3D datasets of museum and freshly fixed specimens. Staining techniques allow visualizing soft tissues despite the use of hard x-rays. (A) Peanut worm (*Themiste* sp.) stained with phosphotungstic acid. (B) Earthworm (*Allolobophora* sp.) stained with phosphotungstic acid. (C) Plates and girdle of a chiton (*Ischnochiton* sp.). (D) Shell of a marine snail (*Cerithium* sp.). (E) Shell of a marine clam (*Cerastoderma* sp.). (F) Cocoon of a ribbon worm (*Amphiporus* sp.) stained with phosphotungstic acid. (G) Lamp shell (*Terebratalia* sp.). (H) Endoskeleton of a sea star (*Pteraster* sp.). (I) Head and thoracical skeleton of a songbird (*Taeniopygia* sp.). Not to scale.

The beginnings of noninvasive 3D specimen analysis using CT techniques can be traced back to their early application in the study of fossil organisms (Sutton (2008) provides a comprehensive review). The acquisition of morphological data is of particular importance in fossil material, as imaging is currently one of the few modalities that allows extraction of biological information from fossilized zoological specimens. The importance of this becomes evident when taking into account that more than 99% of the species that ever inhabited our planet will likely never be available for genetic study (MacLeod 2007). CT and μCT studies have also been successfully extended to fossil specimens preserved in amber (Dierick et al. 2007) as well as to fossil organisms with delicate structures such as daddy longlegs (Garwood et al. 2011).

In order to achieve higher dataset resolutions, SRμCT can be employed to analyze both fossil and living whole specimens (Tafforeau et al. 2006).

This complex imaging modality has also been used for large-scale comparative anatomical projects, for example, involving dozens of representative insect species (Friedrich and Beutel 2010). The particular properties of synchrotron radiation can furthermore be employed to record movement such as tracheal breathing in living insects (Betz et al. 2007) or to visualize soft tissues using phase contrast. However, soft tissue contrast may also be achieved using the absorption contrast protocols typical for μCT scanners by applying different contrast agents to the specimen (Metscher 2009) or through a careful choice of scanning parameters (Dinley et al. 2010, Faulwetter et al. 2013).

10.2.3 Further Useful Applications of Noninvasive Imaging Techniques

Of particular future interest will be the fusion of image data derived from multiple scanning techniques such as MRI (soft tissue information) as well as CT and μCT (hard part information) into a single dataset, a process termed registration. The first steps toward multimodal imaging have been taken in human diagnostics by registering datasets derived from different imaging techniques, and zoo-anatomical research is bound to profit from the pioneering work performed in this field. A few zoological studies have already combined images derived from different noninvasive imaging techniques, in particular in cases where soft tissue information had to be interpreted in the context of complex skeletal structures (see, e.g., Forbes et al. 2006).

Certainly one of the greatest advantages of the previously mentioned imaging techniques is their noninvasive character, which in turn allows tapping the vast resource of preserved specimens that can be found as dry or ethanol-preserved material in natural history museums and zoological institutes. This concept has already been applied to some extent to different vertebrate and invertebrate animal groups such as reptiles (Rieppel et al. 2008), sea urchins (Ziegler 2012), and fish (Chanet et al. 2009, Berquist et al. 2012). Apart from offering a more complete set of samples for comparative experiments than studies using fresh material are likely to provide, the use of museum material in combination with noninvasive imaging techniques constitutes a highly sustainable use of valuable biological resources. This approach is also important from an ethical viewpoint, as it reduces the need for a renewed sacrifice of fresh material. With techniques such as MRI, CT, μCT, and others becoming gradually more available to zoologists, an increase in the importance of natural history collections for comparative anatomical studies can be expected.

10.2.4 Obstacles to Large-Scale Specimen Scanning

The success story of noninvasive imaging techniques in human diagnostics must be interpreted as a paradigm shift that will shape zoological studies in the future. However, adequate access to modern scanning technology is too often still problematic for zoologists. This applies in particular to MRI, an expensive technology that can be performed only at selected facilities. Presumably even less accessible than MRI is SRμCT, since not all of the few existing synchrotrons are available for zoologists that intend to scan animals. Although the situation is more relaxed regarding the access to CT and μCT scanners—they are more abundant, are much cheaper than MRI scanners, and require less specialized support—these instruments are still far from constituting standard equipment in natural history collections and zoological institutes despite their proven usefulness in particular for studies on preserved material.

Furthermore, most scanning equipment is used only during the day, leading to a loss of many hours of valuable scan time on these expensive machines. In the case of experiments performed on preserved material (termed *postmortem* studies in medical research), unsupervised overnight scanning is technically possible and results in a better exploitation of the available resources. For example, because high-resolution 3D MRI scans usually take several hours, overnight scanning of preserved zoological specimens would constitute a logical addition to routine daytime scanning. In case of μCT studies, scan times are much shorter and, therefore, require frequent sample reloading, currently rendering this technique impractical for unsupervised overnight scanning. However, some scanners have been designed to load several samples stacked on top of each other and can then be programmed for consecutive specimen scanning. Another approach is the development of robotic arms on μCT scanners that load samples into the scanning chamber, similar to the automatic loading of samples in modern DNA sequencers. Such robotic devices are also found at some of the synchrotron beamlines, where scanning usually takes less than an hour per sample and frequent reloading would be impractical if performed manually.

10.3 Accelerated Visualization of Zoo-Anatomical Data

The technological improvements in imaging techniques have led to an increased demand for specialized visualization tools, in particular for medical data. Numerous programs are currently available, ranging from proprietary software issued with the purchase of a scanner to open source software available online and free of charge (see Walter et al. (2010) for a comprehensive list). The quality of some of the freely available programs is astounding, and there are numerous examples of scientific articles that show 2D and 3D images created using solely these free tools. However, the application of more complex visualization software is often hampered by the fact that many zoological experts do not possess the necessary programming skills and will therefore shy away from software that requires considerable computational expertise. In our experience, this knowledge gap, in addition to the previously mentioned limited access to scanning equipment, forms the most important impediment to a more widespread use of 3D anatomical data.

10.3.1 3D Visualization Does Not Necessarily Require Cutting-Edge Equipment

A major breakthrough for a more widespread application of digital imaging in zoology can be seen in the possibility to handle 3D datasets on personal computers (PCs) using relatively simple software tools such as ImageJ (http://rsbweb.nih.gov/ij), Fiji (http://fiji.sc/wiki/index.php/Fiji), or VTK (www.vtk.org). Several of the numerous programs intended for 3D visualization in particular in medicine and engineering can be adapted to their application in zoological studies, for example, myVGL (http://www. volumegraphics.com). The advances in imaging software have largely been enabled by hardware improvements such as the availability of more potent graphic cards or more rapid access memory and central processing unit capacity. These hardware improvements in common PC systems can mainly be attributed to an increased use of multimedia files as well as immersive 3D environments that are employed in computer games. As these two domains each represent multibillion dollar industries, improvements in soft- and hardware for 3D visualization can be expected to occur almost constantly.

10.3.2 Local or Remote Visualization?

Although modern PCs support visualization of datasets in the megabyte-size range (e.g., those derived from MRI or CT scanners), datasets in the gigabyte-size range (e.g., those derived from µCT or SRµCT scanners) may be too large to be visualized or manipulated interactively in 3D at full resolution using standard PC systems. One solution for this pressing problem is the creation of datasets with reduced resolution that can be analyzed with the user's PC on a daily basis, before he or she visualizes the original, much larger datasets on a more powerful computing system. In the case of our µCT studies on sea urchins, for example, the full-resolution datasets were each about 25 GB large, but the laptop we intended to use on a daily basis did not provide sufficient hardware resources for visualization of such large datasets. In order to realize the desired constant access to our sea urchin datasets, we therefore reduced the data size of each dataset by a factor of eight by halving its isotropic resolution (e.g., from [15 µm]^3 to [30 µm]^3). This reduction resulted in an individual dataset size of about three GB, which our laptop was able to handle for interactive specimen manipulation in 3D. Although this step evidently resulted in a loss in resolution, the more important aspect was the possibility to rapidly access the datasets in order to qualitatively develop and verify hypotheses on sea urchin hard part anatomy.

Because in such a scenario only the computational resources of the user's PC are utilized, the approach has been termed *local visualization*. Alternatively, datasets can be deposited in full resolution at data centers that provide *remote visualization* environments, which allow access to the more powerful computational resources of the data center itself from the user's computer via a data link. Such dedicated visualization clusters can sometimes be found at data centers on campus (e.g., Leibniz-Rechenzentrum München, Zuse-Institut Berlin). We have successfully employed such a setup for our studies, although the implementation was not trivial and required considerable support from computer scientists and database specialists in addition to the constant availability of state-of-the-art cluster technology (Ziegler et al. 2010b). Unfortunately, such a solution will not be available to every user, and local visualization is therefore bound to be the predominant form of dataset visualization, likely resulting in a permanent tension between dataset size and visualization capacity.

10.3.3 Virtual Dissection of Zoological Specimens

A feature of particular interest to zoologists that are studying anatomical details using digital datasets is the possibility to virtually dissect single or even multiple specimens in parallel. Such an approach can be accomplished by employing interactive dataset slicing tools such as the 3D Viewer or the Volume Viewer that are both available as plug-ins for ImageJ and Fiji. The Volume Viewer plug-in, for example, allows simultaneously opening several 3D datasets and interactively manipulating and rotating these in order to generate comparable 2D virtual sections of the specimens under study (Figure 10.4). Apart from the constant accessibility of specimens for virtual dissection and, consequently, an accelerated investigatory process, digital data permit novel ways of inspecting zoological material in 3D.

In addition to interactive 2D slicing, digital datasets of whole specimens allow creating 3D models that can be used to convey the structural complexity of a scanned specimen in a more plastic manner. 3D modeling is commonly performed as surface (Figures 10.2 and 10.5D) or volume renderings (Figures 10.1C and 10.3). While surface-rendered models constitute discrete objects composed of a mesh of triangles that can be exported into a number of formats, volume-rendered models are constantly computed in real time and are therefore usually not available for export to other, more common file formats (see Russ (2011) for an introduction to surface and volume rendering).

10.3.4 Novel Communication Pathways for 3D Objects

The obvious advances in 3D visualization and modeling confront the classical dissemination pathways of anatomical information (i.e., paper-based publishing) with a considerable dilemma. The advantages that interactive 3D

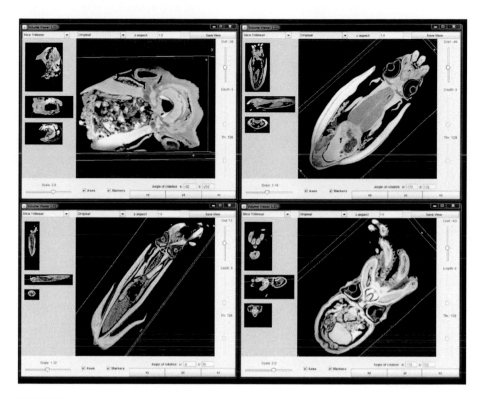

FIGURE 10.4
Digitally acquired 3D datasets can be interactively explored in parallel and in real time using, for example, the ImageJ Volume Viewer plug-in. This screenshot shows four MRI datasets of different cephalopod species that are simultaneously accessed on a standard PC. Rotation of the individual datasets allows virtually dissecting the animals and generating comparable 2D views at arbitrary angles (here, at the level of the eyes). The museum specimens employed in this study were not altered during the imaging process and can be reused for future studies. From top left to bottom right: pearlboat (*Nautilus* sp.), cuttlefish (*Sepia* sp.), squid (*Loligo* sp.), and octopus (*Bathypolypus* sp.). Not to scale.

visualization and modeling entail for the spectator are lost when complex 3D structures must be described in written words or must be portrayed in 2D only. The latter, however, has constituted common practice in zoological publishing for more than 400 years (see Figure 10.1).

Fortunately, some publishing houses (e.g., BioMed Central, Elsevier, Nature Publishing Group, Springer, and Wiley-Liss) have started to embrace novel technologies that enable embedding fully interactive 3D content into scientific publications. This approach requires the author to produce a surface-rendered 3D model, for instance, that of a sea urchin (Ziegler et al. 2010b) or that of a human brain (Ziegler et al. 2011b), and to import these models into a portable document format (PDF) file (the process is described in detail in Ruthensteiner and Hess (2008) and Kumar et al. (2008)). This file

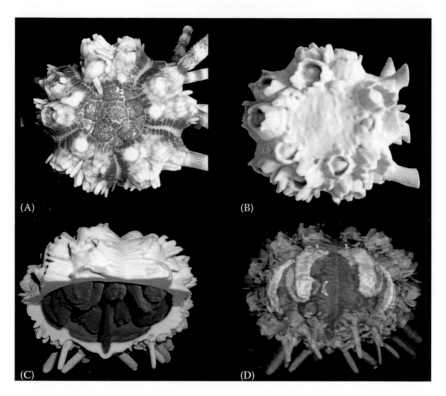

FIGURE 10.5
Communication of 3D objects through rapid prototyping and 3D visualization. (A) Dorsal view of a sea urchin museum specimen (*Eucidaris* sp.). The photograph of this 2.5 cm wide specimen was taken using a digital camera. (B) Corresponding view of a rapid prototyping model of the same specimen. The file used for 3D printing was based on a MRI dataset that was used for manual and automated segmentation followed by 3D modeling. The 3D printout was scaled to 15 cm width. (C) Ventrolateral view of a rapid prototyping model shown with endoskeletal elements partly removed to reveal internal organs. (D) 3D visualization of the surface-rendered model that served as a template for the rapid prototyping models presented here. The digital model is shown in ventrolateral view with semitransparent endoskeleton.

format also offers the possibility to present multiple models next to each other that can then be accessed in parallel for comparative purposes (Kumar et al. 2010, Ziegler et al. 2010a) as well as to some extent the possibility to embed volume-rendered models (Ruthensteiner et al. 2010). Embedded 3D models can thus become an integral part of the publication itself, although obviously only in the electronic version of the article. The fact that PDF files currently constitute the standard file format for the publication of scientific results should guarantee cross-platform accessibility of interactive 3D and other multimedia content, likely for the coming decades (see Ziegler et al. (2011b) for discussion). Further developments in 3D modeling include the increased application of browser-based, interactive 3D models using HTML 5 and WebGL standards, allowing dissemination of 3D content directly over the Internet. Recent examples for this approach include the Zygote Body project (www.zygotebody.com) or Aves 3D (www.aves3d.org).

However, digital 3D models do not necessarily have to be presented only in electronic form. Using rapid prototyping and 3D printing technologies, surface-rendered models can be used to communicate complex 3D structures in a fully tactile form. These techniques permit not only to selectively print those substructures of a specimen that are of interest but also to allow scaling the object as desired (Figure 10.5). A range of possible applications for these techniques in zoology can easily be envisioned, in particular in fields such as functional anatomy and zoology teaching, but also for the replication of valuable museum specimens (Kelley et al. 2007).

10.4 Accelerated Analysis of Zoo-Anatomical Data

The foreseeable drastic increase in 3D anatomical image data makes it imperative to use computational methods for data preparation and image interpretation. Computational tools developed for the processing of medical images, but also in other fields of computer vision, provide the necessary means for machine-based automatization of the analytical processes as well as the reliable quantification and adequate handling of large amounts of high-dimensional data that are routinely generated using high-throughput imaging systems (MacLeod (2007), Shamir et al. (2010), Russ (2011), and Neal and Russ (2012) provide introductions to theoretical and practical shape analysis). Automated image analysis tools will allow zoo-anatomical studies to move beyond the subjective visual inspection of individual samples and the analysis of datasets through a largely qualitative interpretation of individual features.

10.4.1 Automatization of Large-Scale Data Processing

The workload required for preprocessing of image data, parsing their information, and analyzing image content using conventional qualitative means is bound to limit the absolute number of samples that can be considered in a comparative study. In such a situation, computational algorithms already employed in the related disciplines of computer vision, machine learning, and computational statistics can help zoologists to better interact with their data. Computer vision algorithms are capable of automating basic and tedious steps in a data processing pipeline and provide the technical resources necessary for generating large digital sample collections. This applies in particular to data preprocessing, the aforementioned spatial alignment of different datasets (registration), the identification and localization of specific anatomical landmarks (interest point detection) or structures (segmentation), as well as to measurements of basic quantities, such as locations of these structures, distances in between them, or volumes thereof (morphometrics). Computational algorithms can accelerate the processing of these data and increase accuracy by implementing automated quality control through manual or adaptive annotation (Gorlitz et al. 2007, Menze et al. 2008, Donner et al. 2011).

10.4.2 Quantification and Structured Representation

Computational algorithms that identify meaningful descriptors from anatomical scans are capable of transforming complex image content into structured data, that is, into measurements of predefined features in a 3D volume. This provides the basis to search for correlations with other biological information using statistical tools for high-dimensional data processing that are routinely used in bioinformatics. A structured and quantitative representation of image content also constitutes the starting point for implementing and testing formal models describing anatomical changes, such as the evolution of a particular structure over time. A formal representation of image data and a quantitative measure of different anatomical features will furthermore open up new analytical pathways when searching for a correlation between anatomical patterns and genetic or protein data, a process employed in the field of imaging genetics (Hariri et al. 2006, Turner et al. 2006, Whitwell 2009). Computational approaches in this predominantly medical discipline usually involve the regression of an anatomical measure observed in image data to obtain new biomarkers that indicate the progression of a disease. Similar approaches could be used in zoo-anatomical analyses. For example, one may wish to correlate shape variation across a number of species with observed genetic variation in order to identify biological factors that may have an impact on characteristic anatomical features. In a recent large-scale comparison of dog breeds, for example, such correlations of genetic markers with phenotypes have been found in the patterns of ear shape and the frequency of specific alleles (Boyko et al. 2010).

10.4.3 Computational Tools for Pattern Analysis and Image Interpretation

Different tools are available for automatic preprocessing and analysis of 3D image datasets (Shamir et al. (2010) provide a concise list). Most of these tools have been developed in the context of medical image computing, a field that provides ample literature on preprocessing tasks such as 3D image registration and calibration. But also high-level data processing steps including structure identification, segmentation, and image interpretation have been extensively covered (Neal and Russ (2012) provide an introduction). Many of these high-level approaches can be grouped into two classes: on the one hand, *data-driven approaches* that learn from previously processed examples and, on the other hand, *model-driven approaches* that formalize hypotheses, for example, the expected shape and spatial distribution of an anatomical feature or explicit assumptions about noise and variation between observations. These two principal classes will be presented in greater detail in the following two sections.

10.4.4 Data-Driven Approaches for Pattern Analysis

Computational tools that learn from examples with specific properties are often referred to as machine learning or pattern recognition algorithms. Many of these tools are applicable to a wide range of similar tasks with little modification of the underlying algorithm. In order to work well, they require a set of exemplary data that constitute a representative sample. These algorithms have two or three characteristic components: first, a feature representation that is often somewhat problem specific and, second, the choice of the learning algorithm that is employed in addition to this feature representation (Duda et al. 2001, Hastie et al. 2009, Shamir et al. 2010). The optimal choice of features and the adjustment of algorithmic parameters are performed in final predictive tests, which may be considered a third component of these algorithms. Most computer vision approaches rely on such machine learning techniques and use them together with specific features that are capable of describing image content. For example, low-level features that describe magnitude and orientation of edges, that characterize the local image texture, or, at a higher level, that describe the presence and spatial distribution of specific interest points constitute such specific features. For individual objects that are already segmented, integral features such as diameter, area, eccentricity, and similarity with other related shapes can be used to describe the usually rather unstructured image content with just a few numerical descriptors that can then be employed for further mathematical analyses and a statistical comparison.

As a consequence, the feature representation defines which properties of the image are relevant for the analysis and which information can be ignored. Features can be chosen by the user based on experience, or they can be automatically selected from a large set of candidate features during algorithm learning. If, for example, the aim is to discriminate well between two classes, an optimal feature would highlight visible or expected class differences. In image analysis, features can be local, such as the gray-scale value of a particular voxel or the statistics of the gray-scale values surrounding that voxel (Menze et al. 2007a). Local descriptors usually have the same dimensionality as the original input data. It is a particular strength of such low-level image descriptors that they can be easily adapted for the detection of the location of different anatomical landmarks or for the segmentation of particular local structures. For instance, algorithms that perform well in 2D human face and body recognition proved similarly useful in 3D detection and segmentation tasks (Geremia et al. 2010). These approaches may be used for rapid preprocessing of the data, for example, in the extraction of leaf shapes from images (Hearn 2009). However, pattern recognition and classification tasks that assign species names to entire images require a broader high-level feature representation that is more specifically adapted to the problem under study.

While a global description of an anatomical 3D scan can start, in theory, from a high-dimensional feature vector obtained by concatenating all grayscale values of the dataset, in practice, a lower-dimensional representation will be chosen that is guided by knowledge about the problem under study. The image descriptors used could be as basic as a binary indicator of the presence or absence of a specific structure or a numerical value indicating volume, surface, or diameter thereof. In addition, the descriptor may be the location of an anatomical landmark within an appropriate reference frame or the correlation of an observed structure with an expected shape. Of course, most of these high-level representations also require a preprocessing step in order to be automated. This can often be achieved through a computer vision and pattern recognition approach that evaluates local image features in a first step.

To proceed with analysis, all datasets gathered in the course of a study are expressed through this feature representation. This leads to a representation that has a dimensionality equal to the number of features extracted from the anatomical datasets, with the individual samples represented as individual points in this space. In general, two types of algorithms can be used to proceed from here on, and they differ in their requirements for prior knowledge of the feature in question. While *unsupervised methods* can be directly applied to the datasets for revealing structure in an exploratory fashion, *supervised methods* require the user to show the algorithm in a training step which structures should be identified.

Low-dimensional projections and clustering methods are among the most basic approaches for unsupervised analyses. Various linear and nonlinear low-dimensional projection methods exist, and all of them can be applied to shape analysis tasks using 3D anatomical scans (MacLeod 2007). Often, data are represented according to the most general underlying features, for example, in a principal component or an independent component analysis— Figure 10.6 provides an example from our μCT studies on sea urchins (Ziegler et al. 2012b). These methods reduce redundancy when feature entries are chosen in highly correlated representations (Hastie et al. 2009). As an example, the high-dimensional feature vector obtained from concatenating all 10^6 gray-scale values of a 3D image volume could be represented in a set of a few dozen principal components and, hence, by a vector with a few dozen entries. These low-dimensional representations can be used to inspect sample distribution, or they can be used as a new formal representation of the data with highly compressed information. Figure 10.7 depicts a low-dimensional projection that learns a nonlinear metric from the data

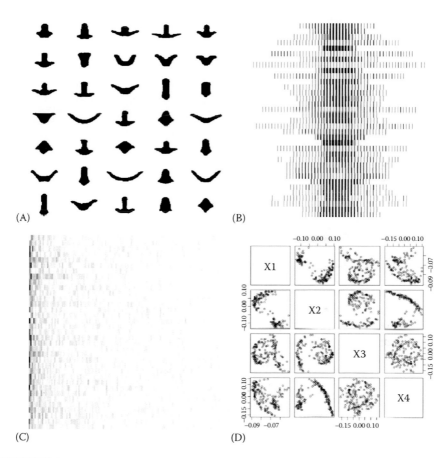

(A) (B) (C) (D)

FIGURE 10.6
Feature representation of binary images used for mathematical shape analysis. (A) Binary segmentations derived from horizontal sections of teeth from 35 sea urchin species. Every single
tooth is represented as a small image patch with 64 × 64 black or white pixels. (B) Representation
of the feature vector after reordering the pixel values obtained before by concatenating all rows
of the sea urchin tooth image patches into one vector with a length of 4096 pixels. Each row
represents one of the 35 binary images. (C) The information presented in (B) is here shown
after the transformation of the 4096-dimensional feature vector using a principal component
analysis. Only the 80 most important features are shown, with informative descriptors being
drawn in dark and unimportant descriptors being drawn in white. The first 10–20 features
contain nearly the complete information from all 4096 original features that were derived from
the image segmentation. (D) Pair plots of the first four principal component features for all
35 samples with every dot representing one sample. Natural groupings between the shapes
of selected sea urchin teeth now become apparent and can be used for further machine-based
analysis.

FIGURE 10.7
Low-dimensional projection of horizontal sections of teeth from 122 different sea urchin species. The image data were acquired using μCT scans of whole museum specimens. The illustration demonstrates how distances between complex 2D shapes can be calculated using supervised machine learning techniques and can then be grouped according to their anatomy using a nonlinear measure.

presented in Figure 10.6 and that may be better adapt to some problems than a principal component analysis.

Such low-dimensional projections are often followed by the application of clustering algorithms that represent a second popular set of unsupervised methods. These algorithms group observations with similar entries in their feature vector and, therefore, provide a glimpse into similarities in high-dimensional feature space. Such groupings can be visualized, for example, in hierarchical trees that reveal general patterns of the data separated into different subgroups. Again, clustering results can be used in the exploratory visualization of the data or as input to automated analysis and, possibly, to supervised methods.

Supervised methods may consider discrete features or continuous features or a set of both (Hastie et al. 2009) and have the advantage that the user can explicitly control which feature must be extracted from a given dataset and what variation within the dataset is safe to ignore. These characteristics make supervised algorithms more robust than those used for exploratory, unsupervised analyses. Moreover, while correlations visible in unsupervised analyses may have occurred by chance within the particular dataset employed, supervised methods allow formalizing empirical hypotheses by teaching the algorithm using a given dataset before testing and verifying it on a fresh set of observations. However, this approach requires annotation, for example, from a semantic understanding of the anatomical feature or through knowledge of the most likely species identifier visible in the scans.

In practice, feature selection and unsupervised and supervised learning algorithms are used jointly in order to form complex algorithms for computer vision (Szeliski 2010). A supervised search for the most discriminative features may help visualize higher-order correlations (Menze et al. 2007b, 2009, 2011a), while pipelines including a clustering or low-dimensional projection of local image features that are later used as input for a supervised learning algorithm represent a successful strategy of many pattern recognition algorithms in computer vision. However, despite their proven usefulness in medical diagnostics and human anatomy, the application of these methods to zoological 3D datasets is still relatively uncommon.

10.4.5 Model-Driven Approaches for Pattern Analysis

As outlined earlier, data-driven machine learning methods can be used to establish and test empirical hypotheses. They require annotating a representative set of data in a first step, then choosing a generic computer vision and machine learning algorithm in a second step, and, finally, validating the results (see, e.g., Boyko et al. (2010)) in a third step. Using such an approach, however, comes at the cost of understanding structural relationships between sets of data. Moreover, choosing an appropriate set of image features may pose a significant problem on its own: if no such single-best feature exists and the user must rely on a rather large set of standard low-level image features, the amount of necessary data to be annotated may grow significantly. The availability of the required datasets, however, may be the limiting factor when dealing with fossil specimens or rare species.

In this case, a viable solution can be the application of functional models of the data as well as a data-generating process instead of data-driven algorithmic tools (Menze et al. 2011b). Such models allow formalizing strong prior assumptions about the data, for example, on shape, the regularity of shape transformations, the type of noise imposed on the observations from the imaging, or the general biological variability (Durrleman et al. 2009, Hearn 2009). Functional models implement specific hypotheses of the data that have fewer parameters requiring estimation and that can be tested on a significantly smaller amount of data. If these hypotheses turn out to sufficiently fit the problem, they also provide better insights into the structure of the problem. This is because different formal models of the data-generating process can be combined and can be used to disentangle the underlying effects. However, for 3D anatomical structures, general shape representations from binary segmentations are usually studied in a rather descriptive fashion and without formalizing the entire data-generating process, for example, in order to evaluate morphological differences in feather shape (Sheets et al. 2006). Here, machine learning algorithms that are capable of learning from a few manually annotated examples (e.g., neural networks, decision trees, or support vector machines) or algorithms that search for structure in the properties of different samples (e.g., nearest-neighbor-clustering or kernel density estimation methods) may help process data at a large scale and provide the means for data preprocessing through data filtering and quality control (Menze et al. 2007b) as well as structure annotation and image segmentation (Geremia et al. 2010). Models can then be used to study the changes of an individual from a selected population along a continuous variable such as time or an arbitrary environmental factor—effectively leading to 4D problems. In clinical applications, this continuous variable may be an index of the severity of a disease or of its progress, but it can also be a time indicator or a semantic index of an anatomical transformation visible at macroscopic scale as well as any other external factor or environmental variable. If multiple 4D series of an entire population are available from a controlled experiment or from a well-stratified dataset with different shapes evolving along an environmental factor, the changes observed in these data may be unmixed into intersubject variation and general evolution of the population (Durrleman et al. 2012). Any continuous transformation can be used to interpolate potential intermediary anatomical stages. For verification, these hypothetical stages should be correlated with known fossil material or with distances inferred from phylogenetic analyzes.

10.4.6 Large-Scale Processing of Anatomical Data: Problems to Overcome

While most of the previously mentioned computational tools are readily available to zoologists because of research performed in computer vision and medical image computing, the biggest obstacle in pushing the computational limits of anatomical data analysis is the lack of actual data. Machine learning tools that rely on data-driven techniques can only be implemented and further developed using large representative sets of anatomical scans and, in case of supervised methods, using annotated data. Model-based approaches do not necessarily require large datasets, but the available data should have been acquired in well-controlled settings and must be provided with detailed annotations in order to stratify these datasets in accordance with external factors. Although it is usually the biologist who acquires the images and defines the research goal of a zoological study, most questions related to computer vision and image interpretation necessitate, at least during the initial processing of datasets for analysis, engineering, and informatics efforts—skills that are frequently wanting in biologists. Consequently, a close interdisciplinary collaboration is imperative.

A number of issues that are relevant to the generation of large-scale anatomical image databases have been discussed intensively in the field of medical image analysis but also in bioinformatics. In zoology, only a few repositories that permit automated species identification exist, for example, the AFORO website for different fish species (Parisi-Baradad et al. 2010). This particular website allows online submission of images of fish otoliths that are then used for automated species identification at a success rate of currently 72%. Other efforts are, for instance, geared toward an automated processing of images of benthic invertebrates for high-throughput analyses of samples that have been acquired for environmental monitoring purposes (Lytle et al. 2010). A major problem encountered in these projects is the natural plasticity of anatomical features as well as a sometimes high degree of convergence between species, often resulting in species misidentification (Hearn 2009).

A number of data repositories that archive and distribute anatomical image data together with annotations from biological experts have been established (Shamir et al. 2010). In the medical field, it has been proposed that repositories should archive the original data—possibly after transformation from proprietary scanner file formats into more common and open data file formats—together with preprocessing protocols and algorithms for registration, calibration, and segmentation. These algorithms in turn should also be uploaded to a repository that is accessible to the public, since such a step will promote the use of these algorithms and render the image processing pipeline more reproducible (Gentleman 2005). This approach would allow the user to work with data derived from any given step of the workflow, as storing the entire processing pipeline would be prohibitively expensive given the exponential growth of possible parameterization in long data preprocessing protocols. Further discussion in the scientific community is bound to shape the optimal protocols for registration, segmentation, and processing as well as the best common standards for treating anatomical digital data. By adhering to such standards, digital specimen data generated using 3D imaging techniques could be transformed into virtual specimens by curatorial staff of natural history collections and consequently be curated in a fashion that would permit query over the Internet. Involving museums in the archival of digital anatomical data would be desirable, as 3D anatomical datasets, similar to physical specimens, require long-term curation after publication of the initial study (see Ziegler et al. (2010b), Rowe and Frank (2011), Berquist et al. (2012), and Faulwetter et al. (2013) for discussion).

Data standardization and public availability of large digital anatomical collections will boost research on algorithms that can process such datasets. In addition, competitions calling for the best algorithm, for example, in a specific species identification problem, could further stimulate research in the computer vision or biomedical image computing communities. Many of the already existing tools in computational biology, biostatistics, and bioinformatics can be adapted to query large anatomical data repositories, analyze features extracted with predefined protocols, and cross-link different resources.

10.5 Conclusions

A prerequisite for a better protection of the environment that humans live in is detailed knowledge of its various components. Using conventional, nondigital approaches, simply too much time would pass to achieve this desirable goal in zoology. Noninvasive imaging techniques allow gathering detailed anatomical data from zoological specimens at relatively high speed and, through their digital output formats, open up a number of analytical pathways that are inaccessible when using analog data formats. The full potential of this approach could be tapped by analyzing digital datasets with advanced 3D visualization and machine learning algorithms. Image analysis constitutes a highly interdisciplinary field and requires constant input from scientists with complementary expertise. In several other biological disciplines, such interdisciplinary cooperation is already commonplace and has resulted in tremendous knowledge gain. Furthermore, the mandatory deposition of digitally acquired anatomical datasets must finally be recognized as an essential step in zoological research—enforced data deposition is standard practice in other scientific fields such as protein science or genetics and has formed the basis for extensive data mining. Digital data in zoo-anatomy will become less expensive to generate, less expensive to handle, and less expensive to archive than physical object data. We are therefore confident that the study of animal anatomy will witness a considerable increase in high-throughput digital research.

Acknowledgments

We would like to thank the staff at the Ernst Mayr Library for making original copies of works by sixteenth-century naturalists accessible to us. The following colleagues generously supplied us with specimens for our imaging experiments: Adam Baldinger, Alexander Gruhl, Gisele Kawauchi, Marta Novo, Andreas Schmidt-Rhaesa, Sergi Taboada-Moreno, and Christopher K. Thompson. We would like to thank Berit Ullrich for discussion of our ideas. Felix Beckmann, Cornelius Faber, Ross Mair, Susanne Mueller, and Malte Ogurreck provided invaluable help with imaging equipment. We are grateful to Gonzalo Giribet for financial support and for his comments on the manuscript. This work was performed in part at the Center for Nanoscale Systems, which is supported by the National Science Foundation under NSF award no. ECS-0335765. Funding for this research was provided by the Deutsche Forschungsgemeinschaft through grant no. ZI-1274/1-1.

References

Aldrovandi U (1606) *De reliquis animalibus exsanguibus*, libri quatuor. Bologna, Italy, B. Bellagamba.

Baker M (2010) The whole picture. *Nature* 463: 977–980.

Belon P (1553) *De aquatilibus*, libri duo. Paris, France, C. Stephane.

Berquist RM, Gledhill KM, Peterson MW, Doan AH, Baxter GT, Yopak KE, Kang N, Walker HJ, Hastings PA, Frank LR (2012) The Digital Fish Library: Using MRI to digitize, database, and document the morphological diversity of fish. *PLoS ONE* 7: e34499.

Betz O, Wegst U, Weide D, Heethoff M, Helfen L, Lee WK, Cloetens P (2007) Imaging applications of synchrotron X-ray phase-contrast microtomography in biological morphology and biomaterials science. I. General aspects of the technique and its advantages in the analysis of millimetre-sized arthropod structure. *Journal of Microscopy* 227: 51–71.

Boistel R, Swoger J, Krzic U, Fernandez V, Gillet B, Reynaud EG (2011) The future of three-dimensional microscopic imaging in marine biology. *Marine Ecology* 32: 438–452.

Boyko AR, Quignon P, Li L, Schoenebeck JJ, Degenhardt JD, Lohmueller KE, Zhao K, Brisbin A et al. (2010) A simple genetic architecture underlies morphological variation in dogs. *PLoS Biology* 8: e1000451.

Budd G, Olsson L (2007) A renaissance for evolutionary morphology. *Acta Zoologica* (Stockholm) 88: 1.

Carbayo F, Marques AC (2011) The costs of describing the entire animal kingdom. *Trends in Ecology and Evolution* 26: 154–155.

Chanet B, Fusellier M, Baudet J, Madec S, Guintard C (2009) No need to open the jar: A comparative study of magnetic resonance imaging results on fresh and alcohol preserved common carps (*Cyprinus carpio* (L. 1758), Cyprinidae, Teleostei). *Comptes Rendus de Biologie* 332: 413–419.

Dierick M, Cnudde V, Masschaele B, Vlassenbroeck J, Van Hoorebeke L, Jacobs P (2007) Micro-CT of fossils preserved in amber. *Nuclear Instruments and Methods in Physics Research A* 580: 641–643.

Dinley J, Hawkins L, Paterson G, Ball AD, Sinclair I, Sinnett-Jones P, Lanham S (2010) Micro-computed X-ray tomography: a new non-destructive method of assessing sectional, fly-through and 3D imaging of a soft-bodied marine worm. *Journal of Microscopy* 238: 123–133.

Donner R, Birngruber E, Steiner H, Bischof H, Langs G (2011) Localization of 3D anatomical structures using random forests and discrete optimization. *Lecture Notes in Computer Science* 6533: 86–95.

Duda R, Hart P, Stock D (2001) *Pattern Classification*. New York, Wiley.

Durrleman S, Pennec X, Trouvé A, Ayache N (2009) Statistical models on sets of curves and surfaces based on currents. *Medical Image Analysis* 13: 793–808.

Durrleman S, Pennec X, Trouvé A, Ayache N, Braga J (2012) Comparison of the endocranial ontogenies between chimpanzees and bonobos via temporal regression and spatiotemporal registration. *Journal of Human Evolution* 62: 74–88.

Faulwetter S, Vasileiadou A, Kouratoras M, Dailianis T, Arvanitidis C (2013) Micro-computed tomography: Introducing new dimensions to taxonomy. *ZooKeys* 263: 1–45.

Forbes JG, Morris HD, Wang K (2006) Multimodal imaging of the sonic organ of *Porichthys notatus*, the singing midshipman fish. *Magnetic Resonance Imaging* 24: 321–331.

Friedrich F, Beutel RG (2010) Goodbye Halteria? The thoracic morphology of Endopterygota (Insecta) and its phylogenetic implications. *Cladistics* 26: 1–34.

Garwood R, Dunlop JA, Giribet G, Sutton MD (2011) Anatomically modern Carboniferous harvestmen demonstrate early cladogenesis and stasis in Opiliones. *Nature Communications* 2: 444.

Gentleman R (2005) *Bioinformatics and Computational Biology Solutions Using R and Bioconductor*. Berlin, Germany, Springer.

Geremia E, Menze BH, Clatz O, Konukoglu E, Criminisi A, Ayache N (2010) Spatial decision forests for MS lesion segmentation in multi-channel magnetic resonance images. *Lecture Notes in Computer Science* 6362: 664–772.

Gesner C (1558) *Historia Animalium, Libri Quatuor*. Zürich, Switzerland, C. Froschower.

Gianolio E, Viale A, Delli Castelli D, Aime S (2011) MR contrast agents. In: Kiessling F, Pichler BJ (eds.), *Small Animal Imaging*. Berlin, Germany, Springer, pp. 165–193.

Giribet G (2010) A new dimension in combining data? The use of morphology and phylogenomic data in metazoan systematics. *Acta Zoologica* (Stockholm) 91: 11–19.

Gorlitz L, Menze BH, Weber MA, Kelm BM, Hamprecht FA (2007) Semi-supervised tumor detection in magnetic resonance spectroscopic images using discriminative random fields. *Lecture Notes in Computer Science* 4713: 224–233.

Hariri AR, Drabant EM, Weinberger DR (2006) Imaging genetics: perspectives from studies of genetically driven variation in serotonin function and corticolimbic affective processing. *Biological Psychiatry* 59: 888–897.

Hastie T, Tibshirani R, Friedman J (2009) *The Elements of Statistical Learning: Data Mining, Inference, and Prediction*. Berlin, Germany, Springer.

Hearn DJ (2009) Shape analysis for the automated identification of plants from images of leaves. *Taxon* 58: 961–981.

Hejnol A, Obst M, Stamatakis A, Ott M, Rouse GW, Edgecombe GD, Martinez P, Baguna J et al. (2009) Assessing the root of bilaterian animals with scalable phylogenomic tools. *Proceedings of the Royal Society B* 276: 4261–4270.

Johnston P (2011) Cross-sectional imaging in comparative vertebrate morphology—the intracranial joint of the coelacanth *Latimeria chalumnae*. In: Saba L (ed.), *Computed Tomography—Special Applications*. Rijeka, Croatia, InTech, pp. 259–274.

Kelley DJ, Farhoud M, Meyerand ME, Nelson DL, Ramirez LF, Dempsey RJ, Wolf AJ, Alexander AL, Davidson RJ (2007) Creating physical 3D stereolithograph models of brain and skull. *PLoS ONE* 10: e1119.

Kumar P, Ziegler A, Grahn A, Hee CS, Ziegler A (2010) Leaving the structural ivory tower, assisted by interactive 3D PDF. *Trends in Biochemical Sciences* 35: 419–422.

Kumar P, Ziegler A, Ziegler J, Uchanska-Ziegler B, Ziegler A (2008) Grasping molecular structures through publication-integrated 3D models. *Trends in Biochemical Sciences* 33: 408–412.

Latorsch C, Imhof H, Sigl R, Settles M, Heß M, Wanninger A (2012) Applications of computational 3D-modeling in organismal biology. In: Alexandru C (ed.), *Modeling and Simulation in Engineering*. Rijeka, Croatia, InTech, pp. 117–142.

Lauridsen H, Hansen K, Wang T, Agger P, Andersen JL, Knudsen PS, Rasmusseun AS, Uhrenholt L, Pedersen M (2011) Inside out: modern imaging techniques to reveal animal anatomy. *PLoS ONE* 6: e17879.

Lytle DA, Martinez-Munoz G, Zhang W, Larios N, Shapiro L, Paasch R, Moldenke A, Mortensen EN, Todorovic S, Dietterich TG (2010) Automated processing and identification of benthic invertebrate samples. *Journal of the North American Benthological Society* 29: 867–874.

MacLeod N (2007) *Automated Taxon Identification in Systematics. Theory, Approaches and Applications*. Boca Raton, FL, CRC Press.

MacLeod N, Benfield M, Culverhouse P (2010) Time to automate identification. *Nature* 467: 154–155.

Menze BH, Kelm BM, Hamprecht FA (2007a) From eigenspots to fisherspots: latent spaces in the nonlinear detection of spot patterns in a highly variable background. In: Lenz HJ, Decker R (eds.), *Studies in Classification, Data Analysis, and Knowledge Organization. Advances in Data Analysis*, Vol. 33. Berlin, Germany, Springer, pp. 255–262.

Menze BH, Kelm BM, Weber MA, Bachert P, Hamprecht FA (2008) Mimicking the human expert: pattern recognition for an automated assessment of data quality in magnetic resonance spectroscopic images. *Magnetic Resonance in Medicine* 59: 1457–1466.

Menze BH, Kelm BM, Masuch R, Himmelreich U, Bachert P, Petrich W, Hamprecht FA (2009) A comparison of random forest and its Gini importance with standard chemometric methods for the feature selection and classification of spectral data. *BMC Bioinformatics* 10: 213.

Menze BH, Kelm BM, Splitthoff DN, Koethe U, Hamprecht F (2011a) On oblique random forests. *Lecture Notes in Computer Science* 6912: 453–469.

Menze BH, Petrich W, Hamprecht FA (2007b) Multivariate feature selection and hierarchical classification for infrared spectroscopy: serum-based detection of bovine spongiform encephalopathy. *Analytical and Bioanalytical Chemistry* 387: 1801–1807.

Menze BH, Van Leemput K, Honkela A, Konukoglu E, Weber MA, Ayache N, Golland P (2011b) A generative approach for image-based modeling of tumor growth. *Lecture Notes in Computer Science* 6801: 735–747.

Metscher BD (2009) MicroCT for comparative morphology: simple staining methods allow high-contrast 3D imaging of diverse non-mineralized animal tissues. *BMC Physiology* 9: 11.

Monaghan MT, Wild R, Elliot M, Fujisawa T, Balke M, Inward DJG, Lees DC et al. (2009) Accelerated species inventory on Madagascar using coalescent-based models of species delineation. *Systematic Biology* 58: 298–311.

Neal FB, Russ JC (2012) *Measuring Shape.* Boca Raton, FL, CRC Press.

Parisi-Baradad V, Manjabacas A, Loombarte A, Olivella R, Chic O, Piera J, Garcia-Ladona E (2010) Automated taxon identification of teleost fishes using an otolith online database—AFORO. *Fisheries Research* 105: 13–20.

Pyron RA, Wiens JJ (2011) A large-scale phylogeny of Amphibia including over 2800 species, and a revised classification of extant frogs, salamanders, and caecilians. *Molecular Phylogenetics and Evolution* 61: 543–583.

Rieppel O, Gauthier J, Maisano J (2008) Comparative morphology of the dermal palate in squamate reptiles, with comments on phylogenetic implications. *Zoological Journal of the Linnean Society* 152: 131–152.

Ritman EL (2011) Current status of developments and applications of micro-CT. *Annual Review of Biomedical Engineering* 13: 531–552.

Rondelet G (1554) *Libri de piscibus marinis, in quibus verae piscium effigies expressae sunt.* Lyon, France, M. Bonhomme.

Rowe T, Frank LR (2011) The disappearing third dimension. *Science* 331: 712–714.

Russ JC (2011) *The Image Processing Handbook.* Boca Raton, FL, CRC Press.

Ruthensteiner B, Baeumler N, Barnes DG (2010) Interactive 3D volume rendering in biomedical publications. *Micron* 41: 886.e1–886.e17.

Ruthensteiner B, Heß M (2008) Embedding 3D models of biological specimens in PDF publications. *Microscopy Research and Technique* 71: 778–786.

Schmidt-Rhaesa A (2009) Morphology and deep metazoan phylogeny. *Zoomorphology* 128: 199–200.

Shamir L, Delaney JD, Orlov N, Eckley DM, Goldberg IG (2010) Pattern recognition software and techniques for biological image analysis. *PLoS Computational Biology* 6: e1000974.

Sheets HD, Covino KM, Panasiewicz JM, Morris SR (2006) Comparison of geometric morphometric outline methods in the discrimination of age-related differences in feather shape. *Frontiers in Zoology* 3: 15.

Sutton MD (2008) Tomographic techniques for the study of exceptionally preserved fossils. *Proceedings of the Royal Society B* 275: 1587–1593.

Szeliski R (2010) *Computer Vision: Algorithms and Applications.* Berlin, Germany, Springer.

Tafforeau P, Boistel R, Boller E, Bravin A, Brunet M, Chaimanee Y, Cloetens P et al. (2006) Applications of X-ray synchrotron microtomography for non-destructive 3D studies of paleontological specimens. *Applied Physics A* 83: 195–202.

Tiedemann F (1816) *Anatomie der Röhren-Holothurie (Holothuria tubulosa), des Pomerantz-farbigen Seesterns (Astropecten aurantiacus) und des Stein-Seeigels (Echinus saxatilis)*. Landshut, Germany, J. Thomann.

Turner JA, Smyth P, Macciardi F, Fallon JH, Kennedy JL, Potkin SG (2006) Imaging phenotypes and genotypes in schizophrenia. *Neuroinformatics* 4: 21–49.

Vaysse A, Ratnakumar A, Derrien T, Axelsson E, Pielberg GR, Sigurdsson S, Fall T et al. (2011) Identification of genomic regions associated with phenotypic variation between dog breeds using selection mapping. *PLoS Genetics* 7: e1002316.

Walter T, Shattuck DW, Baldock R, Bastin ME, Carpenter AE, Duce S, Ellenberg J et al. (2010) Visualization of image data from cells to organisms. *Nature Methods Supplement* 7: S26–S41.

Wheeler QD, Knapp S, Stevenson DW, Stevenson JW, Blum SD, Boom BM, Borisy GG, et al. (2012) Mapping the biosphere: exploring species to understand the origin, organization, and sustainability of biodiversity. *Systematics and Biodiversity* 10: 1–20.

Whitwell JL (2009) Voxel-based morphometry: an automated technique for assessing structural changes in the brain. *The Journal of Neuroscience* 29: 9661–9664.

Ziegler A (2012) Broad application of non-invasive imaging techniques to echinoids and other echinoderm taxa. *Zoosymposia* 7: 53–70.

Ziegler A, Faber C, Bartolomaeus T (2009) Comparative morphology of the axial complex and interdependence of internal organ systems in sea urchins (Echinodermata: Echinoidea). *Frontiers in Zoology* 6: 10.

Ziegler A, Faber C, Mueller S, Bartolomaeus T (2008) Systematic comparison and reconstruction of sea urchin (Echinoidea) internal anatomy: A novel approach using magnetic resonance imaging. *BMC Biology* 6: 33.

Ziegler A, Kunth M, Mueller S, Bock C, Pohmann R, Schröder L, Faber C, Giribet G (2011a) Application of magnetic resonance imaging in zoology. *Zoomorphology* 130: 227–254.

Ziegler A, Mietchen D, Faber C, von Hausen W, Schöbel C, Sellerer M, Ziegler A (2011b) Effectively incorporating selected multimedia content into medical publications. *BMC Medicine* 9: 17.

Ziegler A, Mooi R, Rolet G, De Ridder C (2010a) Origin and evolutionary plasticity of the gastric caecum in sea urchins (Echinodermata: Echinoidea). *BMC Evolutionary Biology* 10: 313.

Ziegler A, Ogurreck M, Steinke T, Beckmann F, Prohaska S, Ziegler A (2010b) Opportunities and challenges for digital morphology. *Biology Direct* 5: 45.

Ziegler A, Schröder L, Ogurreck M, Faber C, Stach T (2012a) Evolution of a novel muscle design in sea urchins (Echinodermata: Echinoidea). *PLoS ONE* 7: e37520.

Ziegler A, Stock SR, Menze BH, Smith AB (2012b) Macro- and microstructural diversity of sea urchin teeth revealed by large-scale micro-computed tomography survey. *Proceedings of SPIE* 8506: 85061G.

11

Quantum Frontier

Joseph F. Fitzsimons, Eleanor G. Rieffel, and Valerio Scarani

CONTENTS

11.1 Summary

The success of the abstract model of computation, in terms of bits, logical operations, programming language constructs, and the like, makes it easy to forget that computation is a physical process. Our cherished notions of computation and information are grounded in classical mechanics, but the physics underlying our world is quantum. In the early 1980s, researchers began to ask how computation would change if we adopted a quantum mechanical, instead of a classical mechanical, view of computation. Slowly, a new picture of computation arose, one that gave rise to a variety of faster algorithms, novel cryptographic mechanisms, and alternative methods of communication. Small quantum information processing devices have been built, and efforts are underway to build larger ones. Even apart from the existence of these devices, the quantum view on information processing has provided significant insight into the nature of computation and information and a deeper understanding of the physics of our universe and its connections with computation.

We start by describing aspects of quantum mechanics that are at the heart of a quantum view of information processing. We give our own idiosyncratic view of a number of these topics in the hopes of correcting common misconceptions and highlighting aspects that are often overlooked. A number of the phenomena described were initially viewed as oddities of quantum mechanics, whose meaning was best left to philosophers, topics that respectable physicists would avoid or, at best, talk about only over a late night beer. It was quantum information processing, first quantum cryptography and then, more dramatically, quantum computing, that turned the tables and showed that these oddities could be put to practical effect. It is these applications we describe next. We conclude with a section describing some of the many questions left for future work, especially the mysteries surrounding where the power of quantum information ultimately comes from.

11.2 Message in a Quantum (or Two)

We begin a deeper excursion into the quantum view of information processing by examining the role of randomness in quantum mechanics, how that role differs from its role in classical mechanics, and the evidence for the deeper role of randomness in quantum mechanics.

11.2.1 Intrinsic Randomness

Randomness, or unpredictability, has been accepted by most human beings throughout history, based on the simple observation that events happen that nobody had been able to predict. The positivist program [32,71] was a daring attempt of getting rid of randomness: as Laplace famously put it, everything would be predictable for a being capable of assessing the actual values of all physical degrees of freedom at a given time. This ultimate physical determinism is protected against the trivial objection that some phenomena remain unpredictable in practice, even in the age of supercomputers. Indeed, such phenomena involve too many degrees of freedom, or require too precise a knowledge of some values, to be predictable with our means. To put it differently, randomness appears as *relative* to our knowledge and computational power: what is effectively random today may become predictable in the future.

Humanity being so familiar with randomness and science having apparently tamed it, the statement that *quantum phenomena entail an element of randomness* hardly stirs any emotion. The trained scientists translate it as "quantum physics deals with complex phenomena," the others as "I have a good excuse not to understand what these people are speaking about." However, what nature is telling us through quantum physics is different: quantum phenomena suggest that there is *intrinsic* randomness in our universe. In other words, some events are unpredictable even for Laplace's being, who knows all that can be known about this physical universe at a given time. It is *absolute* randomness. Coming from scientific quarters, this claim sounds even more daring than positivism. What is the basis for it?

Physicists may well reply that quantum randomness is like the heart in the cup of coffee or the rabbit in the moon: once you have seen it, you see it always. For more convincing evidence, they can point to an outstanding phenomenon: the *observation of the violation of Bell's inequalities*.

11.2.2 Violation of Bell's Inequalities

In order to appreciate the power of Bell's inequalities [14], some notions need to be introduced. In particular, we must be precise about what measurement means, whether in a quantum or classical setting. Once the concepts around measurement are clear, Bell's inequalities follow from a simple statistical argument about probabilities of measurement outcomes.

11.2.2.1 Description of the Measurement Process

It is crucial to begin by providing an *operational description of a measurement process*, which can be decomposed into three steps:

1. First, a physical system to be measured must enter the measurement device. Here, we do not need to know anything about the physical system itself: we just assume that something indicates when the device is *loaded* (actual experiments do not even need this heralding step, but we assume it here for simplicity of the discussion).

2. The device has a knob, or some similar method of selecting settings, with each position of the knob corresponding to a different measurement: the *input* part of the measurement process consists of *choosing a setting*, that is, a position of the knob. Two remarks must be made on this step. First, it is assumed that the process that chooses the setting is uncorrelated from the system to be measured. In short, we say that the choice is *free*. Second, it is *not* assumed that different settings correspond to different measurements within the device: *a priori*, the position of the knob may be uncorrelated with the physical measurement that is actually performed. We are going to sort the results conditioned on the setting, which is the only information about the input we can access.

3. The *output* of a measurement process is readable information. Think of one lamp out of several being lit. For simplicity, in this chapter, we consider the example of binary information: the output of a measurement can be either of two outcomes. Outputs are often *labeled* by numbers for convenience: so, one may associate one lamp with *0* and the other with *1* or, alternatively, with *+1* and *−1*, respectively. But this label is purely conventional and the conclusions should not depend crucially on it.

With respect to the *free* choice of the setting in point 2, bear in mind that one is not requiring the choice to be made by an agent supposedly endowed with *free will*. As will become clear later in this section, what we need is that for two different measuring systems, each consisting of a measuring device and an object being measured, the measurement outcome of system A cannot depend on the setting of system B, and vice versa. If system A has no knowledge of system B's setting, and vice versa, there can be no such dependence. It is impossible to rule out the possibility of such knowledge completely, but steps can be taken to make it appear unlikely. First, communication between the two systems can be ruled out by placing the systems sufficiently far apart that even light cannot travel between them during the duration of the experiment, from measurement setting to reading the measurement outcome. Even without communication, the outcome at system *A* could depend on the setting at system *B* if the setting is predictable. In actual experiments, the choice is made by a physical random process, which is very reasonably assumed to be independent of the quantum systems to be measured. We must be clear about what sort of randomness is required since we will use such a setup to argue for intrinsic randomness. We do not need an intrinsically random process, just a process for choosing a setting for system *A* that is reasonably believed to be unpredictable to system *B* and vice versa. A classical coin flip suffices here, for example, even though the outcome is deterministic given the initial position and momentum, because it is reasonable, though not provable, that system *B* does not have access to the outcome of a coin flip at system *A* and vice versa.

11.2.2.2 Measurement on a Single System

Consider first the characterization of the results of a *single* measurement device. The elementary measurement run (i.e., the sequence *chooses a setting—registers the outcome*) is repeated many times, so that the statistics of the outcomes can be drawn. One observes, for instance, that for setting $x=1$, it holds [Prob $(0|x=1)$, Prob $(1|x=1)$] = [1/2, 1/2]; for setting $x=2$, it holds [Prob $(0|x=2)$, Prob $(1|x=2)$] = [1/3, 2/3]; for setting $x=3$, it holds [Prob $(0|x=3)$, Prob $(1|x=3)$] = [0.99, 0.01]; and so on for as many positions as the knob has. Apart from the recognition *a posteriori* that some positions of the knob do correspond to something different happening in the device, what physics can we learn from this brute observation? Nothing much, and certainly *not* the existence of intrinsic randomness. Indeed, for instance, setting $x=1$ may be associated to the instructions "don't measure any physical property; just choose the outcome by tossing an unbiased coin." This counterexample shows that classical apparent randomness can be the origin of the probabilistic behavior of setting $x=1$. A similar argument can be made for settings $x=2$ and $x=3$, using biased coins.

11.2.2.3 Measurement on Two Separate Systems

However, things change dramatically if we consider *two* measurement devices, if one further assumes that they cannot communicate (and there may be strong reasons to believe this assumption; ultimately, one can put them so far apart that not even light could propagate from one to the other during a measurement run). Then *not all* statistical observations can be deconstructed with two classical processes as we did before. This is the crucial argument, so let us carefully go through it.

We denote by x the input, the setting of the device at location A, and a its outcome; y the input of the device at location B, and b its outcome. Moreover, x and y are assumed to be chosen independently of each other, so that the setting at A is unknown and unpredictable to location B, and vice versa. We restrict to the simplest situation, in which the choice of inputs is binary: so, from now on, in this section, $x, y \in \{0, 1\}$.

First, let us discuss an example of a nontrivial situation, which can nevertheless be explained by classical pseudo-randomness. Suppose that one observes the following statistics for the probabilities Prob $(a, b | x, y)$ of seeing outcomes a and b given settings x and y:

Prob$(0,0|0,0)$=1/2 Prob$(0,1|0,0)$=0   Prob$(1,0|0,0)$=0   Prob$(1,1|0,0)$=1/2

Prob$(0,0|0,1)$=1/4 Prob$(0,1|0,1)$=1/4 Prob$(1,0|0,1)$=1/4 Prob$(1,1|0,1)$=1/4

Prob$(0,0|1,0)$=1/4 Prob$(0,1|1,0)$=1/4 Prob$(1,0|1,0)$=1/4 Prob$(1,1|1,0)$=1/4

Prob$(0,0|1,1)$=1/2 Prob$(0,1|1,1)$=0   Prob$(1,0|1,1)$=0   Prob$(1,1|1,1)$=1/2

In words, this means that $a=b$ when $x=y$, while a and b are uncorrelated when $x \neq y$. The presence of correlations indicates that uncorrelated coins are not a possible explanation. A classical explanation is possible, however. Assume that, in each run, the physical system to be measured in location A carries an instruction specifying that the value of the output should be a_x if the setting is x; and similarly for what happens at B. In other words, from a common source, the physical system sent to A receives instructions to answer a_0 if the setting is 0 and a_1 if the setting is 1, and the system sent to B receives instructions to answer b_0 if the setting is 0 and b_1 if the setting is 1. These instructions are summarized as $\lambda = (a_0, a_1; b_0, b_1)$. The observation statistics earlier require that the source emit only instructions $\lambda = (a_0, a_1; b_0, b_1)$ such that $a_0=b_0$ and $a_1=b_1$. The precise statistics are obtained when the source chooses each of the four possible λ's, $(0, 0; 0, 0)$, $(0, 1; 0, 1)$, $(1, 0; 1, 0)$, and $(1, 1; 1, 1)$, with equal probability, by coin flipping.

Such a strategy is commonly referred to as *preestablished agreement*; the physics jargon has coined the unfortunate name of *local hidden variables* to refer to the λ's. Whether a preestablished agreement can explain a table of probabilities is only interesting if we assume that the output at A cannot depend on the setting at B and vice versa. Without that requirement, any table of probabilities can be obtained by sending out the 16 possible instructions with the probabilities given in the table. This remark illustrates why we insisted that the choice of setting must be made freely and unpredictably. The *local* in *local hidden variables* refers to the requirement that one side does not know the setting on the other side. Before turning to the next example, it is important to stress a point: we are not saying that observed classical correlations *must* be attributed to preestablished agreement (one can observe classical correlations by measuring quantum systems), rather that because classical correlations *can* be attributed to preestablished agreement, it is impossible to use them to provide evidence for the existence of intrinsic randomness.

In order to finally provide such evidence, we consider the following statistics:

Prob(0,0|0,0)=1/2 Prob(0,1|0,0)=0 Prob(1,0|0,0)=0 Prob(1,1|0,0)=1/2

Prob(0,0|0,1)=1/2 Prob(0,1|0,1)=0 Prob(1,0|0,1)=0 Prob(1,1|0,1)=1/2

Prob(0,0|1,0)=1/2 Prob(0,1|1,0)=0 Prob(1,0|1,0)=0 Prob(1,1|1,0)=1/2

Prob(0,0|1,1)=0 Prob(0,1|1,1)=1/2 Prob(1,0|1,1)=1/2 Prob(1,1|1,1)=0

In words, it says that $a=b$ for three out of four choices of settings, while $a \neq b$ for the fourth. Preestablished agreement cannot reproduce this table: it would require the fulfillment of the contradictory set of conditions $a_0 = b_0$, $a_0 = b_1$, $a_1 = b_0$, and $a_1 \neq b_1$. But consider carefully what this means: the outcomes of the measurement process cannot be the result of reading out a preexisting list $\lambda = (a_0, a_1; b_0, b_1)$. Turn the phrase again and we are there: there was an element of unpredictability in the result of the measurements—because, if all the results had been predictable, we could have listed these predictions on a piece of paper, but such a list cannot be written.

We have reached the conclusion that the observation of these statistics implies that the underlying process possesses intrinsic randomness. It is absolutely remarkable that such a conclusion can in principle be reached in a black box scenario, as a consequence of observed statistics, without any discussion of the physics of the process.

One may further guess that the same conclusion can be reached if the statistics are not exactly the ones written earlier, but are not too far from those. This is indeed the case: all the statistics that can be generated by shared randomness must obey a set of linear inequalities; the statistics that violate at least one of those inequalities can be used to deduce the existence of intrinsic randomness. These are the famous *Bell's inequalities*, named after John Bell who first applied these ideas to quantum statistics.[*]

As a matter of fact, the statistics just described cannot be produced by measuring composite quantum systems at a distance: even in the field of randomness, there are some things that mathematics can conceive but physics cannot do for you. Nevertheless, quantum physics can produce statistics with a similar structure, in which 1/2 is replaced by $p = \left(1 + 1/\sqrt{2}\right)/4 \approx 0.43$ and 0 by $1/2 - p = \left(1 - 1/\sqrt{2}\right)/4 \approx 0.07$. This quantum realization still violates Bell's inequalities by a comfortable margin: one would have to go down to $p = 3/8$, for the statistics of this family to be achievable with shared randomness (see

---

[*] As a curiosity, Boole had already listed many such inequalities [19,78], presenting them as conditions that must be trivially obeyed—he could not expect quantum physics!

Chapter 5 of [86]). The bottom line of it all is: *quantum physics violates Bell's inequalities; therefore, there is intrinsic randomness in our universe**.

11.2.3 Quantum Certainties

While randomness is at the heart of quantum mechanics, it does not rule out certainty. Nor is randomness the whole story of the surprise of quantum mechanics: part of the surprise is that certain things that classical physics predicts should happen with certainty and quantum mechanics predicts do not happen, also with certainty. The most famous such example is provided by the Greenberger–Horne–Zeilinger correlations [49,50,75]. These involve the measurement of three separated quantum systems. Given a set of observations, classical physics gives a clear prediction for another measurement: four outcomes can happen, each with equal probability, while the other four never happen. Quantum mechanics predicts, and experiments confirm, that exactly the opposite is the case (see Chapter 6 of [86]).

So, quantum physics is not brute randomness. By making this observation, we take issue with a misconception common in popular accounts of quantum mechanics, and some scholarly articles: that quantum mechanics, at least in the *many-worlds interpretation*, implies that every conceivable event happens in some universe.[†] On the contrary, as we just saw, there are conceivable possibilities (and even ones that a classical bias would call necessities) that cannot happen because of quantum physics. One of us has used this evidence to call for a *fewer-worlds-than-we-might-think* interpretation of quantum mechanics [84].

11.2.4 Think Positive

Because of other quantum properties, such as the inability to precisely measure both position and momentum (see Section 11.3.2), intrinsic randomness

---

* Has nature disproved determinism? This is strictly speaking impossible. Indeed, full determinism is impossible to falsify: one may believe that everything was fully determined by the big bang, so that, among many other things, human beings were programmed to discover quantum physics and thus believe in intrinsic randomness. Others may believe that we are just characters in a computer game played by superior beings, quantum physics being the setting they have chosen to have fun with us (this is not postmodernism: William of Ockham, so often invoked as a paragon of the scientific mindset, held such views in the fifteenth century). The so-called *many-worlds interpretation* of quantum physics saves determinism in yet another way: in short, by multiplying the universes (or the branches of reality) such that all possibilities allowed by quantum physics do happen in this multiversity.

   Hence, it is still possible to uphold determinism as ultimate truth; only, the power of Laplace's being must be suitably enhanced: it should have access to the real code of the universe or to the superior beings that are playing with us or to all the branches of reality. If we human beings are not supposed to have such a power, the observed violation of Bell's inequalities means, at the very least, that some randomness is *absolute for us*. As we wrote in the main text, but now with emphasis: there is intrinsic randomness in *our* universe.

† A typical quote: "There are even universes in which a given object in our universe has no counterpart—including universes in which I was never born and you wrote this article instead." [34].

was rapidly accepted as the orthodox interpretation of quantum phenomena, four decades before the violation of Bell's inequalities was predicted (let alone observed). The dissenting voices, be they as loud as Einstein, Schrödinger, and De Broglie, were basically silenced.

Even so, for more than half a century, physicists seemed to have succumbed to unconscious collective shame when it came to these matters. One perceives an underlying dejection in generations of physicists, and great physicists at that, who ended up associating quantum physics with insurmountable limitations to previous dreams of knowledge of and control over nature. The discourse was something along the lines of: "You can't know position and momentum, because that's how it is and don't ask me why; now, shut up and calculate, your numbers will be very predictive of the few things that we can speak about." Several otherwise excellent manuals are still pervaded by this spirit.

It took a few people trained in both information science and quantum physics to realize that intrinsic randomness is not that bad after all: in fact, it is a *very useful resource for some tasks*. And this is exactly the new, positive attitude: given that our universe is as it is, maybe we can stop complaining and try to do something with it. Moreover, since quantum physics encompasses classical physics and is broader than the latter, surely there must be tasks that are impossible with classical degrees of freedom, which become possible if one moves to quantum ones. Within a few years, this new attitude had triggered the entire field of *quantum information science*.

The epic of the beginning of quantum information science has been told many times and its heroes duly sung. There is Wiesner dreaming of quantum money in the late 1970s and not being taken seriously by anyone [96]. There are Bennett and Brassard writing in 1984 about quantum key distribution (QKD) and quantum bit commitment [16]—the latter to be proved impossible a few years later, the former to be rediscovered by Ekert in 1991 for the benefit of physicists [40]. There is Peter Shor vindicating some previous speculations on quantum computing with a polynomial algorithm for factoring large integers [89].

Before examining some of these topics, however, we pause to introduce the basic concepts underlying quantum information and computation.

11.3 Key Concepts Underlying Quantum Information Processing

Quantum information processing examines the implications of replacing our classical mechanically grounded notions of information and information processing with quantum mechanically grounded ones. It encompasses quantum computing, quantum cryptography, quantum communication protocols, and beyond. The framework of quantum information processing has many similarities to classical information processing, but there are also several striking differences between the two. One difference is that the fundamental unit of computation, the bit, is replaced with the quantum bit, or qubit. In this section, we define qubits and describe a few key properties of multiple qubit systems that are central to differences between classical and quantum information processing: the tensor product structure of quantum systems, entanglement, superposition, and quantum measurement.

11.3.1 One Qubit and Its Measurement

The fundamental unit of quantum computation is the quantum bit, or *qubit*. Just as there are many different physical implementations of a bit (e.g., two voltage levels, toggle switch), there are many possible physical implementations of a qubit (e.g., photon polarization, spin of an electron, excited and ground state of an atom). And, just as in the classical case, information theory can abstract away from the specific physical instantiation and discuss the key properties of qubits in an abstract way.

Classical bits are *two-state* systems: the possible states are 0 and 1. Qubits are *quantum two-level* systems: they can take infinitely many different states. However, crucially, *different does not mean perfectly distinguishable* in quantum physics. In the case of qubits, any given state is perfectly distinguishable from one and only one other state: this is what makes the qubit the simplest, nontrivial quantum system. We cannot indulge here in explaining the mass of experimental evidence that lead physicists to accept such a counterintuitive view, but we can sketch how one describes it mathematically.

Let us first pick a pair of perfectly distinguishable states and label them $|0\rangle$ and $|1\rangle$. These two states can be used to encode a classical bit: as a consequence, all of classical information and computation can be seen as special cases of the quantum ones. In order to describe the other possible quantum states, one has to use *2D complex vectors*:

$$|0\rangle = \begin{pmatrix} 1 \\ 0 \end{pmatrix}, |1\rangle = \begin{pmatrix} 0 \\ 1 \end{pmatrix} \rightarrow |\psi\rangle = \begin{pmatrix} a \\ b \end{pmatrix} = a|0\rangle + b|1\rangle$$

where a and b are complex numbers such that* $|a|^2 + |b|^2 = 1$. One says that $|\psi\rangle$ is a *superposition* of $|0\rangle$ and $|1\rangle$. Of course, superposition is relative to an arbitrary prior choice: physically, $|\psi\rangle$ is just as good a state as $|0\rangle$ and $|1\rangle$. The relationship of distinguishability between two states $|\psi\rangle$ and $|\phi\rangle$

---

* With this normalization and the convention that two vectors $|\psi\rangle$ and $e^{i\theta}|\psi\rangle$ differing by a global constant represent the same state, one is left with two real parameters. It can be shown that the possible states of a qubit are in one-to-one correspondence with the points on the unit sphere in 3D.

is captured by the absolute value of the scalar product, usually denoted $|\langle\psi|\phi\rangle|$: the smaller this number, the more distinguishable the states are. In particular, perfectly distinguishable states are associated to orthogonal vectors.

The existence of infinitely many different states, but of finitely many (two, for qubits) perfectly distinguishable ones, is reflected in the readout of information a.k.a. *measurement*. An elementary desideratum for measurement is that each outcome identifies the physical properties (i.e., the state). In quantum physics, therefore, a measurement consists in choosing a basis of orthogonal vectors as outcomes* so that *after* the measurement, one can say with certainty in which state the system is. However, unless the state is promised to be already in one of the possible outcome states, the information about the state of the system *prior* to the measurement is lost.[†] In general thus, measurement creates a property rather than revealing it: in other words, it is not possible to reliably measure an unknown state without disturbing it.

To give a concrete example, horizontal polarization $|H\rangle$ and vertical polarization $|V\rangle$ of a photon can be perfectly distinguished. Similarly, the two polarizations at 45°, $|\nearrow\rangle = \frac{1}{\sqrt{2}}|H\rangle + \frac{1}{\sqrt{2}}|V\rangle$ and $|\nwarrow\rangle = \frac{1}{\sqrt{2}}|H\rangle - \frac{1}{\sqrt{2}}|V\rangle$ can be perfectly distinguished. The polarization of a photon can be measured in such a way that the two outcomes are horizontal $|H\rangle$ and vertical $|V\rangle$ or in such a way that the two outcomes are polarizations at 45°. A photon with polarization $|\nearrow\rangle$, when measured in a way that distinguishes horizontal polarization $|H\rangle$ from vertical $|V\rangle$ polarization, will become $|H\rangle$ with probability $|\langle\nearrow|H\rangle|^2 = 1/2$ and $|V\rangle$ with probability $|\langle\nearrow|V\rangle|^2 = 1/2$; but if it were measured instead with an apparatus that distinguished $|\nearrow\rangle$ from $|\nwarrow\rangle$, it would be found to be $|\nearrow\rangle$ with certainty.

11.3.2 Uncertainty Relations and No-Cloning Theorem

What we just discussed about measurement has implications that are worth spelling out, since they are well known and will form part of those quantum features that give QKD and quantum computing their power.

First, we can come back to the idea of intrinsic randomness, seen from the angle of quantum measurement. We have said that in general, the measurement prepares a given output state but tells us little about an unknown input state. The reverse of this fact is also true: *if the input state is known, the outcome of a measurement is in general random*. This must be the case in a theory in

---

* More general notions of measurement have been defined, but they can all be put in the framework of a projective measurement (the type of measurement we have just defined) on a larger system.

† A little bit of information on the state prior to the measurement is available: it could not have been orthogonal to the state that was detected, because otherwise, it would have ended up in a different outcome for the measurement.

which there are infinitely many more states than those that can be perfectly discriminated: indeed, all the others cannot be perfectly discriminated and there will be an element of randomness. As we have seen previously, the violation of Bell's inequalities proves that this randomness is an intrinsic feature of our universe and not just a theoretical artifact.

From these observations, it is easy to deduce the famous *uncertainty relations*: given two measurements, most of the time they will project on different bases. Even if an input state gives a certain outcome for the first measurement, it will most probably give only statistical outcomes for the second. The original example of Heisenberg says that a particle cannot have both a well-defined position and a well-defined momentum. While this example is particularly striking for its counterintuitive character, one can define uncertainty relations even for a single qubit. Indeed, uncertainty relations are *derived* notions and certainly not *principles* as the wording often goes: they capture some specific quantitative aspects of the intrinsic randomness through statistical measures such as variances or entropies.

Second, one can prove that it is *impossible to copy reliably an unknown quantum state*, a statement that goes under the name of the *no-cloning theorem*. This is obviously not the case with classical information: one can photocopy a letter without even reading its content. In the quantum case, for each set of fully distinguishable states, there is a mechanism that can copy them. But the mechanism is different for different sets. If the wrong mechanism is used, not only does an incorrect copy emerge but the original is also altered, probabilistically, in a way that is similar to the alteration by measurement. If the state is unknown, it is impossible to reliably choose the correct mechanism. The mathematical proof of the no-cloning theorem is simple (see Chapter 3 of [86] or Chapter 5 of [83]). Here, we can give a proof based on self-consistency of what we said. If it were possible to produce N copies of a given state out of one, one could obtain sufficient statistical information to reconstruct the initial state: this procedure would discriminate all possible states, contrary to the assumption.

11.3.3 Multiple Qubit States: Entanglement

The use of vector spaces to describe physical systems has dramatic consequences when applied to the description of composite systems. To continue with the photon polarization example, consider two photons, propagating in different directions. Both photons can have horizontal polarization, and this we would write $|H, H\rangle$, or both vertical polarization, $|V, V\rangle$. But we are using a vector space: therefore, $a|H, H\rangle + b|V, V\rangle$ should be a possible state of the system too. Only, what does it describe?

Before sketching a reply, let us consider now the polarization of three photons: each photon is a qubit, which (upon choice of a basis) can be described by two complex parameters, so one would expect classically that three qubits are described by $3 \times 2 = 6$ complex parameters. However, it is easy to convince oneself that an orthogonal basis consists of eight vectors, so in fact one needs $2^3 = 8$ complex parameters to describe three qubits.* The proof is simple: the eight states $|H, H, H\rangle$, $|H, H, V\rangle$, $|H, V, H\rangle$, $|H, V, V\rangle$, $|V, H, H\rangle$, $|V, H, V\rangle$, $|V, V, H\rangle$, and $|V, V, V\rangle$ are perfectly distinguishable by measuring each photon in the $H - V$ basis; therefore, they must be described by mutually orthogonal vectors.

A direct generalization of this simple count shows that *the size of the state space grows exponentially with the number of components in the quantum case*, as opposed to only linearly in the classical mechanical case. More specifically, the dimension of two complex vector spaces M and N when combined via the Cartesian product is the sum of the two dimensions: $dim(M \times N) = dim(M) + dim(N)$. The quantum formalism requires rather to combine systems using the *tensor product*, in which case the dimensions multiply: $dim(M \otimes N) = dim(M)dim(N)$, where $\otimes$ denotes the tensor product.

In turn, this means that there are many states that do not have a simple classical counterpart. In particular, while the state of a classical system can be completely characterized by the state of each of its component pieces, most states of a quantum system cannot be described in terms of the states of the system's components: such states are called *entangled states*. The two-photon superposition written earlier is an example of an entangled state for nonzero a and b: $a|H\rangle \otimes |H\rangle + b|V\rangle \otimes |V\rangle \neq |\psi\rangle \otimes |\varphi\rangle$. The statistics violating Bell's inequalities arise from measurements on such states.

In the following, we will be interested particularly in the case in which the system is a multiqubit system and the components of interest are the individual qubits. As it should be clear from the three-photon example, a system of n qubits has a state space of dimension 2^n.

11.3.4 Sophisticated Coda

We want to add two remarks here, aimed at our more sophisticated readers. Other readers can simply skip this section.

In both quantum mechanics and classical mechanics, there are multiple meanings for the word *state*. In classical mechanics, an attractive approach is to take a *state* to mean a probability distribution over all possible configurations rather than as the space of all configurations as we are doing here. Similarly, in quantum mechanics, we have notions of pure states and mixed states, with the mixed states being probability distributions over pure states. We follow the convention that a state means a pure state. In both cases, within the probabilistic framework, the pure states in the quantum mechanical case, or the set of configurations in the classical mechanical

---

* In the main text, we use simplified counts. In fact, as we have seen in a previous footnote, two real parameters are sufficient to parameterize one qubit. Classically, therefore, one would expect that $3 \times 2 = 6$ real parameters are necessary to represent three qubits. But the quantum count gives $2^3 = 8$ complex parameters, -1 real for the normalization and -1 real for a global phase factor, so in total, 14 real parameters.

case, can be identified as the minimally uncertain or maximal knowledge states. In the classical case, such states have no uncertainty; they are the probability distributions with value 1 at one of the configurations and 0 at all others. In the quantum mechanical case, the inherent uncertainty we discussed in Section 11.2.2 means that even minimally uncertain states still have uncertainty; while some measurements of minimally uncertain states may give results with certainty, most measurements of such a state will still have multiple outcomes.

The tensor product structure in quantum mechanics also underlies probability theory and therefore appears in classical mechanics when states are viewed as probability distributions over the set of configurations. Unfortunately, the tensor product structure is not mentioned in most basic accounts of probability theory even though one of the sources of mistaken intuition about probabilities is a tendency to try to impose the more familiar direct product structure on what is actually a tensor product structure. An uncorrelated distribution is the tensor product of its marginals. Correlated distributions cannot be reconstructed from their marginals. For mixed quantum states, it is important to be able to distinguish classical correlations from entanglement, which would require a more sophisticated definition than the one we gave earlier. A major difference between classical probability theory is that minimally uncertain states in the classical case do not contain correlation, whereas in the quantum case, most minimally uncertain states, the pure states, do contain entanglement. In other words, all minimally uncertain states in the classical setting can be written as a tensor product of their marginals, whereas in the quantum setting, most minimally uncertain states cannot be decomposed into tensor factors. For more discussion of the relationship between classical probability theory, classical mechanics, and quantum mechanics, see [59,83,84,97]. For the remainder of the discussion, we return to using *state* to mean *pure state*.

11.4 Quantum Key Distribution

The best known application of quantum mechanics in a cryptographic setting, and one of the earliest examples of quantum information processing relates to the problem of establishing a key, a secret string of bits, shared between two parties (usually called Alice and Bob) that they can use at some later stage as the encryption key for sending secret messages between them. This problem is known as *key distribution*. While the problem is easily solvable if Alice and Bob can meet or if they share some secure channel over which they can communicate, it becomes much harder if all of their communication is potentially subject to eavesdropping.

In order to understand what QKD can and cannot do for you, let us consider the classical scenario of a *trusted courier*. Alice generates a string of bits, burns a copy of it in a DVD, and uses a courier to send it to Bob. Alice and Bob will then share the string of bits (the key), which will be private to them, if everything went well. What could go wrong? One source of concern is the courier: he can read what is on the DVD or allow someone else to read it, while it is on its way. QKD addresses this concern. As with any security protocol, there are threats that a QKD protocol does not deal with. One is authentication. An eavesdropper Eve could convince the courier that she is Bob, thus establishing a shared key between herself and Alice. She can then set up a separate shared key between herself and Bob. This scenario is called a man-in-the-middle attack. Conventional means of authentication, through which Alice can be sure it was Bob who received her string, exist and should be used with QKD to guard against man-in-the-middle attacks. Another concern is that someone might have tapped into Alice's private space (her office, her computer) while she was generating the string, or someone might tap in Bob's private space and read the key. If the private space of the authorized partners is compromised, there cannot be any security, and quantum information processing cannot help.

The crucial feature of QKD is that *if the courier (a quantum communication channel) is corrupted by the intervention of Eve the eavesdropper, Alice and Bob will detect it*. Even more, they will be able to quantify how much information has leaked to Eve. On this basis, they can decide whether the string can be purified with suitable classical information processing and a key be extracted. If too much information has leaked, they will discard the entire string. At any rate, a non-secret key is never used. This eavesdropper-detecting functionality is inextricably linked to the no-cloning theorem and as such could never be achieved using purely classical techniques.

11.4.1 Physical Origin of Security

QKD is a huge research field, encompassing a variety of different QKD protocols, error correction and privacy amplification techniques, and implementation efforts. Here, we focus on the physical origin of security, that is, why Eve's intervention can be detected. For a basic introduction, see Chapter 2 of [86]; more experienced readers can consult a number of excellent review articles [38,45,87].

The security of the first QKD protocol, proposed by Bennett and Brassard in 1984 [16] and therefore called BB84, is based on the combination of the fact that *measurement modifies the quantum state* and the fact that *unknown quantum states cannot be copied* (the no-cloning theorem). Indeed, faced with her desire to learn what Alice is sending to Bob, Eve can try to look directly at the states or she can try to copy them to study at her leisure. Whether she tries to measure or copy, because the information is encoded in a quantum state unknown to her, the measurement or copying mechanism she chooses is almost certain to introduce modification in the quantum state. Because of this modification, Eve not only does not learn the correct state but also does not know what to send to Bob who is expecting to receive a state. Recall that copying with the wrong mechanism disturbs the original as well as the copy, so even in this case, she does not have an unmodified state to send along to Bob. The more Eve gets to know about the key, the more disturbance she causes in the state that reaches Bob. He and Alice can then compare notes publicly on just some of the states he has received to check for modifications and thus detect Eve's interference.

In 1991, Ekert rediscovered QKD [40] using ideas with which we are already familiar: entangled states and Bell's inequalities. If Alice and Bob share entangled states that violate Bell's inequalities, they share strong correlations that they can use to obtain a joint key by measuring these states. Alice's outcome is random for anyone except Bob, and vice versa. In particular, Eve cannot know those outcomes. At the opposite extreme, suppose that Eve knows perfectly the outcomes of Alice and Bob. Then those outcomes are no longer random, and as a consequence, they cannot violate Bell's inequality. In summary, Ekert's protocol exploits a trade-off between the amount of *violation of Bell's inequality* and the information that Eve may have about the outcomes.

11.4.2 Device-Independent QKD

In a very short time, physicists realized that in spite of many superficial differences, BB84 and the Ekert protocol are two versions of the same protocol. Fifteen years later, physicists realized that the two protocols are deeply different after all! In order to understand why, consider a security concern we have not looked at yet. Where are Alice and Bob getting their devices, the ones that create and detect the quantum states that are used in the protocol? Unless they are building the devices themselves, how do they know that the devices are working as advertised? How can they be sure that Eve has not built or modified the devices to enable them to behave in a different way, a way she can attack? They can detect when Eve interferes with a transmission. Can they somehow also detect when something fishy is going on with the devices in the first place?

For BB84, it turns out that a high level of trust in the behavior of the apparatuses is mandatory. In particular, the protocol is secure only if the quantum information is encoded in a *qubit*, that is, a degree of freedom that has only two distinguishable states. If one cannot ensure this, the protocol is insecure: two classical bits suffice to simulate a *perfect* run of the protocol [8]. Ekert's protocol, on the other hand, is based on Bell's inequalities. Referring back to what we wrote earlier, we see that this criterion is based only on conditional statistics: we have described it without having to specify either physical systems (photons, atoms, etc.) or their relevant degrees of freedom (polarization, spin, etc.). In short, one says that the violation of Bell's inequalities is a *device-independent* test [7].

A few remarks are necessary. First of all, we are saying that anyone who can carry out Ekert's protocol can check whether Bell's inequalities are violated. Creating devices that exhibit violations of Bell's inequalities is much more complicated than simply testing for violations: experimentalists, or the producers of QKD apparatuses, must know very well what they are doing. Second, even after establishing violations of Bell's inequalities, Alice and Bob cannot trust their devices blindly: an adversarial provider, for instance, might have inserted a radio that sends out the results of the measurements, a sort of Trojan horse in the private space. This is not a limitation of QKD alone: for any cryptographic protocol, one must trust that there is no radio in any of the devices Alice and Bob are using in their private space. If Alice and Bob's measurement events are spacelike separated, meaning that no signal could travel between them during the time frame of the measurement process, they can be certain that the keys generated are truly random, provided they convincingly violate Bell's inequalities, even if their devices do contain radios or similar means of surreptitiously transmitting information. However, if this communication continues after the key is generated, there is little to stop their devices betraying them and transmitting the key to Eve. Third, if Alice and Bob know quantum physics and find out that their devices are processing qubits, the Ekert protocol becomes equivalent to BB84.

With this understanding, we can go back to our initial topic of randomness and somehow close the loop, before focusing on quantum computation proper.

11.4.3 Back to Randomness: Device-Independent Certification

Even to the one of us who was directly involved in the process, it is a mystery why the possibility of device-independent assessment was noticed only around 2006. Once discovered in the context of QKD, though, it became clear that the notion can be used in other tasks—for instance, certified randomness generation [31,77]. In the preceding text, we discussed how Bell's inequalities can convince anyone of the existence of intrinsic randomness in our universe. Rephrase it all in an industrial context, and one can conclude that the violation of Bell's inequalities can be used to *certify randomness*.

This Bell-based certification has a unique feature: it guarantees that the random numbers are being produced on the spot, by the process itself. Let us explain this important feature in some detail. Consider first the usual statistical tests of randomness, which check for patterns in the produced string. Take a string that passes such a test and copy it on a DVD, then run the test on the copy. Obviously, the copy will pass the test too. Suppose you want to obtain a random string from an untrusted source. How can you check that the string you receive is random? As one example, how do you know that the source is not sending the same string to other customers? Classically, there is no way to check: a copy looks just as random as the original. Quantum mechanics does not provide any additional ability to check for randomness after a string has been obtained, but a string of measurement outcomes from a Bell's experiment that violates Bell's inequalities cannot have been preprogrammed at the source, guaranteeing that the randomness is newly generated and not a copy of a previously generated string. Again, making use of the free choice of settings in the Bell test, it is not possible to use a predetermined string, no matter how random, to specify outcomes in a Bell test and still pass the test. Because the choice of measurement settings changes unpredictably from test to test, it does not matter whether the string passes classical tests for randomness or even came from a previous Bell test. What the analysis of violation of Bell's inequalities guarantees is that any predictable strategy for determining outcomes, even strategies making use of strings certified as random by a previous Bell test, cannot produce outcomes that violate Bell's inequalities. The combination of quantum physics with a test that contains an element of *freedom* is what ultimately allows us to certify randomness as it is generated in this unique way.

11.5 Quantum Computing

The field of quantum computation examines how grounding computation in quantum rather than classical mechanics changes how efficiently computations of various types can be performed. We first present basic notions of quantum computation and then briefly discuss early algorithms before discussing the two most famous quantum algorithms, Shor's factoring algorithm and Grover's search algorithm. After a brief discussion of quantum-computational simulations of quantum systems, we conclude the section with a discussion of known limitations on quantum computation.

11.5.1 Quantum Computing Basics

We start by clarifying what a quantum computer is *not*. Just because a computer makes use of quantum mechanical effects does not mean it is a quantum computer. All modern computers make use of quantum mechanical effects, but they continue to represent information as bits and act on the bits with the same logical operations earlier machines used. The physical way in which the logical operations are carried out may be different, but the logical operations themselves are the same.

Quantum computers process qubits and process them using quantum logic operations, generalizations of classical logic operations that enable, for instance, creation of entanglement between qubits. Mirroring the situation with classical computation, any quantum computation can be broken down into a series of basic quantum logic gates. Indeed, any quantum mechanical transformation of an n qubit system can be obtained by performing a sequence of one and two qubit operations. Unfortunately, most transformations cannot be performed efficiently in this manner, and many of the transformations that can be efficiently performed have no obvious use. Figuring out an efficient sequence of quantum transformations that can solve a useful problem is a hard problem and lies at the heart of quantum algorithm design.

Quantum gates act on quantum states, which means that they can act on superpositions of classical values. Just as a single qubit can be put in a superposition of the two distinguished states corresponding to bit values 0 and 1, a set of n qubits can be placed in a superposition of all 2^n possible values of an n-bit string: 0 ... 00, 0 ... 01, ..., 1 ... 11. Quantum circuits, made up of quantum logic gates, can be applied to such a superposition. For every efficiently computable function f, there is an efficient quantum circuit that carries out the computation of f. When applied to a superposition of all of the 2^n possible input strings, this circuit produces a superposition of all possible input/output pairs for f. Such an application of a circuit to all possible classical inputs is called *quantum parallelism*. Even though properties of quantum measurement mean that only one of these input/output pairs can be obtained from the superposition of all input/output pairs, the idea of *computing over all possible values at once* is the most frequent reason given in the popular press for the effectiveness of quantum computation. We discuss in Section 11.9.1 further reasons why this explanation is misleading.

In this section, we touch on early quantum algorithms, Shor's algorithm and Grover's algorithm, as well as the simulation of quantum systems, the earliest recognized application of quantum computing. After the discovery of Grover's algorithm, there was a 5-year hiatus before a significantly new quantum algorithm was discovered. Not only have a variety of new algorithms emerged since then but also powerful new approaches to quantum algorithm design including those based on quantum random walks, adiabatic quantum computation, topological quantum computation, and one-way or measurement-based computation that we will touch on in Section 11.6.2. For a popular account of more recent algorithms, see [11]. References [72] and [28] provide more technical surveys.

11.5.2 Early Quantum Algorithms

An early result in quantum computation showed that any classical algorithm could be turned into a quantum computation of roughly equivalent complexity. In fact, any reversible classical algorithm can be translated directly into a quantum mechanical one. Any classical computation taking time t and space s can be turned into a reversible one with at most the slight penalty of $O(t^{1+j})$ time* and $O(s \log t)$ space [15]. A classical deterministic computation that returns a result with certainty becomes a deterministic quantum computation that also returns a result with certainty. This fact provides another example of certainty in quantum mechanics as previously discussed in Section 11.2.3. Quantum algorithms can be probabilistic, or they can be deterministic, returning a single final result with probably 1. Just to reemphasize the earlier point that quantum mechanics does not imply that *everything happens*, the obvious deduction from the fact that an algorithm returns one result with certainty is that the other results do not happen at all.

The early 1990s saw the first truly quantum algorithms, algorithms with no classical analog that were provably better than any possible classical algorithm. The first of these, Deutsch's algorithm, was later generalized to the Deutsch–Jozsa algorithm [35]. These initial quantum algorithms were able to solve problems efficiently with certainty that classical techniques can solve efficiently only with high probability. Such a result is of no practical interest since any machine has imperfections so can only solve problems with high probability. Furthermore, the problems solved were highly artificial. Nevertheless, such results were of high theoretical interest since they proved that quantum computation is theoretically more powerful than classical computation.

11.5.3 Shor's Factoring Algorithm and Generalizations

These early results inspired Peter Shor's successful search for a polynomial-time quantum algorithm for factoring integers, a well-studied problem of practical interest. A classical polynomial-time solution has long eluded researchers. Many security protocols base their security entirely on the computational intractability of this problem. At the same time Shor discovered his factoring algorithm, he also found a polynomial-time solution for the discrete logarithm problem, a problem related to factoring that is also heavily used in cryptography. Shor's factoring and discrete log algorithms mean that once scalable quantum computers can be built, all public key encryption algorithms currently in practical use, such as RSA, will be completely insecure regardless of key length.

Shor's results sparked interest in the field, but doubts as to its practical significance remained. Quantum systems are notoriously fragile. Key quantum properties, such as entanglement, are easily disturbed by environmental influences. Properties of quantum mechanics, such as the no-cloning principle, which made a straightforward extension of classical error correction

---

* In the discussion that follows, we will make use of big-O notation, a standard way of characterizing the performance of an algorithm in theoretical computer science. In this notation, we say that a function $g(x)$ is in $O(f(x))$ if and only if for some sufficiently large constant c, $g(x)$ is bounded from above by $cf(x)$ for all values of x.

techniques based on replication impossible, made many fear that error correction techniques for quantum computation would never be found. For these reasons, it seemed unlikely that reliable quantum computers could be built. Luckily, in spite of widespread doubts as to whether quantum information processing could ever be made practical, the theory itself proved so tantalizing that researchers continued to explore it. In 1996, Shor and Calderbank, and independently Steane, discovered quantum error correction techniques that, in John Preskill's words [79], "fight entanglement with entanglement." Today, quantum error correction is arguably the most mature area of quantum information processing.

Both factoring and the discrete logarithm problem are *hidden subgroup problems* [64]. In particular, they are both examples of abelian hidden subgroup problems. Shor's techniques easily extend to all abelian hidden subgroup problems and a variety of hidden subgroup problems over groups that are almost abelian. Two cases of the non-abelian hidden subgroup problem have received a lot of attention: the symmetric group S_n (the full permutation group of n elements) and the dihedral group D_n (the group of symmetries of a regular n-sided polygon). But efficient algorithms have eluded researchers so far. A solution to the hidden subgroup problem over S_n would yield a solution to graph isomorphism, a problem conjectured to be NP-intermediate along with factoring and the discrete log problem. In 2002, Regev showed that an efficient algorithm to the dihedral hidden subgroup problem using Fourier sampling, a generalization of Shor's techniques, would yield an efficient algorithm for the gap shortest vector problem [81]. In 2003, Kuperberg found a subexponential (but still superpolynomial) algorithm for the dihedral group [60] that he has recently improved [61]. Public key cryptographic schemes based on shortest vector problems are among the most promising approaches to finding practical public key cryptographic schemes that are secure against quantum computers.

Efficient algorithms have been obtained for some related problems. Hallgren found an efficient quantum algorithm for solving Pell's equation [54]. Pell's equation, believed to be harder than factoring and the discrete logarithm problem, was the security basis for Buchmann–Williams key exchange and public key cryptosystems [25]. Thus, Buchmann–Williams joins the many public key cryptosystems known to be insecure in a world with quantum computers. Van Dam et al. [91] found an efficient quantum algorithm for the shifted Legendre symbol problem, which means that quantum computers can break certain algebraically homomorphic cryptosystems and can predict certain pseudo-random number generators.

11.5.4 Grover's Algorithm and Generalizations

Grover's search algorithm is the most famous quantum algorithm after Shor's algorithm. It searches an unstructured list of N items in $O(\sqrt{N})$

time. The best possible classical algorithm uses $O(N)$ time. This speed up is only polynomial, but, unlike for Shor's algorithm, it has been proven that Grover's algorithm outperforms any possible classical approach. Although Grover's original algorithm succeeds only with high probability, variations that succeed with certainty are known; Grover's algorithm is not inherently probabilistic.

Generalizations of Grover's algorithm apply to a more restricted class of problems than is generally realized. It is unfortunate that Grover used *database* in the title of his 1997 paper [53]. Databases are generally highly structured and can be searched rapidly classically. Because Grover's algorithm does not take advantage of structure in the data, it does not provide a square root speed up for searching such databases. Childs et al. [27] showed that quantum computation can give at most a constant factor improvement for searches of ordered data such as that of databases. As analysis of Grover's algorithm focuses on query complexity, counting only the number of times a database or function must be queried in order to find a match rather than considering the computational complexity of the process, it is easy to fall into the trap of believing that it must necessarily have better gate complexity, the number of gates required to carry out the computation. This is not always the case, however, since the gate complexity of the query operation potentially scales linearly in N, as is the case for a query of a disordered database. The gate complexity of this operation negates the $O(\sqrt{N})$ benefit of Grover's algorithm, reducing its applications still further, in that the speed up is obtained only for data that has a sufficiently fast generating function.

As a result of the aforementioned restrictions, Grover's algorithm is most useful in the context of constructing algorithms based on black box queries to some efficient function. Extensions of Grover's algorithm provide small speed ups for a variety of problems including approximating the mean of a sequence and other statistics, finding collisions in r-to-1 functions, string matching, and path integration. Grover's algorithm has also been generalized to arbitrary initial conditions, nonbinary labelings, and nested searches.

11.5.5 Simulation

The earliest speculations regarding quantum computation were spurred by the recognition that certain quantum systems could not be simulated efficiently classically [41,65,66]. Simulation of quantum systems is another major application of quantum computing, with small-scale quantum simulations over the past decade providing useful results [24,98]. Simulations run on special-purpose quantum devices provide applications of quantum information processing to fields ranging from chemistry to biology to material science. They also support the design and implementation of yet larger special-purpose quantum devices, a process that ideally leads all the way to the building of scalable general purpose quantum computers.

Many quantum systems can be efficiently simulated classically. After all, we live in a quantum world and have long been able to simulate a wide variety of natural phenomena. Some entangled quantum systems can be efficiently simulated classically, while others cannot. Even on a universal quantum computer, there are limits to what information can be gained from a simulation. Some quantities, like the energy spectra of certain systems, are exponential in quantity, so no algorithm, classical or quantum, can output them efficiently. Algorithmic advances in quantum simulation continue, while the question of which quantum systems can be efficiently simulated classically remains open. New approaches to classical simulation of quantum systems continue to be developed, many benefiting from the quantum information processing viewpoint. The quantum information processing viewpoint has led to improvements in commonly used classical approaches to simulating quantum systems, such as the density matrix renormalization group (DMRG) approach [93] and the related matrix product states (MPS) approach [76].

11.5.6 Limitations of Quantum Computing

Some popular expositions suggest that quantum computers would enable nearly all problems to be solved substantially more efficiently than is possible with classical computers. Such an impression is false. For example, Beals et al. [13] proved for a broad class of problems that quantum computation can provide at most a polynomial speed up. Their results have been extended and other means of establishing lower bounds have also been found, yielding yet more problems for which it is known that quantum computers provide little or no speed up over classical computers. A series of papers established that quantum computers can search ordered data at most a constant factor faster than classical computers and that this constant is small [27]. Grover's search algorithm is known to be optimal in that it is not possible to search an unstructured list of N elements more rapidly than $O(\sqrt{N})$. Most researchers believe that quantum computers cannot solve NP-complete problems in polynomial time, though there is currently no proof of this (a proof would imply $P \neq NP$, a long-standing open problem in computer science).

Other results establish limits on what can be accomplished with specific quantum methods. Grigni et al. [51] showed that for most non-abelian groups and their subgroups, the standard Fourier sampling method, used by Shor and successors, yields exponentially little information about a hidden subgroup. Aaronson showed that quantum approaches could not be used to efficiently solve collision problems [2]. This result means there is no generic quantum attack on cryptographic hash functions that treats the hash function as a black box. By this we mean an attack that does not exploit any structure of the mapping between input and output pairs present in the function. Shor's algorithms break some cryptographic hash functions, and quantum attacks on others may still be discovered, but Aaronson's result says that any attack must use specific properties of the hash function under consideration.

11.6 Quantum Information Processing More Generally

Quantum information processing is a broad field that encompasses, for example, quantum communication and quantum games as well as quantum cryptography and quantum computing. Furthermore, while QKD is the best known quantum cryptographic protocol, many other types of protocols are known, and quantum cryptography remains an active area of research. Here, we briefly survey some quantum cryptographic protocols and touch on their relation to quantum communication and quantum games. We then delve more deeply into blind quantum computation, a recent discovery that combines cryptography and quantum computation.

11.6.1 Quantum Cryptography beyond Key Distribution

While *quantum cryptography* is often used as a synonym for *QKD*, quantum approaches to a wide variety of other cryptographic tasks have been developed. Some of these protocols use quantum means to secure classical information. Others secure quantum information. Many are *unconditionally* secure in that their security is based entirely on properties of quantum mechanics. Others are only quantum computationally secure in that their security depends on a problem being computationally intractable for quantum computers. For example, while *unconditionally* secure bit commitment is known to be impossible to achieve through either classical or quantum means, quantum computationally secure bit commitment schemes exist as long as there are quantum one-way functions [37].

Closely related to QKD schemes are protocols for unclonable encryption [47], a symmetric key encryption scheme that guarantees that an eavesdropper cannot copy an encrypted message without being detected. Unclonable encryption has strong ties with quantum authentication. One type of authentication is digital signatures. Quantum digital signature schemes have been developed [48], but the keys can be used only a limited number of times. In this respect, they resemble classical schemes such as Merkle's one-time signature scheme.

Cleve et al. provide quantum protocols for (k, n) threshold quantum secrets [30]. Gottesman [46] provides protocols for more general quantum secret sharing. Quantum multiparty function evaluation schemes exist [33,55]. Brassard et al. have shown that quantum mechanics allows for perfectly secure anonymous communication [21]. Fingerprinting enables the equality of two strings to be determined efficiently with high probability by comparing their respective fingerprints [10,26]. Classical fingerprints for n-bit strings need to be at least of length $O(\sqrt{n})$. Buhrman et al. [26] show that a quantum fingerprint of classical data can be exponentially smaller.

In 2005, Watrous showed that many classical zero-knowledge interactive protocols are zero knowledge against a quantum adversary [95]. Generally, statistical zero-knowledge protocols are based on candidate NP-intermediate problems, another reason why zero-knowledge protocols are of interest for quantum computation. There is a close connection between quantum interactive protocols and quantum games. Early work by Eisert et al. [39] includes a discussion of a quantum version of the prisoner's dilemma. Meyer has written lively papers discussing other quantum games [69].

11.6.2 Blind Quantum Computation

One area that combines both cryptography and computation is blind quantum computation [23]. Blind computation protocols address a situation that is becoming increasingly common with the advent of cloud computing, namely, how to perform a computation on a powerful remote server in such a way that a person performing the remote computation, the client, can be confident that only she knows which computation was performed (i.e., only she should know the input, output, and algorithm). While classical cryptographic techniques suffice in practice to prevent an eavesdropper from learning the computation if she can only access the communication between the client and the server, this security falls away if the eavesdropper has access to the server. In the case of blind quantum computation, the remote server is considered to be a fully fledged quantum computer, while the client is considered to have access only to classical computation and the ability to prepare certain single qubit states.

This may seem like rather an odd task to focus on, but in a world where we are digitizing our most sensitive information, maintaining the secrecy of sensitive material is more important than ever. Time on supercomputers is often rented, and so it is essentially impossible to ensure that nobody has interfered with the system. The problem becomes even more acute when we consider quantum computers, which will likely appear initially in only very limited numbers.

In 2001, Raussendorf and Briegel proposed a revolutionary new way of performing computation with quantum systems [80]. Rather than using physical interactions between the qubits that make up such a computer to perform computation, as had been done up to that point, they proposed using specially chosen measurements to drive the computation. If the system was initially prepared in a special state, these measurements could be used to implement the basic logic gates that are the fundamental building blocks of any computation. This model of computation is purely quantum: it is impossible to construct a measurement-based computer according to classical physics.

Measurement-based quantum computation supplements classical computation with measurements on a special type of entangled state. The entangled state is universal in that it does not depend on the type of computation being performed. The desired computation is what determines the measurement sequence. The measurements have a time ordering and the exact measurements to be performed depend on the results of previous measurements. Classical computation is used to determine what measurement should be performed next given the measurement results up until that point and to interpret the measurement results to determine the final output of the computation, the answer to the computational problem. The measurements required are very simple: only one qubit is measured at a time. One effect of this restriction to single qubit measurements, as opposed to joint measurements of multiple qubits, is that throughout the computation, the entanglement can only decrease not increase, resulting in an irreversible operation. For this reason, it is sometimes called *one-way* quantum computation. Measurement-based quantum computation utilizing the correct sort of entangled state has been shown to be computationally equivalent in power to the standard model of quantum computation, the circuit model. The outcomes of the measurements, given that they are measurements on an entangled state, exhibit a high degree of randomness. It is a surprising and elegant result that these random measurement outcomes add sufficient power to classical computation that it becomes equivalent in power to quantum computation.

Measurement-based quantum computation provides a particularly clean separation between the classical and quantum parts of a quantum algorithm. It also suggests a fundamental connection between entanglement and the reason for the power of quantum computation. But the issues here are subtler than one might expect at first. In 2009, two groups of researchers [22,52] showed that if a state is too highly entangled, it cannot support quantum computation. Specifically, Gross et al. showed that if the state is too highly entangled, the outcomes of any sequence of measurements can be replaced by random classical coin flips [52]. Thus, if a state is too highly entangled, the resulting outcomes are too random to provide a quantum resource. We will return to this point later when we discuss the mystery surrounding the sources of quantum computing's power. As Gross et al. conclude with respect to entanglement, "As with most good things, it is best consumed in moderation."

This new model of measurement-based quantum computation opens many promising routes for building large-scale quantum computers. Indeed, many researchers are currently working on architectures for distributed quantum computers based on this model that may lead to large-scale quantum computers. However, measurement-based computation is not simply a way to build better computers, but rather a new way to think about computation.

In particular, measurement-based quantum computation provides a convenient lens with which to examine whether or not it is possible to perform a blind computation on a remote computer. The uncertainty principle allows for more information to be encoded in a quantum state than can be accessed through measurements. As measurement-based computation allows quantum computation to be constructed from measurements on quantum states together with a classical rule for adapting subsequent measurements, by using subtly different initial quantum states for the computation, different logic gates can be implemented. Each possible initial state is chosen in such a way that they yield identical results for any possible measurement made by the server, but yet each nudges the computation in a different direction. As a result, it is possible to perform arbitrary calculations blindly.

In fact, quantum properties enable us to take the security one step further. By adapting standard techniques to detect errors in quantum computers, it is possible to detect any interference with the blind computation [42]. Taken together, these results provide us with a way to ensure that our computation remains private and correct without needing to trust the computer or those who have access to it.

Abadi et al. [5] showed that only in the unlikely event that a famous conjecture in complexity theory fails, information theoretically secure blind computation cannot be carried out on a classical computer. If only computational security is required, classical solutions are possible, though it was only in 2009 that the first such scheme was found, Gentry's famous fully homomorphic encryption scheme [43]. Fully homomorphic encryption is most commonly described as enabling universal computation on encrypted data by a party, say a server, that does not have access to a decryption key and learns nothing about the encrypted data values. The server returns the result of the computation on the encrypted data to a party who can decrypt it to obtain meaningful information. Fully homomorphic encryption can also be used to hide the computation being carried out, thus achieving a form of blind computation, but with only a computational security guarantee.

However, there are significant differences between the capabilities of blind quantum computation and classical fully homomorphic encryption. Most importantly, blind quantum computation allows the client to boost their computational power to an entirely different computational complexity class (from P to BQP), unlike known homomorphic encryption schemes. Further, a blind quantum computation can be authenticated, enabling the detection of any deviation from the prescribed computation with overwhelming probability. The security provided by the protocols is also different: While known fully that homomorphic encryption schemes rely on computational assumptions for their security, the security of the blind computation protocol described in [23] can be rigorously proved on information theoretic grounds.

11.7 Classical Lessons from Quantum Information Processing

The quantum information processing viewpoint provides insight into complexity issues in classical computer science and has yielded novel classical algorithmic results and methods. The usefulness of the complex perspective for evaluating real valued integrals is often used as an analogy to explain this phenomenon. Classical algorithmic results stemming from the insights of quantum information processing include lower bounds for problems involving locally decodable codes, local search, lattices, reversible circuits, and matrix rigidity. Drucker and de Wolf [36] survey a wealth of purely classical computational results, in such diverse fields as polynomial approximations, matrix theory, and computational complexity, that resulted from taking a quantum-computational view.

In two cases, quantum arguments have been used to establish security guarantees for purely classical cryptographic protocols. Cryptographic protocols usually rely on the empirical hardness of a problem for their security; it is rare to be able to prove complete, information theoretic security. When a cryptographic protocol is designed based on a new problem, the difficulty of the problem must be established before the security of the protocol can be understood. Empirical testing of a problem takes a long time. Instead, whenever possible, *reduction* proofs are given that show that if the new problem were solved, it would imply a solution to a known hard problem. Regev designed a novel, purely classical cryptographic system based on a certain lattice problem [82]. He was able to reduce a known hard problem to this problem, but only by using a quantum step as part of the reduction proof. Gentry, for his celebrated fully homomorphic encryption scheme [43], provides multiple reductions; one of which requires a quantum step.

11.8 Implementation Efforts

Over the past two decades since the discovery of Shor's and Grover's algorithms, progress in realizing a scalable quantum computer has begun to gather pace. Technologies based on liquid-state nuclear magnetic resonance techniques (NMR) provided a test bed for many proof of concept implementations of quantum algorithms and other quantum information processing tasks. However, because of problems of cooling, liquid-state NMR is not considered a viable route to a scalable quantum computer. The leading candidates for viable routes to scalable quantum computers have long been ion trap and optical quantum computing. Recently, however, progress in superconducting qubits has shown significant promise for scalable quantum computing. Superconducting quantum processors could be constructed using techniques and facilities similar to today's semiconductor-based processors. Recently, IBM has demonstrated gate fidelities approaching the threshold necessary for fault-tolerant quantum computation [29].

While scalable quantum computing has not yet been achieved, QKD has already been developed into a viable technology. Today, commercial QKD systems are already available from a number of manufacturers including id Quantique and MagiQ Technologies. Other quantum cryptographic techniques have not yet matured to this level, but many, including blind quantum computation [12], have been demonstrated in a laboratory setting.

11.9 Where Does the Power of Quantum Computing Come from?

In contrast to the case for QKD, the source of the power of quantum computation remains elusive. Here, we review some of the explanations commonly given, explaining both the limitations of and the insights provided by each explanation.

11.9.1 Quantum Parallelism?

As discussed in Section 11.5, the most common reason given in the popular press for the power of quantum computation is *quantum parallelism*. However, quantum parallelism is less powerful than it may initially appear. We only gain information by measuring, but measuring results in a single input/output pair, and a random one at that. By itself, quantum parallelism is useless. This limitation leaves open the possibility that quantum parallelism can help in cases where only a single output, or a small number of outputs, is desired. While it suggests a potential exponential speed up for all such problems, as we saw in Section 11.5.6, for many problems, it is known that no such speed up is possible.

Certain quantum algorithms that were initially phrased in terms of quantum parallelism, when viewed in a clearer light, have little to do with quantum parallelism. Mermin's explanation of the Bernstein–Vazirani algorithm, originally published in his paper *Copenhagen Computation: How I Learned to Stop Worrying and Love Bohr* [68], contributed to this enlightenment. He was the first to see that, without changing the algorithm at all, just viewing it in a different light, the algorithm goes from one phrased in terms of quantum parallelism in which a calculation is needed to see that it gives the desired result to one in which the outcome is evident. The Bernstein–Vazirani algorithm [17] and Mermin's argument in particular deserve to be better known because of the insight they give as to how best to view quantum computation.

11.9.2 Exponential Size of Quantum-State Space?

A second popular explanation is the exponential size of the state space. This explanation is also flawed. To begin with, as we have seen, exponential spaces also arise in classical probability theory. Furthermore, what would it mean for an efficient algorithm to take advantage of the exponential size of a space? Even a superposition of the exponentially many possible values of an n-bit string is only a single state of the quantum-state space. The vast majority of states cannot even be approximated by an efficient quantum algorithm [58]. As an efficient quantum algorithm cannot even come close to most states in the state space, quantum parallelism does not, and efficient quantum algorithms cannot, make use of the full state space.

11.9.3 Quantum Fourier Transforms?

Most quantum algorithms use quantum Fourier transforms (QFTs). The Hadamard transformation, a QFT over the group Z_2, is frequently used to create a superposition of 2^n input values. In addition, the heart of most quantum algorithms makes use of QFTs. Shor and Grover both use QFTs. Many researchers speculated that QFTs were a key to the power of quantum computation, so it came as a surprise when Aharonov et al. [9] showed that QFTs are classically simulable. Given the ubiquity of QFTs in quantum algorithms, researchers continue to consider QFTs as one of the main tools of quantum computation, but in themselves, they are not sufficient.

As any quantum computation can be constructed out of a series of gates consisting of QFTs and transformations that preserve the computational basis, it has been suggested that the minimum number of layers of Fourier transforms required for an efficient implementation of a particular quantum transformation gives rise to a hierarchy (known as the Fourier Hierarchy) containing an infinite number of levels, which cannot be collapsed while maintaining polynomial circuit size [88]. The zeroth and first levels of such a hierarchy correspond to the classical complexity classes P and BPP, respectively, while many interesting quantum algorithms, such as Shor's factoring algorithm, occupy the second level. Nonetheless, the truth of the Fourier Hierarchy conjecture that the levels cannot be collapsed remains an open problem.

11.9.4 Entanglement?

Jozsa and Linden [57] show that any quantum algorithm involving only pure states that achieve exponential speed up over classical algorithms must entangle a large numbers of qubits. While entanglement is necessary for an exponential speed up, the existence of entanglement is far from sufficient to guarantee a speed up, and it may turn out that another property better characterizes what enables a speed up. Many entangled systems have been shown to be classically simulable [67,94]. Indeed, the Gottesman–Knill theorem [1], as well as results on the classical simulation of match gates [90], has shown that there exist nonclassical computational models that allow for highly entangled states that are efficiently classically simulable. Furthermore, if one looks at query complexity instead of algorithmic complexity, improvements can be obtained with no entanglement whatsoever. Meyer [70] shows that in the course of the Bernstein–Vazirani algorithm, which achieves an N to 1 reduction in the number of queries required, no qubits become entangled. Going beyond quantum computation, it becomes more obvious that entanglement is not required to reap benefits. For example, the BB84 QKD protocol makes no use of entanglement. While measurement-based quantum computation, discussed in Section 11.6.2, graphically illustrates the use of entanglement as a resource for quantum computation, it turns out that if states are too highly entangled, they are useless for measurement-based quantum computation [22,52]. In the same paper in which they showed that entanglement is necessary, Jozsa and Linden end their abstract with *we argue that it is nevertheless misleading to view entanglement as a key resource for quantum-computational power* [57]. The reasons for quantum information processing's power remains mysterious; Vedral refers to "the elusive source of quantum effectiveness" [92].

11.10 What if Quantum Mechanics Is Not Correct?

Physicists do not understand how to reconcile quantum mechanics with general relativity. A complete physical theory would require modifications to general relativity, quantum mechanics, or both. Any modifications to quantum mechanics would have to be subtle as the predictions of quantum mechanics hold to great accuracy, and most predictions of quantum mechanics will continue to hold, at least approximately, once a more complete theory is found. Since no one yet knows how to reconcile the two theories, no one knows what, if any, modifications would be necessary or whether they would affect the feasibility or the power of quantum computation.

Once the new physical theory is known, its computational power can be analyzed. In the meantime, theorists have looked at what computational power would be possible if certain changes in quantum mechanics were made. So far, these changes imply greater computational power rather than less. Abrams and Lloyd [6] showed that if quantum mechanics were nonlinear, even slightly, all problems in the class #P, a class that contains all NP problems and more, would be solvable in polynomial time. Aaronson [3] showed that any change to one of the exponents in the axioms of quantum mechanics would yield polynomial-time solutions to all PP problems, another class containing NP. These two results are closely related, in that a classical computer augmented with the power to solve either class of problems efficiently would have identical power. With these results in mind, Aaronson [4] suggests that limits on computational power should be considered a fundamental principle guiding physical theories, much like the laws of thermodynamics.

11.11 Conclusions

We hope this glimpse of quantum information processing has intrigued you. If so, there are many excellent resources for learning more, from books on quantum computation [74,83] to arxiv.org/archive/quant-ph where researchers post papers with their most recent results.

Advances in quantum information processing are also driving the development of other technologies beyond computation and communication. Quantum information techniques have led to advances in lithography, providing a means to affect material at scales below the classical wavelength limit [20]. Quantum information processing has motivated significant strides in our ability to control quantum systems [97]. Further, quantum mechanics allows for significant improvements in the performance of a variety of sensors. Theoretical improvements have been demonstrated in a number of settings, initially restricted to simple parameter estimation [18,44,99] but later extended to imaging and other complex tasks [63]. Experimentally, such quantum techniques have been demonstrated to provide increased accuracy in estimating phase shifts induced by optical materials [73], spectroscopy [62,85], and in estimating magnetic field strengths [56].

Many open problems remain. Some are of a fundamental nature. What does nature allow us to compute efficiently? What does nature allow us to make secure? Others are of a more practical nature. How will we build scalable quantum computers? For what problems are there effective quantum algorithms? How broad an impact will quantum information processing have? At the very least, quantum computation, and quantum information processing more generally, has changed forever how humanity thinks about and works with physics, computation, and information.

References

1. S. Aaronson and D. Gottesman. Improved simulation of stabilizer circuits. *Physical Review A*, 70(5):052328, 2004.
2. S. Aaronson. Quantum lower bounds for the collision problem. In *Proceedings of STOC'02*, pp. 635–642, 2002.

3. S. Aaronson. Quantum computing, postselection, and probabilistic polynomial-time. *Proceedings of the Royal Society A*, 461:3473–3482, 2005.
4. S. Aaronson. The limits of quantum computers. *Scientific American*, 298(3):62–69, March 2008.
5. M. Abadi, J. Feigenbaum, and J. Kilian. On hiding information from an oracle. *Journal of Computer and System Sciences*, 39(1):21–50, 1989.
6. D. S. Abrams and S. Lloyd. Nonlinear quantum mechanics implies polynomial-time solution for NP-complete and #P problems. *Physical Review Letters*, 81:3992–3995, 1998.
7. A. Acin, N. Brunner, N. Gisin, S. Massar, S. Pironio, and V. Scarani. Device-independent security of quantum cryptography against collective attacks. *Physical Review Letters*, 98:230501, 2007.
8. A. Acin, N. Gisin, and L. Masanes. From Bell's theorem to secure quantum key distribution. *Physical Review Letters*, 97(12):120405, 2006.
9. D. Aharonov, Z. Landau, and J. Makowsky. The quantum FFT can be classically simulated. Los Alamos Physics Preprint Archive, http://xxx.lanl.gov/abs/quant-ph/0611156, 2006.
10. A. Ambainis and Y. Shi. Distributed construction of quantum fingerprints. http://www.citebase.org/cgi-bin/citations?id=oai:arXiv.org:quant-ph/0305022, 2003.
11. D. Bacon and W. van Dam. Recent progress in quantum algorithms. *Communication of the ACM*, 53(2):84–93, 2010.
12. S. Barz, E. Kashefi, A. Broadbent, J. F. Fitzsimons, A. Zeilinger, and P. Walther. Demonstration of blind quantum computing. *Science*, 335(6066):303–308, 2012.
13. R. Beals, H. Buhrman, R. Cleve, M. Mosca, and R. de Wolf. Quantum lower bounds by polynomials. *Journal of the ACM*, 48(4):778–797, 2001.
14. J. S. Bell. On the Einstein–Podolsky–Rosen paradox. *Physics*, 1:195–200, 1964.
15. C. H. Bennett. Time/space trade-offs for reversible computation. *SIAM Journal on Computing*, 18(4):766–776, 1989.
16. C. H. Bennett and G. Brassard. Quantum cryptography: Public key distribution and coin tossing. In *Proceedings of IEEE International Conference on Computers, Systems, and Signal Processing*, Adelaide, South Australia, Australia, pp. 175–179, 1984.
17. E. Bernstein and U. V. Vazirani. Quantum complexity theory. *SIAM Journal on Computing*, 26(5):1411–1473, 1997. A preliminary version of this paper appeared in the *Proceedings of the 25th Association for Computing Machinery Symposium on the Theory of Computing*.
18. S. Boixo, S. T. Flammia, C. M. Caves, and J. M. Geremia. Generalized limits for single-parameter quantum estimation. *Physical Review Letters*, 98(9):90401, 2007.
19. G. Boole. On the theory of probabilities. *Philosophical Transactions of the Royal Society of London*, 152:225–252, 1862.
20. A. N. Boto, P. Kok, D. S. Abrams, S. L. Braunstein, C. P. Williams, and J. P. Dowling. Quantum interferometric optical lithography: Exploiting entanglement to beat the diffraction limit. *Physical Review Letters*, 85:2733–2736, September 2000.
21. G. Brassard, A. Broadbent, J. Fitzsimons, S. Gambs, and A. Tapp. Anonymous quantum communication. In *Proceedings of the Advances in Cryptology 13th International Conference on Theory and Application of Cryptology and Information Security*, Kuching, Malaysia, pp. 460–473, Springer-Verlag, 2007.
22. M. J. Bremner, C. Mora, and A. Winter. Are random pure states useful for quantum computation? *Physical Review Letters*, 102:190502, 2009.

23. A. Broadbent, J. Fitzsimons, and E. Kashefi. Universal blind quantum computation. In *Proceedings of the 50th Annual IEEE Symposium on Foundations of Computer Science (FOCS'09)*, Atlanta, GA, pp. 517–526, IEEE, 2009.

24. K. L. Brown, W. J. Munro, and V. M. Kendon. Using quantum computers for quantum simulation. *Entropy*, 12(11):2268–2307, 2010.

25. J. Buchmann and H. C. Williams. A key-exchange system based on imaginary quadratic fields. *Journal of Cryptology*, 1(2):107–118, 1988.

26. H. Buhrman, R. Cleve, J. Watrous, and R. de Wolf. Quantum fingerprinting. *Physical Review Letters*, 87(16):167902, 2001.

27. A. M. Childs and T. Lee. Optimal quantum adversary lower bounds for ordered search. *Lecture Notes in Computer Science*, 5125:869–880, 2008.

28. A. M. Childs and W. van Dam. Quantum algorithms for algebraic problems. *Reviews of Modern Physics*, 82:1–52, January 2010.

29. J. M. Chow, J. M. Gambetta, A. D. Corcoles, S. T. Merkel, J. A. Smolin, C. Rigetti, S. Poletto et al. Complete universal quantum gate set approaching fault-tolerant thresholds with superconducting qubits. arXiv preprint arXiv:1202.5344, 2012.

30. R. Cleve, D. Gottesman, and H.-K. Lo. How to share a quantum secret. *Physical Review Letters*, 83(3):648–651, 1999.

31. R. Colbeck and A. P. A. Kent. Private randomness expansion with untrusted devices. *Journal of Physics A: Mathematical and Theoretical*, 44:095305, 2011.

32. A. Comte. Cours de philosophie positive, Bachelier, 1835.

33. C. Crépeau, D. Gottesman, and A. Smith. Secure multi-party quantum computation. In *Proceedings of STOC'02*, Montreal, Quebec, Canada, pp. 643–652, 2002.

34. D. Deutsch. David Deutsch's many worlds. *Frontiers*, December 1998.

35. D. Deutsch and R. Jozsa. Rapid solution of problems by quantum computation. *Proceedings of the Royal Society of London Series A*, A439:553–558, 1992.

36. A. Drucker and R. de Wolf. Quantum proofs for classical theorems. arXiv: 0910.3376, 2009.

37. P. Dumais, D. Mayers, and L. Salvail. Perfectly concealing quantum bit commitment from any quantum one-way permutation. *Lecture Notes in Computer Science*, 1807:300–315, 2000.

38. M. Dušek, N. Lütkenhaus, and M. Hendrych. *Progress in Optics*, 49:381, 2006.

39. J. Eisert, M. Wilkens, and M. Lewenstein. Quantum games and quantum strategies. *Physical Review Letters*, 83, 1999.

40. A. K. Ekert. Quantum cryptography based on Bell's theorem. *Physical Review Letters*, 67(6):661–663, August 1991.

41. R. Feynman. *Feynman Lectures on Computation*. Addison-Wesley, Reading, MA, 1996.

42. J. F. Fitzsimons and E. Kashefi. Unconditionally verifiable blind computation. arXiv preprint arXiv:1203.5217, 2012.

43. C. Gentry. Fully homomorphic encryption using ideal lattices. In *Proceedings of the 41st Annual ACM Symposium on Theory of Computing*, Bethesda, MD, pp. 169–178, ACM, 2009.

44. V. Giovannetti, S. Lloyd, and L. Maccone. Quantum-enhanced measurements: Beating the standard quantum limit. *Science*, 306(5700):1330–1336, 2004.

45. N. Gisin, G. Ribordy, W. Tittel, and H. Zbinden. Quantum cryptography. *Reviews of Modern Physics*, 74:145, 2002.

46. D. Gottesman. On the theory of quantum secret sharing. *Physical Review A*, 61:042311, 2000.

47. D. Gottesman. Uncloneable encryption. *Quantum Information and Computation*, 3:581–602, 2003. Preprint at Los Alamos Physics Preprint Archive, http://xxx.lanl.gov/abs/quant-ph/0210062.

48. D. Gottesman and I. L. Chuang. Quantum digital signatures. arXiv:quant-ph/0105032, November 2001.

49. D. M. Greenberger, M. A. Horne, A. Shimony, and A. Zeilinger. A Bell's theorem without inequalities. *American Journal of Physics*, 58:1131–1143, 1990.

50. D. M. Greenberger, M. A. Horne, and A. Zeilinger. Going beyond Bell's theorem. In *Bell's Theorem, Quantum Theory and Conceptions of the Universe*, pp. 73–76. Kluwer, Dordrecht, the Netherlands, 1989.

51. M. Grigni, L. Schulman, M. Vazirani, and U. V. Vazirani. Quantum mechanical algorithms for the nonabelian hidden subgroup problem. In *Proceedings of STOC'01*, Crete, Greece, pp. 68–74, 2001.

52. D. Gross, S. T. Flammia, and J. Eisert. Most quantum states are too entangled to be useful as computational resources. *Physical Review Letters*, 102:190501, 2009.

53. L. K. Grover. Quantum computers can search arbitrarily large databases by a single query. *Physical Review Letters*, 79(23):4709–4712, 1997.

54. S. Hallgren. Polynomial-time quantum algorithms for Pell's equation and the principal ideal problem. In *Proceedings of STOC'02*, Montreal, Quebec, Canada, pp. 653–658, 2002.

55. P. Hayden, D. Leung, and G. Smith. Multiparty data hiding of quantum information. *Physical Review A*, 71, 062339, 2005.

56. J. A. Jones, S. D. Karlen, J. Fitzsimons, A. Ardavan, S. C. Benjamin, G. A. D. Briggs, and J. J. L. Morton. Magnetic field sensing beyond the standard quantum limit using 10-spin NOON states. *Science*, 324(5931):1166–1168, 2009.

57. R. Jozsa and N. Linden. On the role of entanglement in quantum computational speed-up. *Proceedings of the Royal Society of London Series A*, 459:2011–2032, 2003.

58. E. Knill. Approximation by quantum circuits. arXiv:quant-ph/9508006, 1995.

59. G. Kuperberg. A concise introduction to quantum probability, quantum mechanics, and quantum computation. Unpublished notes, available at https://www.math.ucdavis.edu/~greg/intro-2005.pdf, 2005. Accessed on May 25, 2013.

60. G. Kuperberg. A subexponential-time quantum algorithm for the dihedral hidden subgroup problem. *SIAM Journal on Computing*, 35(1):170–188, 2005.

61. G. Kuperberg. Another subexponential-time quantum algorithm for the dihedral hidden subgroup problem. arXiv:1112.3333, 2011.

62. D. Leibfried, M. D. Barrett, T. Schaetz, J. Britton, J. Chiaverini, W. M. Itano, J. D. Jost, C. Langer, and D. J. Wineland. Toward Heisenberg-limited spectroscopy with multiparticle entangled states. *Science*, 304(5676):1476–1478, 2004.

63. S. Lloyd. Enhanced sensitivity of photodetection via quantum illumination. *Science*, 321(5895):1463–1465, 2008.

64. C. Lomont. The hidden subgroup problem: review and open problems. arXiv:quant-ph/0411037, 2004.

65. Y. I. Manin. *Computable and Uncomputable*. Sovetskoye Radio, Moscow, Russia, (in Russian), 1980.

66. Y. I. Manin. *Mathematics as Metaphor: Selected Essays of Yuri I. Manin*. American Mathematical Society, Providence, RI, 2007.

67. I. Markov and Y. Shi. Simulating quantum computation by contracting tensor networks. arXiv:quant-ph/0511069, 2005.

68. N. D. Mermin. Copenhagen computation: How I learned to stop worrying and love Bohr. *IBM Journal of Research and Development*, 48:53, 2004.
69. D. A. Meyer. Quantum strategies. *Physical Review Letters*, 82:1052–1055, 1999.
70. D. A. Meyer. Sophisticated quantum search without entanglement. *Physical Review Letters*, 85:2014–2017, 2000.
71. J. S. Mill. Auguste Comte and positivism, Trubner, 1866.
72. M. Mosca. Quantum algorithms. arXiv:0808.0369, 2008.
73. T. Nagata, R. Okamoto, J. L. O'Brien, K. Sasaki, and S. Takeuchi. Beating the standard quantum limit with four-entangled photons. *Science*, 316(5825):726–729, 2007.
74. M. Nielsen and I. L. Chuang. *Quantum Computing and Quantum Information*. Cambridge University Press, Cambridge, U.K., 2001.
75. J. W. Pan, D. Bouwmeester, M. Daniell, H. Weinfurter, and A. Zeilinger. Experimental test of quantum nonlocality in three-photon Greenberger-Horne-Zeilinger entanglement. *Nature*, 403:515–518, 2000.
76. D. Perez-Garcia, F. Verstraete, M. M. Wolf, and J. I. Cirac. Matrix product state representations. *Quantum Information & Computation*, 7(5):401–430, 2007.
77. S. Pironio, A. Acin, S. Massar, A. Boyer de la Giroday, D. N. Matsukevich, P. Maunz, S. Olmschenk et al. Random numbers certified by Bell's theorem. *Nature*, 464:1021, 2010.
78. I. Pitowsky. George Boole's' conditions of possible experience' and the quantum puzzle. *British Journal of the Philosophy of Science*, 45(1):95–125, 1994.
79. J. Preskill. Reliable quantum computers. *Royal Society of London Proceedings Series A*, 454:385, January 1998.
80. R. Raussendorf and H. J. Briegel. A one-way quantum computer. *Physical Review Letters*, 86(22):5188–5191, 2001.
81. O. Regev. Quantum computation and lattice problems. In *Proceedings of FOCS'02*, Vancouver, British Columbia, Canada, pp. 520–529, 2002.
82. O. Regev. On lattices, learning with errors, random linear codes, and cryptography. In *Proceedings of STOC'05*, Baltimore, MD, pp. 84–93, 2005.
83. E. G. Rieffel and W. Polak. *A Gentle Introduction to Quantum Computing*. MIT Press, Cambridge, MA, 2010.
84. E. Rieffel. Certainty and uncertainty in quantum information processing. In *Proceedings of the AAAI Spring Symposium 2007*, Stanford, CA, pp. 134–141, 2007.
85. C. F. Roos, M. Chwalla, K. Kim, M. Riebe, and R. Blatt. 'Designer atoms' for quantum metrology. *Nature*, 443(7109):316–319, 2006.
86. V. Scarani, L. Chua, and S. Y. Liu. *Six Quantum Pieces*. World Scientific, Singapore, 2010.
87. V. Scarani, H. Bechmann-Pasquinucci, N. J. Cerf, M. Dusek, N. Lutkenhaus, and M. Peev. The security of practical quantum key distribution. *Reviews of Modern Physics*, 81(3):1301–1350, 2009.
88. Y. Shi. Quantum and classical tradeoffs. *Theoretical Computer Science*, 344(2): 335–345, 2005.
89. P. W. Shor. Algorithms for quantum computation: Discrete log and factoring. In *Proceedings of FOCS'94*, Santa Fe, NM, pp. 124–134, 1994. ftp://netlib.att.com/netlib/att/math/shor/quantum.algorithms.ps.Z.
90. L. G. Valiant. Quantum computers that can be simulated classically in polynomial time. In *Proceedings of the Thirty-Third Annual ACM Symposium on Theory of Computing*, Crete, Greece, pp. 114–123, ACM, 2001.

91. W. van Dam, S. Hallgren, and L. Ip. Quantum algorithms for some hidden shift problems. In *Proceedings of SODA'03*, Baltimore, MD, pp. 489–498, 2003.

92. V. Vedral. The elusive source of quantum effectiveness. arXiv:0906.2656, 2009.

93. F. Verstraete, D. Porras, and J. I. Cirac. Density matrix renormalization group and periodic boundary conditions: A quantum information perspective. *Physical Review Letters*, 93(22):227205, 2004.

94. G. Vidal. Efficient classical simulation of slightly entangled quantum computations. *Physical Review Letters*, 91:147902, 2003.

95. J. Watrous. Zero-knowledge against quantum attacks. In *Proceedings of STOC'06*, Seattle, WA, pp. 296–305, 2006.

96. S. Wiesner. Conjugate coding. *SIGACT News*, 15:78–88, 1983.

97. H. M. Wiseman and G. J. Milburn. *Quantum Measurement and Control*. Cambridge University Press, Cambridge, U.K., 2010.

98. M.-H. Yung, J. D. Whitfield, S. Boixo, D. G. Tempel, and A. Aspuru-Guzik. Introduction to quantum algorithms for physics and chemistry. arXiv:1203.1331, *to appear in Advances in Chemical Physics*, page 44, March 2012.

99. B. Yurke. Input states for enhancement of fermion interferometer sensitivity. *Physical Review Letters*, 56(15):1515–1517, 1986.

12

DYVERSE: From Formal Verification to Biologically Inspired Real-Time Self-Organizing Systems

Eva M. Navarro-López*

CONTENTS

* This work has been supported by the Engineering and Physical Sciences Research Council
(EPSRC) of the UK under the framework of the project *DYVERSE: A New Kind of Control for
Hybrid Systems* (EP/I001689/1). The author is also grateful for the support of the Research
Councils United Kingdom under the grant EP/E50048/1.

12.1 Motivation and Outline of the Chapter

Scientists, similar to artists, find their *raison d'être* in the willingness to make the world a better place or at least to understand it better—both stem from the same generosity. By better understanding the world around us, we raise consciousness and question the existing order. There are many examples of this in science, engineering, and technology. Two prominent ones are Leonardo da Vinci in the fifteenth and sixteenth centuries and Alan Turing in the first half of the twentieth century. Leonardo, the *artist engineer*, transcended the boundaries of the arts, anatomy, engineering, architecture, mechanics, and mathematics. His designs of automated machines resembling human or animal bodies (*anathomia artificialis*) would nowadays belong within the fields of robotics, mechatronics, or bioengineering, as developed in the past 30 years. Leonardo da Vinci was a protocontrol engineer.

Alan Turing, on the other hand, was not an engineer; he was a scientist par excellence. He is considered one of the fathers of computer science and brought together concepts of computation, biology, mathematics, logics, chemistry, and botanics at a time when interdisciplinary research was rare. Not only did he contribute to the design of the world's first commercially available digital computer, but he also established the foundations of computability theory, anticipated artificial intelligence, machine learning, and the idea of genetic algorithms or neural networks and contributed to cryptanalysis and dynamical system analysis. In his last years in Manchester, he embarked upon research into the formation of spatiotemporal patterns in living organisms and cells. His reaction–diffusion model of morphogenesis is still causing controversy among biologists.

These two men of ideas were similarly controversial, visionary, endlessly curious, and roused the jealousy and skepticism of their contemporaries. They were also so far ahead of their time that they did not see all their ideas come to fruition. Today, as engineers and scientists, we consider polymaths such as da Vinci and Turing sources of inspiration in our own work, and we wonder: who are the Leonardos and Alan Turings of the future?

This chapter traces three threads of thought. Firstly, the idea of learning from nature to improve engineering and technological designs. Nature is a never-ending source of inspiration for science as well as for engineering. There are many examples in the study of the collective behavior of cells, animals, and human behavior. This is an integral part of the theory of complex networks and multiagent systems, consisting of a large number of interacting and interdependent systems (nodes) connected in a nontrivial and nonregular manner (links) (Mesbahi and Egerstedt, 2010; Newman, 2010). How could we design a robust and resilient control system that could reorganize its structure and functions to cope with critical situations in the same manner as our brain?

Secondly, we wish to invert the thinking and use engineering methodologies to influence natural processes. Engineering is more than observation and interpretation; it is about transforming, building, and making things work. In short, engineering is the art of applying practical ideas—the simpler the better—to solve problems. A recent growth area of research has been the application of control engineering paradigms in systems biology (Sontag, 2004; Doyle et al., 2011). However, the fact that the combination of control theory, dynamical system analysis, and systems biology has the potential to reveal important patterns in diseases such as cancer or diabetes has not been widely acknowledged.

The third thread of thought is to incorporate the algorithmic and logical thinking of computer science into complex systems' design. Probably, the most significant example is the use of formal verification to check, in an automated manner, the correct behavior of dynamical systems and, most recently, hybrid dynamical systems (van der Schaft and Schumacher, 2000), that is, systems that combine continuous and discrete and smooth and abrupt dynamics.

These three threads of thought are unified within the novel framework of DYVERSE. DYVERSE stands for the **DY**namically driven **VER**ification of Systems with Energy considerations. Here, energy refers to the abstract energy of dynamical systems, which is studied within the dissipativity theory (Willems, 1972). DYVERSE is a computational–dynamical framework for the discrete abstraction and formal verification of complex dynamical behaviors of hybrid dynamical systems. The intent is to ensure the correct behavior of devices and to eliminate all the factors that degrade performance by using appropriate control schemes. Although the hybrid system paradigm has made great advances in the past 10 years in the modeling, analysis, and control of complex systems, there is still a gap in the effective integration and application of the theory. DYVERSE is aimed toward overcoming this problem by bridging the theories of formal verification, bifurcations and stability, dissipativity, and switching control.

This chapter will trace the main strands of the theories that underlie the ideas mentioned earlier and paint a bigger picture of their potential future applications. Inspired by Alan Turing's morphogenesis, the concept of real-time self-organizing systems is introduced, and ideas to expand the DYVERSE framework to networked and distributed systems are sketched. For this purpose, *reconfigurable hybrid automata* are introduced.

The work presented in this chapter does not intend to be comprehensive. It sketches future possibilities and opens up new avenues for the integration of theoretical computer science, control engineering, applied mathematics, and network science. The implications are vast and wide ranging: from electric power networks to electromechanical systems with impacts and friction and from the understanding of memory recovery or plasticity in our brain to the modeling of tumor growth in patients with cancer.

12.2 Nature: The Never-Ending Source of Inspiration

The first thread of thought of this chapter is to understand how important advances in science and engineering have been motivated by the rules dictating natural and physical processes. The laws and phenomena in nature are formalized and approximated by means of mathematical, physical, and computational models or a combination of these.

Typically, natural phenomena are described by means of differential or difference equations, which are approximations of the actual world that allow us to define the behavior of the past and future states of a system by using the current state. In this section, we go beyond classical mathematical models and show the most recent approaches to describe the self-organization and emergence of collective behavior in biological, physical, and social systems. Inspired by the behavioral patterns of ant colonies, schools of fish, flocks of birds, cell sorting, or social human relationships, several models, analysis techniques, and control schemes have been devised in order to improve a wide variety of engineering and technological applications that facilitate our welfare and day-to-day life. Complex networks and multiagent system models are highlighted.

Roughly speaking, a complex network is a group of systems (a large number of them) interconnected in a nontrivial and nonregular manner. Typically, each system is considered as a dynamical system with continuous states that is located within the nodes of a graph. In the case of multiagent systems, normally, no dynamics are considered in each node, and the links between nodes and the communication between them are mostly considered in order to study their evolution.

The chief characteristic of these networked systems is the complex nature of their interconnection topology, which defines the behavior of the overall structure and entails the onset of complex dynamical phenomena not present in the individual systems. Two key ideas are pivotal to propose a model for these systems: (1) Individual elements following elementary behavioral rules can produce complex behavioral patterns, and (2) in many cases, individual elements achieve a global goal with minimal communication with other elements in the network and without having a complete picture of the overall structure.

This section will start explaining the idea of morphogenesis, in particular, Turing's model of morphogenesis, which is considered as one of the first models that intended to describe self-organizing mechanisms and the emergence of global behavior from local interactions in natural and physical systems.

12.2.1 *Natural Wonders Every Child Should Know* and Turing's Morphogenesis

As scientists, we think that most phenomena we observe have an explanation. Our endless curiosity challenges the existing order, and we wonder: is nature predictable? Can we know with certainty how natural, physical, and engineering systems evolve? Does unexpected complexity arise from simplicity? We are within a solar system whose planetary motion may be unpredictable. We build electronic circuits that can exhibit, for some operation conditions, chaotic behavior, for instance, Chua's circuit (Chutinan, 1992). We can observe the behavior of a pendulum that evolves as we expected until an event occurs, and after that, its behavior becomes unpredictable (Strogatz, 1994).

In 1952, Alan Turing, one of the fathers of computer science, proposed a mathematical model that later on has been employed to explain the formation of structures and patterns in living organisms (for instance, the patterns of spots and stripes on animals' coats). This model was originally proposed for explaining the phenomenon of morphogenesis—from the Greek, origin of shapes—in cells and is well known as Turing's morphogenesis.

The book *Natural Wonders That Every Child Should Know* (Brewster, 1912) was an important source of inspiration for Turing in his childhood, as it is explained in his biography (Hodges, 1992). This book brought to Turing (as to many other children) the spirit of scientific enquiry, the discovery of the why's and how's of the world around us, and the challenge to the existing order. Paraphrasing D'Arcy W. Thompson, "the ultimate problems of biology are as inscrutable as of old" (Hodges, 1992). D'Arcy Thompson's seminal book *On Growth and Form* (Thompson, 1917) was another important influence for Turing. Thompson's book is considered as the foundation of what is now referred to as biomathematics, that is, bringing mathematics to biology. D'Arcy Thompson proposed to understand the form of plants and animals (i.e., morphology) by means of pure mathematics.

After devising an abstract model, referred to as the universal machine (Turing machine) that could simulate any computer algorithm (Turing, 1937) following striking simple rules, Alan Turing introduced what was later to become the basis of artificial intelligence (Turing, 1948). He was inspired by the human brain and asked himself "can machines think?" (Turing, 1950): "I propose to investigate the question as to whether it is possible for machinery to show intelligent behaviour [...] The analogy with the human brain is used as a guiding principle" (Turing, 1948).

His aspirations of approximating natural processes by mathematical rules went beyond computer science foundations. He was one of the first to propose a formal model that could explain the self-organization phenomena present in a wide variety of biological systems. And he did so with an impressive clarity of thought, as all great thinkers (see Figure 12.1, where the first two lines of his seminal paper are shown).

"The Chemical Basis of Morphogenesis" (Turing, 1952) proposes a reaction–diffusion model of spatial pattern formation. This was the only paper that Turing published on this topic, which was further explored in 1953 by the MSc thesis of his student Bernard Richards (Richards, 2005).

The basic elements of Turing's morphogenesis are the morphogens. The morphogens are the "form producers," chemical substances diffusing

---

THE CHEMICAL BASIS OF MORPHOGENESIS

B Y A. M. TURING, F.R.S. *University of Manchester*

(Received 9 November 1951— Revised 15 March 1952)

It is suggested that a system of chemical substances, called morphogens, reacting together and diffusing through a tissue, is adequate to account for the main phenomena of morphogenesis.

FIGURE 12.1
The first two lines of Turing's paper about the reaction–diffusion model of morphogenesis. (From Turing, A.M., *Philos. Trans. R. Soc. Lon. Ser. B Biol. Sci.* 237(641), 37, 1952.)

through a mass of tissue of given geometrical form and reacting together within it. They can be proteins, molecules, genes, or hormones. "The Chemical Basis of Morphogenesis" showed theoretically that a system subject to infinitesimal perturbations, with chemicals reacting and diffusing (activators and inhibitors), could spontaneously evolve to spatial patterns from an initially uniform state, where no prior pattern existed. These emerged patterns are now well known as Turing's patterns or Turing's structures. The transitions and behavior patterns associated with these phenomena are exhibited by nonlinear dynamical systems. They are typically related to the loss of stability of an equilibrium point. This loss of stability usually has its origin in what in mathematical terms is known as a bifurcation, that is, the point that marks a structural change. Figure 12.2 sums up the main mechanisms involved in Turing's reaction–diffusion model of morphogenesis.

Turing predicted the mathematical basis for oscillating nonlinear chemical reactions such as the Belousov–Zhabotinsky reaction. This is a chemical reaction that, under some conditions and far from equilibrium, *magically* produces self-organized spiral waves with beautiful patterns. This phenomenon was firstly discovered by Boris Belousov in 1951. He was contemporary with Turing, although it seems they were not aware of each other's work. Belousov's work was followed by Zhabotinsky in the 1960s, but it was not until the 1990s that chemical reactions similar to the one discovered by Belousov in 1951 were accepted as examples of Turing-like patterns.

Finally, Turing, through his unfinished work on phyllotaxis, attempted to explain mathematically the process of arrangement of leaves on plants ["Morphogen Theory of Phyllotaxis" (Saunders, 1992)], the first application of his morphogenesis theory. This is a problem that has not been solved yet.

Turing pioneered computational biology, which posits the idea that nature can be viewed as a system following mathematical and computational rules. He brought together computing, biology, and mathematics at a time when science was dominated by physics. Turing's ideas of the mathematical formalization of the transition between different structures in biological systems

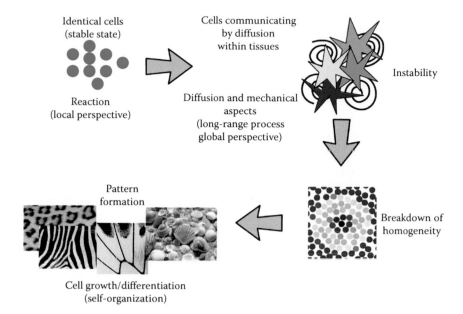

Identical cells
(stable state)

Reaction
(local perspective)

Cells communicating
by diffusion
within tissues

Diffusion and mechanical
aspects
(long-range process
global perspective)

Instability

Pattern
formation

Breakdown of
homogeneity

Cell growth/differentiation
(self-organization)

FIGURE 12.2
A particular interpretation of Turing's reaction–diffusion model. Interdependences of bio-chemical and mechanical processes and the importance of the existence of an instability (a switching process, a discontinuity) breaking the homogeneity and leading to heterogeneous patterns.

have inspired many innovations in engineering and mathematics: from the computational morphogenesis in optimization problems to evolvable software systems, from computer graphics to building design, from ecology to material sciences, and from hydrodynamics to astrophysics. In chemistry and biology, the application of Turing's model is still controversial, yet his reaction–diffusion mechanism has become one of the standard models of theoretical biology. For a more recent perspective of Turing's morphogenesis, the reader is referred to Howard et al. (2011) and Maini et al. (1997), where several applications of the reaction–diffusion model are reviewed.

12.2.2 Self-Organization and Collective Behavior

The pursuit of models for describing the collective behavior of animals or insects can be very useful for solving specific problems encountered in some emerging engineering applications. The synchronized patterns discernible in an ant colony, a school of fish, and a flock of birds yield valuable insights. The study of swarm intelligence illumines a wide range of complex engineering systems, such as cooperative robotics (Martínez et al., 2007), computer animation (Reynolds, 1987), multiagent control design incorporated in what is referred to as multiagent networks (Uhrmacher and Weyns, 2009;

Mesbahi and Egerstedt, 2010), distributed computer systems, software engineering (Jennings and Bussmann, 2003), multivehicle cooperative control (Ren et al., 2007), distributed control (Paley et al., 2007; Tanner et al., 2007), and social network modeling (Mikhailov and Calenbuhr, 2002; Barabási, 2005a; Barabási, 2007).

Intriguingly, consensus decision making in animals or insects does not require verbal communication, and simple local rules produce complex collective behavior, dispensing with the necessity of a leader (see Figure 12.3).

Underpinning all this is the phenomenon of self-organization in biological systems (Schweitzer, 1997; Camazine et al., 2003), that is, the emergence of behavioral patterns with unexpected properties and regularities. The activities of ants, bees, birds, and fish have inspired numerous swarming models, flocking algorithms, and consensus protocols to describe collective dynamical patterns.

The analysis of the emergence of dynamical behaviors and patterns in groups and in systems in general is not new. Empedocles, a Greek philosopher of the fifth century BC, articulated a poetic conception of collective behavior. He conceived changes as the result of mixture and separation of unalterable substances (fire, air, water, earth). That is, the combination of the actions of cooperation (aggregation) and competition (segregation) explains how the world is formed and is the basis of all changes (see Figure 12.4 for an example).

(1)   (2)

FIGURE 12.3
Swarm intelligence. (1) Monarch butterflies in a forest of Michoacán in Central Mexico and (2) bee swarm. A single butterfly or bee is not smart, but their colonies are. (Picture (1) was taken by the author. Picture (2) is by Karen Ridenour, who has kindly granted permission for its use in this chapter. The picture was discovered at http://www.texasento.net/bee-swarm.html, a webpage by Mike Quinn.)

(1)   (2)

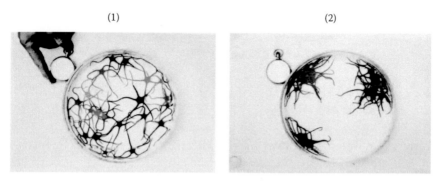

FIGURE 12.4
Aggregation and segregation in nature. (1) Brittle starfish aggregate readily when put into a clean vessel of sea water, (2) 10 min after (1) was taken. (Extracted from Allee, W.C., *The Social Life of Animals*, New York, W.W. Norton & Company Inc., p. 45, 1938, Available from the Biodiversity Heritage Library at: http://www.archive.org/details/sociallifeofanim00alle.)

There are two levels in the phenomenon of self-organization in groups: a local level of the individual and a global level of the group (Figure 12.5). At the local level, the individual:

- Has a local view, informed only by its closest neighbors
- Follows local elementary behavioral rules, such as sticking together, avoiding collisions, or moving in the same direction as its neighbors
- Has minimal communication with other individuals

(1)   (2)

FIGURE 12.5
The paradox between unthinking individuality and collective intelligence. (1) Many individuals with a local view and (2) emergence of collective intelligence in group behavior. (Picture (1) is a flock of pelicans in Puerto Arista, Tonalá, Chiapas, México, and was taken by the author. Picture (2) is by Diana Hale, who has kindly granted permission for its use in this chapter. This picture is a flock of birds in Norfolk, U.K.)

On the other hand, at the global level,

- Collective behavior emerges.
- The group solves complex tasks.
- The operation is distributed, not centralized.
- The group adapts to changes.

The interplay of locality and universality gives rise to two key features: (1) Collective behavior emerges in the absence of centralized control or leadership, and (2) local rules for individuals lead to global patterns, transcending the detailed nature of the components. Furthermore, individual elements following elementary behavioral rules may produce complex behavioral patterns.

There are many textbooks and papers related to this phenomenon. For instance, examples of self-organizing systems are given in the books (Schweitzer, 1997; Mikhailov and Calenbuhr, 2002; Camazine et al., 2003). A description of self-organized and self-assembled structures from the perspective of physics is given in Desai and Kapral (2009).

Typically, the study of self-organization in groups is carried out under the framework of multiagent systems, and agent-based models are employed to abstract these systems. An agent is an autonomous entity acting upon an environment by using sensors and actuators. Multiagent systems consist of agents that are connected through a network topology and interact with each other by a communication protocol (Uhrmacher and Weyns, 2009). The ultimate goal is to carry out a common complex task.

The main dynamical behaviors exhibited in multiagent systems are swarming and flocking. Swarming poses a modeling-like problem and is the action of moving or gathering in groups (Vicsek et al., 1995; Couzin et al., 2005; Moreau, 2005). Flocking is the action of congregating or traveling in flock and entails a control-like problem, that is, how to force all individuals to move as a group. The archetypal flocking algorithm is the boids flocking algorithm proposed by Reynolds (1987), which is a computer model of coordinated animal motion and was originally proposed to visually simulate flocking and schooling behaviors for the animation industry.

Finally, consensus protocols are protocols for the coordination of agents to reach agreement as a group. Consensus protocols pose a stability-type problem (in terms of dynamical systems) and establish a common solution for the entire group (Olfati-Saber and Murray, 2004). The reader is referred to the following surveys on collective behavior and consensus for more details on group cooperative algorithms (Coradt and Roper, 2005; Sumpter, 2006; Couzin, 2008; Vicsek, 2008).

I conclude this section with a final thought: how can we learn from these natural processes to improve engineering systems and technological designs? It is possible to mimic biological systems' mechanisms to better understand human behavior and design engineering and computing systems that were inconceivable not so long ago. Three very recent examples of this are highlighted as follows:

1. *Brain2Grid*: The use of neuronal network paradigms to analyze large-scale electric power networks integrating renewable energy sources (Venayagamoorthy, 2009, 2011).

2. *SpiNNaker*: The design of the next generation of massive parallel computers that mimic the human brain (Furber and Brown, 2011) and the onset of what is referred to as neural system engineering (Furber and Temple, 2007).

3. *The next generation of evolving social networks*: Although human interaction in groups is much more complex than the interactions of cells and animals, it follows similar rules, and so pattern-formation rules detected in cells of living tissues can be employed to explain human behavior patterns. Recent work (Navarro-López and Poliakov, 2011) proposes a novel framework for the modeling and analysis of human interaction and interests. It is based on Alexei Poliakov's work on spatial pattern formations in mixtures of cells (Poliakov et al., 2008; Taylor et al., 2011). He explained how living *agents* organize according to their similarity to each other. The key aspect of this discovery is the link between the position of the *agents* and their behavior. Poliakov et al. go further and establish linkages of people from well-proven linkages of cells.

12.2.3 Complex Networks

Does a natural order underlie the manner our society, human relationships, and our bodies work? In order to answer that question, scientists in systems require an abstract model to understand complex structures and complex behaviors. The result is the theory of complex networks or network science: the coupling of heart cells; the synchronization patterns of beta-cells in the pancreas, which is an important factor in diabetes; the spread of diseases; and the ceaseless interdependencies and interlockings that power those micro and macro-universes, our brains—complex networks are everywhere!

In 10 years since the establishment of complex networks as a field of study, there has been great progress in the understanding of large-scale complex systems. Various models have been proposed in order to study the complex topology of networks. In addition, crucial collective dynamical processes have been uncovered: mainly, synchronization and other self-organizing behavioral patterns (cascading or spreading dynamics, among others) (Barrat et al., 2010). Most of these studies are based on graph theory tools, as the complex network is interpreted as a large number of interacting and inter-dependent systems (nodes) coupled in a nontrivial and nonregular manner (links). Typically, node dynamics are not considered or just simplified.

The models devised to understand the connectivity and structure of these systems are referred to as small-world networks (Watts and Strogatz, 1998; Watts, 1999), random networks (Erdös and Rényi, 1960), scale-free networks (Barabási and Albert, 1999; Bianconi and Barabási, 2001; Albert and Barabási, 2000, 2002; Barabási and Bonabeau, 2003), and, recently, evolving or adaptive networks (Dorogovtsev and Mendes, 2003; Gross and Sayama, 2009). In the past 5 years, there have been many books published on the topic of complex networks (Newman et al., 2006; Caldarelli and Vespignani, 2007; Barrat et al., 2010; Cohen and Havlin, 2010; Dorogovtsev, 2010; Newman, 2010). Besides, two very good introductory texts are Strogatz (2003) and Watts (1999).

The tendency has been to use mathematical measurements to describe the topological properties of the network (especially, related to size, density, and connectivity). As a consequence, the dominating perspective in the study of complex networks is more statistical than behavioral. Survey papers explaining the basic topologies and topological features or measures of networks include the following: Strogatz (2001), Albert and Barabási (2002), Newman (2003), Wang and Chen (2003), Boccaletti et al. (2006), and Da Fontoura et al. (2007).

Complex is more than complicated. With this, we mean that a complex system is more than the sum of many parts. The complexity stems from the fact that global behaviors emerge in the system and they cannot be explained just from the behaviors and characteristics of the individual parts. Complex networks consist of a large number of elements that interact with each other and their environment in order to organize in specific emergent structures. Therefore, the whole is not the sum of the parts. The complexity in complex networks arises on three fronts:

1. Structural complexity
    a. The connections between nodes are nontrivial and nonregular (see Figure 12.6).
    b. Complex topological features, such as hierarchies, communities, or giant components; the latter related to the phenomenon of percolation (Achlioptas et al., 2009).
    c. Network evolution: The topology and the number of nodes can change over time. For example, in the World Wide Web, new pages and links are created and lost every second.
    d. Connection diversity: The links between nodes can be of a different nature, having different weights, directions, and types. For example, synapses in the nervous systems are stronger or weaker and inhibitory or excitatory.

(1) (2)

FIGURE 12.6
Referring to the topology: regular versus complex. (1) Spider's webs in the canals of Manchester: identically repeating patterns and (2) school friendship network (nodes: children, links: friendship): nontrivial and nonregular patterns, from www.visualcomplexity.com. Complex is more than complicated. The study of complex networks goes beyond the traditional thinking of networks as simple and steady graphs. (Picture (1) was taken by the author. Picture (2) was generated by James Moody with the material published in Moody, J., *Am. J. Sociol.*, 107(3), 679, 2001. With permission.)

2. Node dynamical complexity
   a. Nodes can be nonlinear dynamical systems exhibiting dynamical behaviors, such as bifurcations, oscillations, and chaotic behavior.
   b. Node diversity: A network may have many different types of nodes. For example, a power grid integrates many types of nodes, such as electric generators (of many types), loads, or storages.
3. Interdependencies and mutual interactions
   a. A network is affected by many internal and external complex factors.
   b. There are intradependencies and interdependencies between networks and subnetworks. This is well understood when we think of networks of networks. For example, a blackout in a power grid may lead to chain reactions in our day-to-day life, in industrial production, or affect other networks, such as transportation and traffic, communication and information, and economic and financial networks.

As explained in the previous section, flocking or swarming models inspired by insect and animal group behavior have been proposed to model collective behavioral patterns in networked systems. But many collective behavioral patterns are still a mystery: how can we lose a neuron every second and not only still be functional but actually flourish as creative and intelligent beings? Flexibility, adaptation, and self-organization make us stronger. Could any engineering system cope with the loss of components? Resilience—that is, the ability of recovering from shock—is one of the main features of complex networks. They are adaptive and evolving systems that resist random removal of components. Many other characteristics put complex networks apart from other complex systems, mainly

- They are decentralized systems with no central authority.
- The larger the size of the complex network, the larger its heterogeneity and the variability of its properties.
- Heavy-tail properties, that is, local events, can trigger global rearrangements across the entire system at all levels (Csermely, 2006).
- Small-world properties: Although there are large numbers of nodes, the number of links between two arbitrary nodes is small. Here, it is useful to recall the 6° of separation between you and everyone else on the planet. This idea proposes a complex network formed by every person on our planet, connected by the link of *knowing each other*. There are, according to this theory, on average and at most, only six links between you and anyone else (Watts, 1999).
- The strength of weak ties: A long-range connection may lead to a stronger interaction between two nodes than short-range connections of neighboring nodes. For example, people not in a network of close friends may be more useful to help find a new job, as Granovetter's experiment established in the late 1960s. For more details on this property, the reader is referred to the book by Csermely (2006).

With respect to dynamics and behaviors, the consideration of dynamical behaviors in networks is made under the framework of dynamical networks, which are just networks of coupled dynamical oscillators. The simplest collective behavior in dynamical networks is synchronization. Some examples of synchronization are fireflies flashing in unison, pacemaker cells in the heart oscillating in harmony, the synchronized pedestrians that made the London's Millennium Footbridge wobble, and the synchronized clapping of an audience. An introductory book of synchronization is Strogatz (2001); two other books about synchronization are Boccaletti (2008) and Wu (2007); the latter follows an algebraic approach. Very well-written papers showing the ideas of synchronization are Jordan (2003), Strogatz (2003), Wang and Chen (2003), and Arenas et al. (2008).

Synchronization is the emergence of coherent behavior in which the elements of the system follow the same dynamical pattern. In the case of a dynamical network, this means that all the nodes tend toward the same asymptotic evolution, which is unknown a priori and emerges from the negotiation process between neighboring nodes. In addition, it cannot be explained in terms of the individual dynamics of each single node in the network. In order to model the synchronization phenomenon, three main elements are necessary:

1. A mathematical description of the dynamical behavior of each of the nodes in the network. Typically, a collection of coupled oscillators.
2. An interaction or coupling protocol for the nodes to communicate with each other.
3. A connectivity pattern or network topology describing the interconnections between neighboring nodes.

A great number of existing models for synchronization comprise identical oscillators that are linearly coupled. Many biological oscillators, however, are not coupled by smooth interactions; on the contrary, they communicate by firing sudden impulses. This is the case of firing neurons, flashing fireflies, or chirping crickets. They can be modeled by pulse-coupled oscillators. A widely employed model of this type is Peskin's model for the self-synchronization of the cardiac pacemaker, known as the integrate-and-fire model of oscillators (Mirollo and Strogatz, 1990). Finally, dynamical networks of nonidentical oscillators are Winfree's model (Winfree, 1967) or Kuramoto's model (Strogatz, 2000).

There are more complex behaviors than synchronization, swarming, or flocking in complex networks. Some very interesting work (Nakao and Mikhailov, 2010) studies the emergence of Turing-like patterns in the context of complex networks.

There is an unending list of examples and application domains for complex networks; the reader is referred to the books (Newman et al., 2006; Caldarelli and Vespignani, 2007; Barrat et al., 2010; Cohen and Havlin, 2010; Dorogovtsev, 2010; Newman, 2010) or the introductory paper (Barabási, 2007). Now, I shall proceed to enumerate some of the most recent applications of complex network paradigms in the study of some large-scale engineering systems and critical infrastructures: in particular, electricity transmission networks (Rosas i Casals, 2009), electric transportation networks (Scirè et al., 2005), electric power networks (Arianos et al., 2009; Brummitt et al., 2012), or water distribution systems (Yazdani and Jeffrey, 2011). New advances are being made in what is now referred to as networks of networks, interacting networks, or interdependent dynamical networks (Gao et al., 2012).

Analyzing these emerging complex applications requires not just the use of new technological advances but also a better understanding of social behavior. In the satellite event of *networks of networks* within the most recent *International Conference on Network Science* in Budapest in June 2011, terms such as social computational science, computational epidemiology, or science of science were employed.

12.3 The Other Way Round: From Engineering and Systems Theory to Nature

The second thread of thought to follow is the use of engineering, computational, and systems thinking to influence and better understand natural processes. Biological and complex engineering systems seem, on the surface, to be very different entities, but they pose similar challenges in the development of new analysis tools and models. We will explore the new role of dynamical systems theory, network science, control engineering and theory, and theoretical computer science in biological systems and medical applications. Only the most recent advances will be highlighted. In the case of computer science at the service of life sciences, only formal methods of computer science and new computational-oriented models for dynamical systems—especially the hybrid system framework of hybrid automata—will be revisited.

12.3.1 Role of Mathematics and Physics

A new multidisciplinary approach is applying dynamical systems thinking to biology and is known as systems biology (Alon, 2007). Systems biology, as established by Peter Wellstead in his paper (Wellstead, 2010), marks a radical shift in mathematical biology (Murray, 2002) and goes beyond the systematic and mathematical analysis of physical and engineering systems.

The beauty of mathematics lies in its clarity and rigor; mathematics brings order to complexity in an analytical manner. Over the past decade, much has been written about the interplay of mathematics and biology; some examples are Cohen (2004), Hoy (2004), May (2004), and Nanjundiah (2005). A recent report, published by the Foresight Horizon Scanning Centre within the Government Office for Science of the United Kingdom,* draws attention to the fact that interaction between biology and mathematics will be a matter of course in the near future.

Mathematics and biology mutually enhance one another (Cohen, 2004). There have been many mathematical advances arising from biological problems—some significant examples have been presented in Section 12.2. On the other hand, mathematical challenges have also transformed biology.

Although the application of mathematical tools has facilitated great progress in genomics, ecology, evolution, epidemiology, or bioinformatics, the application of mathematics in life sciences is not as well developed as in physical sciences or engineering. In Nanjundiah (2005), it is argued that one reason is that "it is difficult to conceive a living creature, taken as whole, as being reducible to mathematical formalisms such as those that embody

---

* *Growing Synergy between Biology and Mathematics,* May 2011, available at www.sigmascan.org.

the laws of physics." And "mathematics is unlikely to be effective in biology because of evolution."

Robert May in "Uses and Abuses of Mathematics in Biology" (May, 2004) elucidates the problem of applying analytical methods to a science fundamentally based on experimental validation as biology. It is emphasized that many simulation or statistical software packages—initially designed for linear systems—are applied to highly nonlinear systems, even with discontinuities. As a consequence, the onset of complex dynamical behaviors because of these nonlinearities is ignored (Hoy, 2004; May, 2004). Another problem is the tendency of devising overly complex models with more emphasis on computational simulation than the abstraction and analysis of the prime dynamical behaviors. Perhaps this is a sign of the division between biologists and dynamical system mathematicians. There are signs, however, that things are changing; applied mathematicians and biologists are already coming together, and soon there may be a complete integration of the disciplines. Some of the modeling and simulation challenges to be tackled are explained in Sauro et al. (2006).

Applied mathematicians interpret the world through mathematical models to approximate and predict the evolution of dynamical systems. There has been some success in the use of dynamical models to describe biological and medical applications (Demongeot et al., 2009) and different processes in human and life sciences (Bellomo et al., 2010). A great variety of in silico models have been proposed, that is, mathematical–computational models that complement the traditional in vivo or in vitro models, and can be simulated to quantitatively study the emergent interacting behavior of a biological system consisting of complex processes and many components. Many examples are found in neuroscience, where different neurological diseases are modeled: in particular, Parkinson's disease (Wellstead, 2010; Liu et al., 2011a), Alzheimer's disease (Bhattacharya et al., 2011), epilepsy (Alamir et al., 2011), or neurological processes involved in addictions (Metin and Sengör, 2010). A wide variety of neural models are studied, as spiking neurons (Izhikevich, 2007), and the phenomenon of synchronization in neuronal networks has attracted special interest in the past few years (Tabareau et al., 2010; Sun et al., 2011; Yu et al., 2011).

Diabetes is another disease whose mechanism has been revealed by dynamical system analysis. The synchronization of pancreatic beta-cells has been shown to play a key role in the development of the disease (Barajas-Ramírez et al., 2011). There has also been success in identifying common features in different types of cancer at various stages of the disease. Most of these results are related to intracellular properties and genetic mutations that lead to the onset of replicating malignant cells provoking tumor growth and to the analysis of the effects of therapies such as chemotherapy (see, e.g., Spencer et al., 2004; Aguda et al., 2008; Tanaka et al., 2008). A recent survey on advances of the application of mathematics in cancer over the past 50 years is in Byrne (2010). There are also many models that study biochemical signaling network processes, key to understanding the onset of chronic inflammatory diseases (Ashall et al., 2009). Gene regulatory networks have also been studied through the prism of different mathematical models and dynamical analysis (Polynikis et al., 2009).

But not all contributions to biology or life sciences stem from mathematics and computational mathematics. Formal methods in computer science, control theory, and network science are also shaping the future. Some comments on network science are given in the succeeding text, and the roles of control theory and theoretical computer science are further explored in the next two subsections.

Network science, mainly dominated by physicists, is based on experimental data. Most statistical physics methods are applied to understand the topological properties of complex networks, namely, size, density, and connectivity. There are many examples of complex network theory uncovering the behavior of biological networks and medical applications (Feng et al., 2007). The term *network biology*, initially applied to understand cell-to-cell interactions through the theory of complex networks (Barabási and Oltvai, 2004), is becoming widely employed. Furthermore, the most recent term *network medicine* (Barabási et al., 2011) highlights the role of complex network paradigms to better understand numerous diseases, for instance, neurological diseases (Villoslada et al., 2009), metabolic networks (Jeong et al., 2000), and, finally, cancer metastasis (Chen et al., 2009) or cancer growth (Galvão and Miranda, 2008). The applications are vast and promising.

12.3.2 Role of Control Engineering

A control system is based on measuring an output, comparing to a desired output, and generating an error that is employed to make the necessary corrections in the input so that the system output reaches a desired value. Depending on the type of model considered for the system or the type of control goal defined, different types of control techniques are applied, namely, linear, nonlinear, discontinuous, switched, hybrid, networked, cognitive, cooperative, optimal, adaptive, robust, or supervisory.

Control engineers, unlike mathematicians, put theory into practice. Not only do engineers observe, but they also transform and build with a prime goal: the simpler the better. This perspective suits biology to perfection, which is why control systems thinking will continue to play a key role in the analysis of dynamical processes in living organisms, exposing the limitations of mathematical approaches, as established in Nanjundiah (2005) and commented in the previous section.

There is growing interest in the control community to apply control theory and control engineering methodologies to transform the course of biological or natural processes by means of the design of feedback mechanisms. An illuminating report recently published by the IEEE Control

Systems Society (Samad and Annaswamy, 2011) and a subsequent editorial article (Middleton, 2011) highlight the important role that control results may have in systems biology in the near future. The document "Control in Biological Systems" (Doyle et al., 2011) explains why the application of control-system paradigms to medical applications is very different from other areas, and "it is much less mature, has far wider impact on human life and is much less established." There are many barriers to cross. One of the challenges is to find a common language within the parameters of these disciplines.

Two recent books demonstrate the emerging interest in bringing together control theory and systems biology (Iglesias and Ingalls, 2010; Cosentino and Bates, 2012). On the one hand, Cosentino and Bates (2012) provide an excellent summary of classical control techniques written in an understandable language for biologists. On the other hand, Iglesias and Ingalls (2010) describe recent trends of this emerging multidisciplinary research and how prominent control-system researchers use their expertise to solve some problems in different life science applications.

It is gratifying that some of the best international nonlinear control theorists are breaking the barricades and combining their efforts with biologists, physicists, and life scientists (Cho et al., 2005; Arcak and Sontag, 2008; Danzl et al., 2009; Scardovi et al., 2010; Tabareau et al., 2010; Alamir et al., 2011; Barajas-Ramírez et al., 2011; Liu et al., 2011a,b). Promising new directions in control have been directly inspired by biology (Sontag, 2004; Wellstead, 2010). It is significant that electric and control engineering departments of international repute are currently studying biological systems, and one of the most prestigious journals in control systems *Automatica* recently published a complete issue on systems biology.*

Although this combination of control theory, dynamical systems analysis, and systems biology has vast potential to reveal important patterns in the evolution of different diseases, it is not widely exploited. The idea of *feedback medicine*—that is, the use of control techniques to monitor and predict patients' problems—has been recently introduced in Wellstead et al. (2008).

Great progress has been made mainly in neurodegenerative diseases (Wellstead, 2010; Alamir et al., 2011; Liu et al., 2011a) and the study of diseases such as cancer. Optimal control techniques have been applied to evaluate the effects of chemotherapy or other therapies in cancer (Swan, 1990; Matveev and Savkin, 2002; Krabs and Pickl, 2010). Applications in cardiovascular systems and endocrinology are highlighted in Doyle et al. (2011), or body weight models are provided (Christiansen et al., 2008) to tackle obesity, a high-priority problem because of our unhealthy lifestyle.

Control in complex networks is practically nonexistent. There is more work done related to synchronization in dynamical networks of coupled

---

* *Automatica*, 47(6), 1095–1278, June 2011.

oscillators (DeLellis et al., 2010), which can be interpreted as a stabilization problem. Only very recent work (Liu et al., 2011b) starts defining complex network concepts in terms derived from control theory, in particular, controllability of linear systems. This is further explained in Section 12.4.3.

12.3.3 Role of Theoretical Computer Science

The field of biomedical or biological theoretical computer science is relatively young compared to mathematical biology or biomedical control. Applications of formal methods of computer science in life sciences are increasingly found in the specialized literature. Hybrid system models and formal verification techniques are flourishing as novel and promising approaches to uncover the complex mechanisms in living organisms.

Hybrid dynamical systems are systems combining continuous dynamics (either continuous time or discrete time) and discrete input or event-driven discrete state transitions (van der Schaft and Schumacher, 2000). There has been increasing interest in using the hybrid system framework to model and analyze systems in life sciences; see, for example, the special issue of the *Philosophical Transactions of the Royal Society A* dedicated to the applications of hybrid dynamical systems to biological and medical applications (Aihara and Suzuki, 2010) or the recent workshop of *Hybrid Modeling in Systems Biology* held in November 2011 within the framework of the *IEEE International Conference on Bioinformatics and Biomedicine* to be followed by a special issue of the journal *Theoretical Computer Science* in 2012. The important role that hybrid system models and approaches have in life sciences is also discussed in Mailleret and Lemesle (2009).

Several modeling frameworks exist for hybrid dynamical systems, whose foundations were established in the 1960s with Witsenhausen (1966). From here on in this chapter, only illustrations of the computational-oriented modeling framework of hybrid automata are provided (Henzinger, 1996).

An automaton is a computational abstraction of the transitions of a system between discrete states (on and off, for instance). A hybrid automaton, additionally, considers dynamical evolution in each state. It is an adaptation of finite automata and is therefore related to computer science discrete representations and symbolic dynamics.

Examples of the use of hybrid automata for modeling biological processes are few. The most recent and relevant include modeling excitable-cell behavior (Ye et al., 2008) and intracellular dynamics (Cho et al., 2005); hybrid input–output automata to describe complex behavioral patterns in cardiac tissue, such as fibrillation (Bartocci et al., 2009); and the use of hybrid automata and reachability analysis for biological protein regulatory networks, particularly, protein signaling processes (Ghosh and Tomlin, 2004; Campagna and Piazza, 2010) to establish the initial conditions from which a particular biologically significant steady state is attainable. Other works related to reachability analysis of linear biological systems are proposed in Dang et al. (2011), where a technique of hybridization of systems—transformation of a smooth system into a class of hybrid automata—is proposed.

The process of checking in an automated manner that a system behaves correctly is referred to as *formal verification* in the theory of computer science; the elimination of negative behaviors falls into the field of control engineering. Formal verification originated within computer science to show that software or hardware systems had been correctly designed. Recently, it has been applied to prove properties about dynamical systems (see, e.g., Tomlin et al., 2003, and Carter and Navarro-López, 2012).

For formal verification, new discrete abstractions are required in order to represent key information of biological systems and their large number of variables and complex dynamical behaviors (Collins et al., 2011; Cook et al., 2011). The term *symbolic systems biology* has been proposed (Lincoln and Tiwari, 2004). This is a very young field of research, though recent examples include Collins et al. (2011) in which piecewise-affine hybrid systems—that is, a system that switches between linear systems—are employed to abstract key features of biochemical reaction systems. Besides, Bartocci et al. (2010) propose a logic referred to as *biological oscillators synchronization logic* in order to formally specify the phenomenon of synchronization in a network of biological coupled oscillators based on Kuramoto's model.

Generally speaking, there are two approaches to formal verification. The first is *model checking* (Clarke, 2008), where the model of the system is checked using algorithms to automatically test that it satisfies a desired specification (property). The second approach is *deductive verification* or *theorem proving* where known axioms and rules are employed to logically prove that a formal specification holds (Akbarpour and Paulson, 2010). The idea is to mimic how humans carry out mathematical proofs.

Model checking is more widely employed for the formal verification of biological systems than theorem proving, for instance, in biochemical reaction systems (Collins et al., 2011) or in the synchronization of biological coupled oscillators (Bartocci et al., 2009, 2010). For a view on the application of model checking and hybrid automata in systems biology, the reader is referred to Mysore (2006) and Mishra (2009). The probabilistic model checker PRISM is increasingly being employed; one recent application is in the understanding of biological pathways (Kwiatkowska et al., 2008). The application of formal verification techniques can be useful in discovering more reliable and effective treatments for diseases. Recently, model checking techniques have been employed to uncover the signaling pathways in pancreatic cancer (Gong et al., 2011).

Key questions arise: Are the results presented in Sections 12.3.1 through 12.3.3 really applicable and clinically feasible? Do these approaches usher a new age of scientific publications that fail to make a substantial advance in life sciences? There is a long way to go before we can answer these questions, but the fact of joining efforts from different disciplines is a necessary and very good start that can offer a new approach to understand crucial processes in living organisms.

12.4 DYVERSE: Putting the Pieces of the Jigsaw Together

The third thread of thought is the combination of computational models, formal verification, network science, control engineering tools, and systems thinking. The proposed framework for this is referred to as DYVERSE. DYVERSE meshes together the approaches and disciplines presented in Sections 12.2 and 12.3 and uses the hybrid system framework to model complex systems.

12.4.1 Toward a Computational Dynamical Systems Theory

There are many subclasses in hybrid systems. This is, in general, a nomenclature issue, and labels for hybrid systems that on the surface seem different actually represent the same idea. They typically arise in control systems' literature and include systems such as heterogeneous, multimodal, multicontroller, logic-based switching control, discrete event, variable structure, discontinuous/switched, complementarity, nonsmooth, reset, jump, piecewise smooth, mixed logical dynamical, or impulsive. Broadly speaking, a hybrid system is a system that continually changes and has unpredictable behavior. They combine continuous and discrete, smooth, and abrupt dynamics—discontinuities or discrete events, for example.

Hybrid dynamical systems have been studied over the past 15 years as the 2-C science—computation and control. They were considered to be the control theory of tomorrow (Matveev and Savkin, 2000) 12 years ago and represent a very active research theme with many unsolved problems in theory and practice. Since the first special issue dedicated to hybrid systems in 1995,[*] several special issues have appeared in top computer science and control-related journals. A new journal was launched in 2006, *Nonlinear Analysis: Hybrid Systems*, and two international conferences have been consolidated over the past years, *Hybrid Systems: Computation and Control* (HSCC) and *Analysis and Design of Hybrid Systems* (ADHS).

Recently, hybrid dynamical systems have evolved to *cyber-physical systems*—advanced embedded systems combining computation, control, and communication—and can be regarded as the 3-C science (computation, control, and communication). Cyber-physical systems build on the technology of networked embedded control systems. Advances in hybrid dynamical systems and cyber-physical systems have a potential impact on our life and welfare. They underpin the theory beneath emerging complex applications, mainly high-confidence health care, security, and transportation. The U.S. National Science Foundation identified cyber-physical systems as a high-priority research topic: one that represents an information

---

[*] In *Theoretical Computer Science*, Vol. 138(1), February 1995.

technology revolution.* In this chapter, we prefer the term *complex network systems* to *cyber-physical systems*.

Every mathematical model is an approximation of the actual world and is full of limitations. However, some models are better than others at describing the evolution of certain systems. DYVERSE is a computational dynamical framework for the modeling, analysis, and control of complex control systems. DYVERSE stands for the **DY**namically driven **VER**ification of **S**ystems with **E**nergy considerations. Here, energy refers to the abstract energy of dynamical systems, which may not have a physical interpretation, and is studied by means of the dissipativity theory. Dissipativity theory analyzes dynamical system behavior by means of the exchange of energy with the environment (Willems, 1972). The study of the behavior of a system in terms of the energy it can store or dissipate has an extraordinary value, since it provides a rather physical and intuitive interpretation of dynamical properties, such as stability (Navarro-López, 2002). A dissipative system is a system that cannot store all the energy that has been given, that is, it dissipates energy in some way.

But DYVERSE is not only about formal verification and dissipativity analysis: DYVERSE uses the verification results to modify and improve the system behavior. DYVERSE represents a fresh perspective within the theory of hybrid systems and dynamical systems in general and provides new insights into the modeling, analysis, and control of systems with switching dynamics and complex behaviors.

Although the hybrid system paradigm has made great advances in the past 15 years in the modeling, analysis, and control of many complex systems, there is still a gap in the effective integration and application of the theory. DYVERSE bridges the theories of bifurcation analysis and stability, dissipativity, formal verification, formal languages, and switching control, which has never been performed before. Preliminary ideas on DYVERSE were presented in previous work (Navarro-López and Carter, 2010).

The modeling framework that DYVERSE uses is the hybrid automaton framework (Henzinger, 1996). It is an adaptation of finite automata and is therefore related to computer science discrete representations and symbolic dynamics.

In Figure 12.7, the main elements of a general hybrid automaton with inputs are shown. A hybrid automaton is a finite-state automaton including a dynamical system—a set of ordinary differential equations—in each discrete location q_i. The continuous state of the system is represented by x. In each discrete location, the state x belongs to a different domain (represented by the set $Dom(q_i)$ for each q_i). The continuous inputs to the system are represented by the vector u. The transitions between discrete states are represented by edges with guards and resets. The trajectory of the system will

---

* Networking and information technology research and development program, NFS report, CPS week, San Francisco, CA, April 2009.

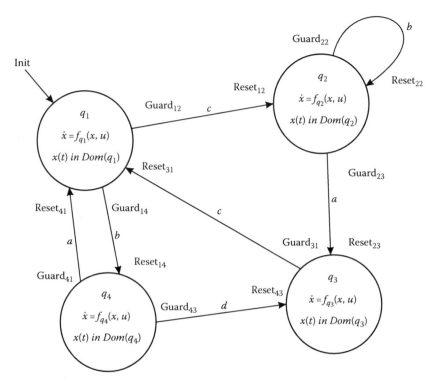

FIGURE 12.7
Graphical representation of a hybrid automaton with its main elements.

change its discrete location q_i, moving to another location q_j, only if a guard condition is satisfied. Before entering the new location q_j, the continuous state x can be reset to a new value. The combination of this guard and reset in each transition is the main cause of discontinuous changes in the system evolution. Moreover, a symbol (in the form of a letter in the figure), which belongs to an alphabet, is associated with each edge or transition between the discrete locations. These symbols can be considered as the discrete outputs or events in the system.

The basic hybrid automaton model employed in our approach is adapted from the hybrid model given in Johansson et al. (1999) and Lygeros et al. (2003) and has been recently proposed in Navarro-López (2009) and Navarro-López and Carter (2010a) as the discontinuous dynamical system (DDS) hybrid automaton with inputs and outputs. The DDS hybrid automaton is a general hybrid automaton that provides a suitable mathematical model for DDSs with discontinuous state derivatives (e.g., systems with dry friction). We could understand in a better way the potential applications of the DDS hybrid automaton by means of a study case that is a simplified model of the torsional behavior of a conventional vertical oil well drill string (Navarro-López and Cortés, 2007). This system with discontinuous

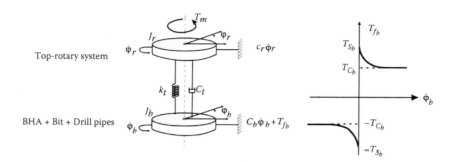

FIGURE 12.8
Two-degree-of-freedom model of a simplified vertical conventional oil well drill string used to model bit sticking-related behaviors (Navarro-López and Suárez, 2004). The drive torque T_m comes from the electric motor at the surface. The Coulomb and static friction torques are represented by T_{C_b} and T_{S_b}, respectively. (From Navarro-López, E.M. and Suárez, R., Practical approach to modelling and controlling stick-slip oscillations in oilwell drillstrings, in *IEEE Conference on Control Applications*, Taipei, Taiwan, pp. 1454–1460, September 2004.)

friction for the case of two degrees of freedom (two torsional inertias, J_r and J_b) is shown in Figure 12.8.

This is a substantially simplified model of an actual drill string. However, it captures all the main behaviors that we are interested in when studying the harmful stick–slip oscillations present at the bit and the bottom-hole assembly (BHA). One of the consequences of stick–slip oscillations is that the top of the drill string rotates with a constant rotary speed, whereas the bit and the BHA rotary speeds vary between zero and up to six times the rotary speed measured at the surface. For this example, the system state is $x = (x_1, x_2, x_3)^T$, with x_1 and x_3 the angular velocities of the top-rotary system and the bit ($\dot{\phi}_r$ and $\dot{\phi}_b$), respectively, and x_2 is the difference between the two angular displacements ($\phi_r - \phi_b$). The torque $u > 0$ is applied at the top-rotary system, and the weight on the bit is represented by $W_{ob} > 0$. Both, u and W_{ob}, are considered as two varying input parameters. The discontinuous nature of the drill string is because of the discontinuous friction (T_{f_b}), which models the bit–rock interaction.[*]

The discontinuous system presented in Figure 12.8 can be reinterpreted in the computational framework of hybrid systems by means of the DDS hybrid automaton presented in Figure 12.9.

The DDS hybrid automaton is being modified to include jumps in the states (a discontinuity usually associated with systems with impacts), and we have recently proposed the MRB hybrid automaton—that is, the multirigid hybrid automaton—in other works (O'Toole and Navarro-López, 2012). These modifications are inspired by the idea of mythical modes in discontinuous systems, originally proposed in Nishida and Doshita (1987) and further expanded to hybrid systems by Mosterman and Biswas (1998) and Mosterman and Biswas

---

[*] Details on this system are given in different papers published by Navarro-López, which are available at http://staff.cs.manchester.ac.uk/~navarroe/pubs/discontinuous/

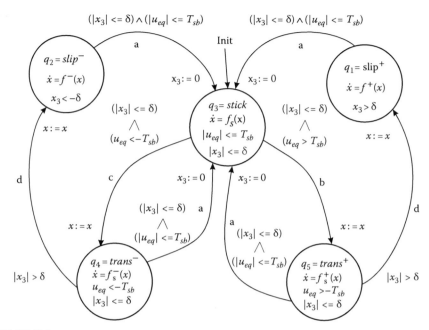

FIGURE 12.9

DDS hybrid automaton modeling the switching behavior of the simplified representation of an oil well drill string shown in Figure 12.8. (Reproduced from Navarro-López, E.M. and Carter, C., Hybrid automata: An insight into the discrete abstraction of discontinuous systems, *Int. J. Syst. Sci.*, 42(11), 1883-1898, 2011. Journal's web site http://www.tandfonline.com. With permission from Taylor & Francis Group.)

(2000). The chief characteristic of these hybrid automaton-based models is that, following the computational divide-and-conquer approach, a system with multiple discontinuous elements can be represented by the composition of several hybrid automata (Navarro-López and Carter, 2011).

Modifications in the already-existing DDS hybrid automaton must be considered to model complex network systems. The extension of the DDS hybrid automaton for complex control network systems is the *reconfigurable hybrid automaton*, whose key ideas will be introduced in Section 12.4.3.

12.4.2 Formal Verification Driven by Dynamical Properties and Control

The controlled motion of swarm satellites, the cooperation of several robotic systems to meet a common goal, the operation of large-scale electric power networks with renewable generation sources, the automation of public transport scheduling systems in metropolitan areas, the synchronization patterns of beta-cells in your pancreas, the self-organization of cells, and the interconnections in brain networks are examples of interdependent, interacting, distributed, and networked systems. Their operation is usually safety critical and demands the preservation of some properties despite uncertainties and the recovery from contingencies minimizing losses. Furthermore, they are highly nonlinear and exhibit different transitions and switching behavior patterns. They can be interpreted as hybrid dynamical systems.

Controlling these complex network systems requires the application of methods beyond classical control engineering procedures or stability analysis techniques. To this end, the combination of formal verification methods from computer science and dynamical analysis techniques is a promising development, and the theory of hybrid dynamical systems has a pivotal role to play in this.

But what makes the behavior of hybrid systems so complex? Where does the unpredictable behavior of these systems originate: from the system itself or from its environment? Because of the discontinuous changes in states, hybrid systems entail complex behaviors not present in systems that are purely discrete or continuous. These *complex behaviors* usually lead to faults (e.g., genetic mutations triggering mechanisms that may lead to cancer) or dynamics degrading performance (e.g., heart arrhythmias). The challenge is to ensure that systems behave correctly, eliminating all negative behavior.

There has been much success in applying formal verification methods to validate dynamical properties in particular classes of discontinuous and hybrid systems (Henzinger et al., 1998; Chutinan and Krogh, 2003; Tomlin et al., 2003; Guéguen et al., 2009). However, it is still a challenge to provide satisfactory solutions for the verification of complex dynamical behaviors of hybrid systems. One of the main obstacles has been the difficulty of obtaining a computational representation of these complex dynamical properties. Besides, most of the results have been focused on safety and reachability analysis. Consideration of the changes of the energy of the system, as the DYVERSE approach proposes, may be the answer to this problem.

There are three main parts of the novel approach DYVERSE, inspired by the three-step control cycle: model, check, and act (Figure 12.10):

1. *Discrete abstraction of dynamical behaviors*: Dynamical behaviors and stability patterns are specified by means of formal languages plus the evolution of the stored and supplied energies of the system. Specifically, a set of formal languages is associated with each key behavior. Each word of the languages represents a sequence of events (transitions) generated by the dynamical evolution of the hybrid automaton for an initial hybrid state. We use ω-regular languages—that is, infinite word languages (Muller, 1963)—in order to properly describe oscillatory behaviors (infinite behaviors). The languages obtained can be used in the verification process.

   For the hybrid automaton shown in Figure 12.9, we have obtained the set of formal languages for each key long-term dynamical

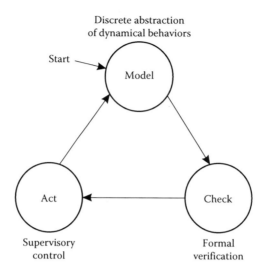

FIGURE 12.10
The three-step control cycle of DYVERSE.

behavior and for each initial discrete location (Carter and Navarro-López, 2011). Specific examples of behaviors present in this drilling mechanism with their corresponding language are presented in Figure 12.11. Considering q_3 as the initial discrete location, we can therefore associate the languages that represent the three behavior patterns present in the system:

a. Positive velocity equilibrium (q_1): $[(b+c)(\varepsilon+d)a]^*bd$
b. Stuck bit equilibrium (q_3): $[(b+c)(\varepsilon+d)a]^*$
c. Positive stick–slip (periodic behavior that infinitely repeats the loop $q_3 \rightarrow q_5 \rightarrow q_1 \rightarrow q_3 \ldots$): $[(b+c)(\varepsilon+d)a]^*(bda)^\omega$

In the formal languages, ε is the empty string, the operation * is the Kleene closure of a string—the string is repeated as many times as we want, including none—and ω is the ω-closure of a string—the string is repeated an infinite number of times.

The way we have abstracted dynamical behaviors from a computer science viewpoint by means of formal languages may be inconclusive and ambiguous for some behaviors. Additional dynamical considerations should be used. In our case, the key to break this ambiguity will be the use of energy-related properties of the system together with its discrete abstraction. Consequently, the dynamical behaviors will also be specified by means of the type of evolution of the stored energy (V) of the hybrid automaton. Following the terminology and rationale of dissipative systems, the dynamical behaviors will be specified in terms of dissipation,

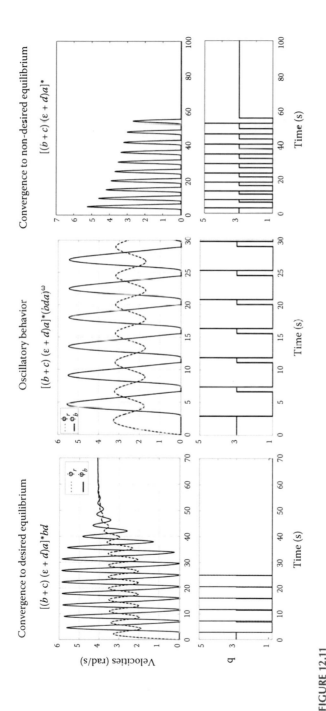

FIGURE 12.11

Regular languages associated with each key long-term dynamical behavior in a simplified model of an oil well drill string obtained for the DDS hybrid automaton of Figure 12.9 with q_3 the initial discrete location. The convergence to a desired equilibrium is the convergence to a positive desired velocity at the top-rotary system as well as at the BHA. The oscillatory behavior corresponds to bit stick–slip oscillations (the bit velocity oscillates between zero and a positive velocity). Finally, the convergence to a nondesired equilibrium corresponds to the situation of a permanent stuck bit (the bit stops rotating after some period of time and never starts again).

injection, or conservation of energy; furthermore, the stored energy *V* acts as a *common storage function*. We distinguish three types of evolution of *V*: decreasing, nondecreasing, and conservative. This energy-related specification of behaviors is taken from dissipativity theory of discrete-time and continuous-time systems (Navarro-López, 2002).

These ideas can be applied to the abstraction of self-oscillating and stable behaviors of the oil well drill string. For this example, we can propose as the stored energy *V* the sum of the kinetic and the potential energy:

$$V = \frac{1}{2}\left[k_t(x_2 - \bar{x}_2)^2 + J_r(x_1 - \bar{x}_1)^2 + J_b(x_3 - \bar{x}_3)^2 \right],$$

with $\bar{x}$ the unique desired equilibrium point. From Figure 12.12, it can be appreciated that for the convergence to the equilibrium, the storage function *V* is decreasing, whereas for the stick–slip situation, *V* is nondecreasing, and for the stuck situation, *V* is conservative.

2. *Formal verification*: The second part of DYVERSE is formal verification.

We propose a dynamically driven and behavioral-oriented verification procedure to establish that a system is free of undesirable oscillations and instabilities and meets the design specifications.

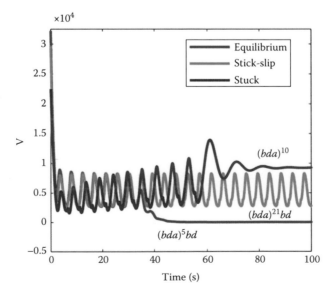

FIGURE 12.12
Evolution of the stored energy for the system shown in Figure 12.9 and associated words for the time interval shown in the figure.

Verification in hybrid dynamical systems is usually based on computing the reachable and safe state space of a hybrid automaton and requires the approximation of the continuous space regions. In many hybrid systems, and because of the nature of the complex behaviors involved, to define the properties to verify in terms of a reachability problem can be unfeasible. Our alternative approach is to specify the properties of the system in terms of *goal behaviors* and *forbidden behaviors*. These behaviors will be words (elements of a formal language) generated by the discrete transitions of the hybrid automaton plus the type of evolution of the stored energy of the system, as presented earlier. Consequently, the computational verification method is guided by dynamical properties of the system. Roughly speaking, DYVERSE can be interpreted as a *compiler* of dynamical behaviors and energy properties. This is a novel approach for the verification of dynamical properties of hybrid automata.

3. *Control synthesis*: A step further is to use all the mentioned theory in order to propose a control scheme that can steer the hybrid system trajectories to the desired dynamics by considering the verification results. We propose a supervisory control architecture for this purpose, which is shown in Figure 12.13 and consists of four main blocks:

   a. An *events+status block* storing the sequence of discrete events generated by the hybrid automaton (H) and the information about the most recent behavior

   b. An *energy evaluation block* generating the type of V

   c. A *verification block* that receives as inputs the word (sequence of discrete outputs) generated by H, the type of V, and the desired specification and provides as output the behavior of the system

   d. A *hybrid controller block*, including a switching controller and a finite-state automaton

The novel contribution of this control scheme is that we include, in the control loop, an additional verification block that combined with the hybrid controller will be able to modify the hybrid automaton behavior. Depending on the language generated by the hybrid automaton and on the evolution of the common storage function V, an appropriate controller will be selected. Typically, the hybrid automaton will be a controlled system for which we have to check whether or not the proposed control design achieves the control goals. In other words, the verification block can be considered as a monitoring system.

Some relevant elements of the control scheme are

   a. A *behavior B* of the hybrid automaton is a tuple $(w, Type_V)$, where w is a word within the language associated with each long-term dynamical behavior and $Type_V$ is a set defining the types of evolution of the function V.

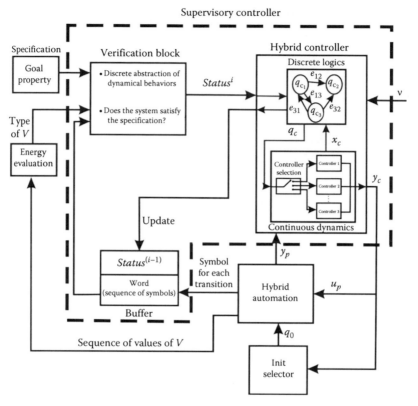

FIGURE 12.13
Supervisory control architecture. The reference input v represents the target; it may consist of discrete and continuous inputs.

b. Finite sets of *forbidden behaviors* (B_F) and *goal behaviors* (B_G), such that $B_F \cap B_G = \varnothing$.

c. The *Status* is a tuple $Status = (w, Type_V, safe)$, with $safe = 1$ if $B \in B_G$, $safe = 0$ if $B \in B_F$.

d. A sequence of discrete outputs of the hybrid automaton, w, corresponding to the symbols associated to each edge or transition between discrete locations (see Figure 12.7).

The hybrid controller in Figure 12.13 is continuously receiving the continuous output of the hybrid automaton (y_p), and depending on y_p, an appropriate controller is applied. The continuous output of the hybrid controller (y_c) will act as the continuous input to the plant (u_p). In addition, the control strategy can be changed when the verification block is activated. The output of the verification block is the *Status* of the system. Roughly speaking, the verification block produces the *Status* of the system that will be processed by a discrete-logic controller within the hybrid controller. Finally, the *Status* is stored in the buffer where the sequence of discrete outputs (symbols) is. The verification block will not be activated if the hybrid automaton remains in the same discrete location. That is, there is no verification if no transition has been generated by the hybrid automaton. Moreover, we have to take into account that the *Init* state of the hybrid automaton changes with time and may also vary depending on the controller output. We have considered here the verification block as a black box, and no details are given on its implementation.

The idea of bringing together computer science paradigms and control engineering techniques is not new (Asarin et al., 2000; Maler, 2002) and has usually been solved through supervisory control schemes and under the framework of discrete-event systems. A recent work that bridges computer science abstraction and control-system design, similar to the idea of DYVERSE, is the work by Tabuada (2009).

12.4.3 Biologically Inspired Real-Time Self-Organizing Systems: The Morphogenesis of the Twenty-First Century

Memory is often recovered after a brain hemorrhage. In other words, the brain can reorganize its structure and functions to cope with critical situations. This is an example of the kind of self-organizing mechanisms discussed in Section 12.2. Novel computational–dynamical models—to capture evolving network phenomena and analyze the emergence of collective dynamical behaviors in complex network systems—can be the key to the better understanding of such adaptive processes. Hybrid system methodologies hold out the promise of this.

The approach employed in complex networks so far is more statistical than behavioral and consequently has important limitations in the analysis of collective emergent dynamical properties and network evolution and adaptation. This reinforces the necessity to enhance complex network theory by complementing the traditional analysis with new perspectives. We believe that the theory of complex networks requires the integration of powerful tools from theoretical computer science and dynamical system theory. The union of computer science and network science is our grand ambition. Going further, the imperative is to bring together computer science paradigms, control, dynamical systems analysis, and complex networks theory, in short, to combine the still infant theory of hybrid systems with network science.

Complex network theory and its new type of applications require knowledge transfer using one method in one domain/discipline, and transfer it into another. It represents a new Renaissance: the pursuit of beauty in design, multidisciplinarity, and creative thinking. László Barabási, one of the most prominent network scientists, puts the finger on it in the paper "Taming complexity" (Barabási, 2005b): "The science of networks is experiencing a boom. But despite the necessary multidisciplinary approach to tackle the theory of complexity, scientists remain largely compartmentalized in their separate disciplines. Can they find a common voice?"

The expansion of DYVERSE to complex networks and distributed systems represents a new perspective in order to find this common voice. It combines the theory of hybrid automata, dynamical systems and control theory, formal verification, and complex networks. The consideration of network science paradigms within the field of control hybrid systems is the most significant novelty of this research. Moreover, the inclusion of hybrid systems methodologies within complex network models will facilitate the design of adaptive mechanisms to modify and control the topology and dynamics of the network to achieve a predefined goal in an efficient manner.

We propose a new generation of complex network models to analyze the pattern in different evolving and adaptive dynamical processes, in which the topology is directly dependent on the dynamics of the nodes. Nature is our source of inspiration, particularly the social behavior of animals, biological cooperation of organisms, and the different patterns of self-organization of cells.

Special attention will be paid to the modularity of the network. That is, the network will be divided into units (closely connected nodes) that perform a common goal almost independently. Hence, we consider the functionality or changing dynamics and goals of the nodes. This follows the biological and engineering intuition that designs with modularity have higher adaptability and better resist the random removal of their components or the changes of environment. This evolving self-organizing dynamical approach may provide answers to questions that are still open: from greater understanding of the brain's plasticity to an explanation of how modern cities evolve and from the operation of beta-cells in our pancreas to unpredictable patterns of people mobility.

Preexisting models of complex networks are not appropriate for this purpose. Network science has made great progress in the past 10 years. New models for evolving and adaptive networks have been proposed (Dorogovtsev and Mendes, 2003; Gross and Sayama, 2009; Gros, 2011). The idea of stabilization or controllability in networks is recent (Csermely, 2006; Liu et al., 2011b). However, there are still many gaps in the modeling of the adaptive, continually changing, and switching dynamical processes of networks. Furthermore, they fail to effectively reproduce a key aspect of complex networks: switching and discontinuous dynamical processes and modularly varying goals.

Hybrid system models, especially hybrid automata, are very good for capturing the evolution of switching and discontinuous dynamical processes.

But preexisting hybrid automata models are not sufficiently flexible to be applied to evolving complex networks: we have to introduce adaptive changing conditions to trigger the transitions between discrete states. The mechanism of adaptation will depend on the topology of the connections (edges) between discrete states and on the dynamical behavior of the subsystems within the hybrid automata. The result will be a *reconfigurable hybrid automaton*: a new hybrid system model incorporating complex network thinking.

The reconfigurable hybrid automaton will break the fixed connections of the classical hybrid automaton model. Our reconfigurable hybrid automaton can recognize evolving and adaptive transitions between the dynamical subsystems of the hybrid automaton. The reconfigurable hybrid automaton, in addition to the elements shown in Figure 12.7, has evolving and adaptive transitions between the dynamical subsystems and discrete locations.

Imagine the hybrid automaton shown in Figure 12.7, but now, the edges or transitions and their associated guards and resets can change with time following some adaptive rules. These rules would depend on the evolution of the continuous state x of the subsystems, the discrete locations q, and the topology of the connections between the discrete locations.

The reconfigurable hybrid automaton uses a general model to capture the coupled behavior of nodes or subsystems and the switching processes present in the overall network. In addition, it provides a unique framework to incorporate control laws to drive the network to a desired state or goal or sequence of states. In this manner, under this framework, advanced computational and control-system techniques can be applied to complex networks. The idea of driver nodes (terminology inherited from biology) proposed in the paper (Liu et al., 2011b) can be successfully exploited in this framework. In the case of hybrid automata, this means that the control is not applied to all the locations; it is selectively applied, and not all the dynamical systems in the hybrid automata are influenced by control inputs.

Inspired by cell-to-cell interaction patterns, the balance of energy in the systems may play a key role to design the evolving and adaptive mechanisms in the transitions of the reconfigurable hybrid automaton. Hence, control methodologies coming from dissipativity systems and passivity-based controllers could be employed. Moreover, the rationale of Turing's morphogenesis is also interesting to adapt for our framework. The interplay of reaction and diffusion processes that facilitate the emergence of self-organizing behaviors is something to mimic and incorporate in the adaptive mechanisms of the transitions in the reconfigurable hybrid automaton. For these reasons, the systems modeled by the reconfigurable hybrid automaton are referred to as *biologically inspired real-time self-organizing systems*.

To sum up, from the perspective of control theory, two control mechanisms are introduced into the reconfigurable hybrid automaton:

1. The change of the topology by means of an adaptive scheme. This will imply the change of the edges between the discrete states of the reconfigurable hybrid automaton, in addition to the guards and resets associated to each edge.
2. The change of the continuous dynamics of each subsystem by means of controllers applied in selected discrete states (nodes) of the reconfigurable hybrid automaton.

In our approach, networks are considered as self-organizing and adaptive structures, and the topology itself can be employed as an integral part of the control system. The reconfigurable hybrid automaton consists of interacting and interdependent subsystems and connections between them; the evolution of one cannot be understood without the evolution of the others, and changes in one will affect the performance of the others. The dynamical implications of the evolving topology of a network of systems as this will play a key role in the design of efficient and robust architectures for large-scale systems in general.

The framework proposed here is general and applicable to a broad class of physical, engineering, or biological systems. We believe that the framework of DYVERSE networks can be successfully applied to enhance a better understanding of self-organizing and adaptive biological systems such as tumors. We could, for example, study the extracellular collective relationships that lead to the formation of tumors when patients have developed cancer. By uncovering these relationships, we may be able to propose more efficient treatments and detect a cancer profile earlier. We could also uncover the mysterious behavioral and coupling patterns in the networks of our brain. This will be key to the healing of unforeseen damage and assist doctors in treatment.

12.5 Conclusions: Looking Ahead

László Barabási highlights in his most recent opinion writing (Barabási, 2012): "Who owns the science of complexity? Physics? Engineering? Biology, mathematics, computer science? All of the above? Anyone?" The answer is as follows: the complexity is in all of them and none of them.

After all our observations, we return to the question posed in Section 12.1: who are the Leonardos and Alan Turings of the future? Our hopes for the future reside in scientists and engineers who are on the boundaries of disciplines, with open minds to challenge the existing order.

Acknowledgments

Probably my work would have never been linked to Turing's morphogenesis without Jenny Trigg, an artist and independent spirit who first alerted me to the possibilities of Turing's model of morphogenesis.

I have constant support and inspiration from my team of wonderful colleagues as Rebekah Carter, Mike O'Toole, Dave Lester (at the heart of the DYVERSE team in Manchester), and Lawrence Paulson (from the Computer Laboratory in Cambridge). Thanks a lot to Alexei Poliakov (from the MRC National Institute for Medical Research in London), who introduced me to the world of cell interactions. I also wish to thank Pietro Liò (from the Computer Laboratory in Cambridge), who is helping me understand the wonders of systems biology and tumor growth, and Ben Small (from the Manchester Interdisciplinary Biocentre), who is helping me discover the mysteries of biochemical signaling networks. I am especially grateful to Professor Bernard Richards (from the School of Computer Science in Manchester), who has shared with me his work with Alan Turing and has encouraged me to study the fascinating topic of morphogenesis.

References

Achlioptas, D., R.M. D'Souza, and J. Spencer. 2009. Explosive percolation in random networks. *Science* 323: 1453–1455.

Aguda, B.D., Y. Kim, M.G. Piper-Hunter, A. Friedman, and C.B. Marsh. 2008. MicroRNA regulation of a cancer network: Consequences of the feedback loops involving miR-17-92, E2F, and Myc. *PNAS* 105(50): 19678–19683.

Aihara, K. and H. Suzuki. 2010. Special issue: Theory of hybrid dynamical systems and its applications to biological and medical systems. *Philosophical Transactions of the Royal Society A. Mathematical, Physical & Engineering Sciences* 368: 4893–4914.

Akbarpour, B. and L.C. Paulson. 2010. MetiTarski: An automatic theorem prover for real-valued special functions. *Journal of Automated Reasoning* 44(3): 175–205.

Alamir, M., J.S. Welsh, and G.C. Goodwin. 2011. Synaptic plasticity based model for epileptic seizures. *Automatica* 47: 1183–1192.

Albert, R. and A.-L. Barabási. 2000. Topology of evolving networks: Local events and universality. *Physical Review Letters* 85(24): 5234–5237.

Albert, R. and A.-L. Barabási. 2002. Statistical mechanics of complex networks. *Reviews of Modern Physics* 74: 47–97.

Allee, W.C. 1938. *The Social Life of Animals.* New York: W.W. Norton & Company Inc. Available from the Biodiversity Heritage Library at: http://www.archive.org/details/sociallifeofanim00alle.

Alon, U. 2007. *An Introduction to Systems Biology. Design Principles of Biological Circuits.* Boca Raton, FL: Taylor & Francis Group.

Arcak, M. and E.D. Sontag. 2008. A passivity-based stability criterion for a class of interconnected systems and applications to biochemical reaction networks. *Mathematical Biosciences and Engineering* 5: 1–19.

Arenas, A., A. Díaz-Guilera, J. Kurths, Y. Moreno, and C. Zhou. 2008. Synchronization in complex networks. *Physics Reports* 469: 93–153.

Arianos, S., E. Bompard, A. Carbone, and F. Xue. 2009. Power grid vulnerability: A complex network approach. *Chaos* 19: 013119.

Asarin, E., O. Bournez, T. Dang, O. Maler, and A. Pnueli. 2000. Effective synthesis of switching controllers for linear systems. *Proceedings of the IEEE* 88(7): 1011–1025.

Ashall, L., C.A. Horton, D.E. Nelson, P. Paszek, C.V. Harper, K. Sillitoe, S. Ryan et al. 2009. Pulsatile stimulation determines timing and specificity of NF-κB-dependent transcription. *Science* 324: 242–246. See also supporting online material.

Barabási, A.-L. and R. Albert. 1999. Emergence of scaling in random networks. *Science* 286: 509–512.

Barabási, A.-L. 2005a. The origin of bursts and heavy tails in human dynamics. *Nature* 435(May): 207–211.

Barabási, A.-L. 2005b. Taming complexity. *Nature Physics* 1(November): 68–70.

Barabási, A.-L. 2007. The architecture of complexity. From network structure to human dynamics. *IEEE Control Systems Magazine* 27(4): 33–42.

Barabási, A.-L. 2012. The network takeover. *Nature Physics* 8: 14–16.

Barabási, A.-L. and E. Bonabeau. 2003. Scale-free networks. *Scientific American* May: 50–59.

Barabási, A.-L., N. Gulbahce, and J. Loscalzo. 2011. Network medicine: A network-based approach to human disease. *Nature Review Genetics* 12: 56–68.

Barabási, A.-L. and Z.N. Oltvai. 2004. Network biology: Understanding the cell's functional organization. *Nature Reviews, Genetics* 5(February): 101–114.

Barajas-Ramírez, J.G., E. Steur, R. Femat, and H. Nijmeijer. 2011. Synchronization and activation in a model of a network of β-cells. *Automatica* 47(6): 1243–1248.

Barrat, A., M. Barthélemy, and A. Vespignani. 2010. *Dynamical Processes on Complex Networks*. Cambridge, U.K.: Cambridge University Press.

Bartocci, E., F. Corradini, M.R. di Berardini, E. Entcheva, S.A. Smolka, and R. Grosu. 2009. Modeling and simulation of cardiac tissue using hybrid I/O automata. *Theoretical Computer Science* 410: 3149–3165.

Bartocci, E., F. Corradini, E. Merelli, and L. Tesei. 2010. Detecting synchronisation of biological oscillators by model checking. *Theoretical Computer Science* 411: 1999–2018.

Bellomo, N., H. Berestycki, F. Brezzi, and J.-P. Nadal. 2010. Mathematics and complexity in human and life sciences. Introduction to the special issue of *Mathematical Models and Methods in Applied Sciences* 20(1): 1391–1395.

Bhattacharya, B.S., D. Coyle, and L.P. Maguire. 2011. A thalamo-cortico-thalamic neural mass model to study alpha rhythms in Alzheimer's disease. *Neural Networks* 24: 631–645.

Bianconi, G. and A.-L. Barabási. 2001. Competition and multiscaling in evolving networks. *Europhysics Letters* 54: 436–442.

Boccaletti, S. 2008. *The Synchronized Dynamics of Complex Systems*. Amsterdam, the Netherlands: Elsevier.

Boccaletti, S., V. Latora, Y. Moreno, M. Chávez, and D.U. Hwang. 2006. Structure and dynamics of complex networks. *Physics Reports* 424: 175–308.

Brewster, E.T. 1912. *Natural Wonders Every Child Should Know*, original edition entitled *A Child's Guide to Living Things*. New York: Doubleday, Page & Co.

Brummitt, C.D., R.M. D'Souza, and E.A. Leicht. 2012. Suppressing cascades of load in interdependent networks. *Proceedings of the National Academy of Sciences of United States of America* 109(12): E680-E689.

Byrne, H.M. 2010. Dissecting cancer through mathematics: From the cell to the animal model. *Nature Review Cancer* 10: 221–230.

Caldarelli, G. and A. Vespignani, eds. 2007. *Large Scale Structure and Dynamics of Complex Networks: From Information Technology to Finance and Natural Science*. Singapore: World Scientific.

Camazine, S., J.-L. Deneubourg, N.R. Franks, J. Sneyd, G. Theraulaz, and E. Bonabeau. 2003. *Self-Organization in Biological Systems*. Princeton studies in complexity. Princeton, NJ: Princeton University Press.

Campagna, D. and C. Piazza. 2010. Hybrid automata, reachability, and systems biology. *Theoretical Computer Science* 411: 2037–2051.

Carter, R. and E.M. Navarro-López. 2011. Abstractions of hybrid systems: Formal languages to describe dynamical behaviour. In: *Proceedings of the 18th IFAC Triennial World Congress* (Milano, Italy, August 31–September 2), pp. 4552–4557.

Carter, R. and E.M. Navarro-López. 2012. Dynamically-driven timed automaton abstractions for proving liveness of continuous systems. In: *Proceedings of the 10th International Conference on Formal Modelling and Analysis of Timed Systems (FORMATS)* (London, U.K., September 18–20).

Chen, L.L., N. Blumm, N.A. Christakis, A.-L. Barabási, and T.S. Deisboeck. 2009. Cancer metastasis networks and the prediction of progression patterns. *British Journal of Cancer* 101: 749–758.

Cho, K.-H., K.H. Johansson, and O. Wolkenhauer. 2005. A hybrid systems framework for cellular processes. *BioSystems* 80: 273–282.

Christiansen, E., A. Swann, and T.I.A. Sorensen. 2008. Feedback models allowing estimation of thresholds for self-promoting body weight gain. *Journal of Theoretical Biology* 254: 731–736.

Chutinan, L.O. 1992. The genesis of Chua's circuit. *Archiv Elektronic Übertransgungstechnik* 46(4): 250–257.

Chutinan, A. and B. Krogh. 2003. Computational techniques for hybrid system verification. *IEEE Transactions on Automatic Control* 48(1): 64–75.

Clarke, E.M. 2008. Birth of model checking. In: *25 Years of Model Checking*, Lecture notes on computer science, Vol. 5000, pp. 1–26.

Cohen, J.E. 2004. Mathematics is biology's next microscope, only better; biology is mathematics' next physics, only better. *PLoS Biology* 2(12): e439.

Cohen, R. and S. Havlin. 2010. *Complex Networks: Structure, Robustness and Function*. Cambridge, U.K.: Cambridge University Press.

Collins, P.J., L. Habets, J.H. van Schuppen, I. Černá, J. Fabriková, and D. Šafránek. 2011. Abstraction of biochemical reaction systems on polytopes. In: *Proceedings of the 108th IFAC World Congress* (Milano, Italy, August 28–September 2), pp. 14869–14875.

Cook, B., J. Fisher, E. Krepska, and N. Piterman. 2011. Proving stabilization of biological systems. In: *Proceedings of the Conference on Verification, Model Checking, and Abstract Interpretation* (Austin, TX, January 23–25).

Coradt, L. and T.J. Roper. 2005. Consensus decision making in animals. *Trends in Ecology and Evolution* 2(8): 449–456.

Cosentino, C. and D. Bates. 2012. *Feedback Control in Systems Biology*. Boca Raton, FL: Taylor & Francis Group.

Couzin, I.D. 2008. Collective cognition in animal groups. *Trends in Cognitive Science* 13(1): 36–43.

Couzin, I.D., J. Krause, N.R. Franks, and S.A. Levin. 2005. Effective leadership and decision-making in animal groups on the move. *Nature* 433: 513–516.

Csermely, P. 2006. *Weak Links: Stabilizers of Complex Systems from Proteins to Social Networks*. Heidelberg, Germany: Springer-Verlag.

Da Fontoura Costa, L., F.A. Rodrigues, G. Travieso, and P.R. Villas Boas. 2007. Characterization of complex networks: A survey of measurements. *Advances in Physics* 56: 167–242.

Dang, T., C. Le Guernic, and O. Maler. 2011. Computing reachable states for nonlinear biological models. *Theoretical Computer Science* 412(21): 2095–2107.

Danzl, P., J. Hespanha, and J. Moehlis. 2009. Event-based minimum-time control of oscillatory neuron models. Phase randomization, maximal spike rate increase, and desynchronization. *Biological Cybernetics* 101: 387–399.

DeLellis, P., M. di Bernardo, T.E. Gorochowski, and G. Russo. 2010. Synchronization and control of complex networks via contraction, adaptation and evolution. *IEEE Circuits and Systems Magazine* Third Quarter: 64–82.

Demongeot, J., J.-P. Françoise, and D. Nerini. 2009. Introduction to the special issue From biological and clinical experiments to mathematical models. *Philosophical Transactions of the Royal Society A* 367: 4657–4663.

Desai, R.C. and R. Kapral. 2009. *Dynamics of Self-Organized and Self-Assembled Structures*. Cambridge, U.K.: Cambridge University Press.

Dorogovtsev, S. 2010. *Lectures on Complex Networks*. Oxford, U.K.: Oxford University Press.

Dorogovtsev, S. and J.F.F. Mendes. 2003. *Evolution of Networks. From Biological Nets to the Internet and WWW*. Oxford, U.K.: Oxford University Press.

Doyle III, F.J., W. Bequette, R. Middleton, B. Ogunnaike, B. Paden, R.S. Parker, and M. Vidyasagar. 2011. Control in biological systems. In: *The Impact of Control Technology: Overview, Success Stories, and Research Challenges*, pp. 57–67.

Erdös, P. and A. Rényi. 1960. On the evolution of random graphs. *Public Mathematical Institute of Hungarian Academy of Sciences* 5: 167–256.

Feng, J., J. Jost, and M. Qian, eds. 2007. *Networks: From Biology to Theory*. New York: Springer-Verlag.

Furber, S. and A. Brown. 2011. Biologically-inspired massively-parallel architectures— Computing beyond a million processors. Whitepaper on the EPSRC-funded project SpiNNaker. Manchester, U.K.: The University of Manchester.

Furber, S. and S. Temple. 2007. Neural systems engineering. *Journal of the Royal Society Interface* 4(13): 193–206.

Galvão, V. and J.G.V. Miranda. 2008. A computational model for cancer growth by using complex networks. *Physica A* 387: 5279–5286.

Gao, J., S.V. Buldyrev, H.E. Stanley, and S. Havlin. 2012. Networks formed from interdependent networks. *Nature Physics* 8: 40–48.

Ghosh, R. and C. Tomlin. 2004. Symbolic reachable set computation of piecewise affine hybrid automata and its application to biological modelling: Delta-Notch protein signalling. *IEEE Transactions on Systems Biology* 1(1): 170–183.

Gong, H., Q. Wang, P. Zuliani, J. Faeder, M. Lotze, and E. Clarke. 2011. Symbolic model checking of signaling pathways in pancreatic cancer. In: *Proceedings of 3rd International Conference on Bioinformatics and Computational Biology, BICoB 2011* (New Orleans, LA, March 23–25).

Gros, C. 2011. *Complex and Adaptive Dynamical Systems: A Primer*, 2nd edn. Springer-Verlag.

Gross, T. and H. Sayama, eds. 2009. *Adaptive Networks: Theory, Models and Applications*. New York: Springer-Verlag.

Guéguen, H., M.-A. Lefebvre, J. Zaytoon, and O. Nasri. 2009. Safety verification and reachability analysis of hybrid systems. *Annual Reviews in Control* 33: 25–36.

Henzinger, T. 1996. The theory of hybrid automata. In: *Proceedings of the 11th IEEE Symposium of Logic in Computer Science* (New Brunswick, NJ, July 27–30), pp. 278–292.

Henzinger, T., P.-H. Ho, and H. Wong-Toi. 1998. Algorithmic analysis of nonlinear hybrid systems. *IEEE Transactions on Automatic Control* 43(4): 540–554.

Hodges, A. 1992. *Alan Turing: The Enigma*. London, U.K.: Vintage Books.

Howard, J., S.W. Grill, and J.S. Bois. 2011. Turing's next steps: The mechanochemical basis of morphogenesis. *Nature Reviews Molecular Cell Biology* 12: 392–398.

Hoy, R. 2004. New math for biology is the old new math. *Cell Biology Education* 3(Summer): 90–92.

Iglesias, P.A. and B.P. Ingalls, eds. 2010. *Control Theory and Systems Biology*. Cambridge, MA: The MIT Press.

Izhikevich, E.M. 2007. *Dynamical Systems in Neuroscience: The Geometry of Excitability and Bursting*. Cambridge, MA: The MIT Press.

Jennings, N.R. and S. Bussmann. 2003. Agent-based control systems. Why are they suited to engineering complex systems? *IEEE Control Systems Magazine* June: 61–73.

Jeong, H., B. Tombor, R. Albert, Z.N. Oltvai, and A.-L. Barabási. 2000. The large-scale organization of metabolic networks. *Nature* 407(October): 651–654.

Johansson, K.H., M. Egerstedt, J. Lygeros, and S. Sastry. 1999. On the regularization of Zeno hybrid automata. *Systems & Control Letters* 38: 141–150.

Jordan, P. 2003. All together now. *Nature* 421: 780–782.

Krabs, W. and S. Pickl. 2010. An optimal control problem in cancer chemotherapy. *Applied Mathematics and Computation* 217: 1117–1124.

Kwiatkowska, M., G. Norman, and D. Parker. 2008. Using probabilistic model checking in systems biology. *ACM SIGMETRICS Performance Evaluation Review* 35(4): 14–21.

Lincoln, P. and A. Tiwari. 2004. Symbolic systems biology: Hybrid modeling and analysis of biological networks. In: *Proceedings of the 7th International Conference on Hybrid Systems: Computation and Control*, Lecture notes in computer science, Vol. 2993 (R. Alur and G. Pappas, Eds.), Berlin, Germany, Springer-Verlag, pp. 660–672.

Liu, J., H.K. Khalil, and K.G. Oweiss. 2011a. Model-based analysis and control of a network of basal ganglia spiking neurons in the normal and Parkinsonian states. *Journal of Neural Engineering* 8: 045002.

Liu, Y.-Y., J.-J. Slotine, and A.-L. Barabási. 2011b. Controllability of complex networks. *Nature* 473: 167–173. See also supplementary information.

Lygeros, J., K.H. Johansson, S.N. Simić, J. Zhang, and S. Sastry. 2003. Dynamical properties of hybrid automata. *IEEE Transactions on Automatic Control* 48(1): 2–17.

Mailleret, L. and V. Lemesle. 2009. A note on semi-discrete modelling in the life sciences. *Philosophical Transactions of the Royal Society A. Mathematical, Physical & Engineering Sciences* 367: 4779–4799.

Maini, P.K., K.J. Paintera, and H.N.P. Chaub. 1997. Spatial pattern formation in chemical and biological systems. *Journal of the Chemical Society, Faraday Transactions* 93(20): 3601–3610.

Maler, O. 2002. Control from computer science. *Annual Reviews in Control* 26: 175–187.

Martínez, S., J. Cortés, and F. Bullo. 2007. Motion coordination with distributed information: Obtaining global behavior from local interaction. *IEEE Control Systems Magazine* 27(4): 75–88.

Matveev, A.S. and A.V. Savkin. 2000. *Qualitative Theory of Hybrid Dynamical Systems (Control and Engineering)*. Birkhäuser, Basel.

Matveev, A.S. and A.V. Savkin. 2002. Application of optimal control theory to analysis of cancer chemotherapy regimens. *Systems & Control Letters* 46: 311–321.

May, R.M. 2004. Uses and abuses of mathematics in biology. *Science* 303: 790–793.

Mesbahi, M. and M. Egerstedt. 2010. *Graph Theoretic Methods in Multiagent Networks*. Princeton, NJ: Princeton University Press.

Metin, S. and N.S. Sengör. 2010. Dynamical system approach in modelling addiction. In: *Brain Inspired Cognitive Systems, 4th International Symposium on Cognitive Neuroscience* (Madrid, Spain, July 14–16).

Middleton, R. 2011. Challenges and future directions. *IEEE Control Systems Magazine* August: 13–14.

Mikhailov, A.S. and V. Calenbuhr. 2002. *From Cells to Societies. Models of Complex Coherent Action*. New York: Springer-Verlag.

Mirollo, R.E. and S.H. Strogatz. 1990. Synchronization of pulse-coupled biological oscillators. *SIAM Journal of Applied Mathematics* 50(6): 1645–1662.

Mishra, B. 2009. Intelligently deciphering unintelligible designs: Algorithmic algebraic model checking in systems biology. *Journal of the Royal Society Interface* doi:10.1098/rsif.2008.0546. Published online.

Moody, J. 2001. Race, school integration, and friendship segregation in America. *American Journal of Sociology* 107(3): 679–716.

Moreau, L. 2005. Stability of multiagent systems with time-dependent communication links. *IEEE Transactions on Automatic Control*, 52(2): 169–182.

Mosterman, P.J. and G. Biswas. 1998. A theory of discontinuities in dynamic physical systems. *Journal of the Franklin Institute* 335B(3): 401–439.

Mosterman, P.J. and G. Biswas. 2000. A comprehensive methodology for building hybrid models of physical systems. *Artificial Intelligence* 121: 171–209.

Muller, D.E. 1963. Infinite sequences and finite machines. In: *Proceedings of the Fourth Annual Symposium on Switching Circuit Theory and Logical Design* (Chicago, IL, October 28–30), pp. 3–16.

Murray, J.D. 2002. *Mathematical Biology*. New York: Springer-Verlag.

Mysore, V.P. 2006. Algorithmic algebraic model checking: Hybrid automata and systems biology. PhD thesis. Department of Computer Science, New York University, New York.

Nakao, H. and A.S. Mikhailov. 2010. Turing patterns in network-organized activator-inhibitor systems. *Nature Physics* 6: 544–550.

Nanjundiah, V. 2005. Mathematics and biology. *Current Science* 88(3): 388–393.

Navarro-López, E.M. 2002. *Dissipativity and Passivity-Related Properties in Nonlinear Discrete-Time Systems*. Barcelona, Spain: Universitat Politècnica de Catalunya (UPC), Institute of Industrial and Control Engineering (IOC). ISBN: 84-688-1941-7.

Navarro-López, E.M. 2009. Hybrid-automaton models for simulating systems with sliding motion: Still a challenge. In: *3rd IFAC Conference on Analysis and Design of Hybrid Systems, ADHS 2009* (Zaragoza, Spain, September), pp. 322–327.

Navarro-López, E.M. and C. Carter. 2011. Hybrid automata: An insight into the discrete abstraction of discontinuous systems. *International Journal of Systems Science*, 42(11), 1883-1898, 2011, 1–16.

Navarro-López, E.M. and C. Carter. 2010. Languages spoken by dynamical behaviours: A new approach to hybrid control systems? In: *13th ACM International Conference on Hybrid Systems: Computation and Control, Cyber-Physical Systems Week 2010, WIP-Session* (Stockholm, Sweden, April 11–16).

Navarro-López, E.M. and D. Cortés. 2007. Avoiding harmful oscillations in a drillstring through dynamical analysis. *Journal of Sound and Vibration*, 307(1–2): 152–171.

Navarro-López, E.M. and A. Poliakov. 2011. Geo-behavioural interest networks. In: *International Conference on Network Science NetSci 2011* (Budapest, Hungary, June 6–10, 2011).

Navarro-López, E.M. and R. Suárez. 2004. Practical approach to modelling and controlling stick-slip oscillations in oilwell drillstrings. In: *IEEE Conference on Control Applications* (Taipei, Taiwan, September), pp. 1454–1460.

Newman, M.E.J. 2003. The structure and function of complex networks. *SIAM Review* 45(2): 167–256.

Newman, M.E.J. 2010. *Networks: An Introduction*. Oxford, U.K.: Oxford University Press.

Newman, M.E.J., A.-L. Barabási, and D.J. Watts. 2006. *The Structure and Dynamics of Networks*. Princeton, NJ: Princeton University Press.

Nishida, T. and S. Doshita. 1987. Reasoning about discontinuous change. In: *6th National Conference on Artificial Intelligence (AAAI-87)* (Seattle, WA, July), pp. 643–648.

Olfati-Saber, R. and R.M. Murray. 2004. Consensus problems in networks of agents with switching topology and time-delays. *IEEE Transactions on Automatic Control* 49(9): 1520–1533.

O'Toole, M. and E.M. Navarro-López. 2012. A hybrid automaton for a class of multi-Contact rigid-body systems with friction and impacts. In *4th IFAC Conference on Analysis and Design of Hybrid Systems* (Eindhoven, the Netherlands, June 6–8).

Paley, D.A., N.E. Leonard, R. Sepulchre, D. Grünbaum, and J.K. Parrish. 2007. Oscillator models and collective motion: Spatial patterns in the dynamics of engineered and biological networks. *IEEE Control Systems Magazine* 27(4): 89–105.

Poliakov, A., M. Cotrina, A. Pasini, and D. Wilkinson. 2008. Regulation of EphB2 activation and cell repulsion by feedback control of the MAPK pathway. *The Journal of Cell Biology* 183(5): 933–947.

Polynikis, A., S.J. Hogan, and M. di Bernardo. 2009. Comparing different ODE modelling approaches for gene regulatory networks. *Journal of Theoretical Biology* 261: 511–530.

Ren, W., R.W. Beard, and E.M. Atkins. 2007. Information consensus in multivehicle cooperative control. *IEEE Control Systems Magazine* April: 71–82.

Reynolds, C.W. 1987. Flocks, herds, and schools: A distributed behavioral model. *Computer Graphics* 21(4): 15–24.

Richards, B. 2005. Turing, Richards and morphogenesis. *The Rutherford Journal. The New Zealand Journal for the History and Philosophy of Science and Technology* 1. Available at http://www.rutherfordjournal.org/article010109.html.

Rosas i Casals, M. 2009. Topological complexity of the electricity transmission network. Implications in the sustainability paradigm. PhD thesis, Universitat Politècnica de Catalunya, Barcelona, Spain.

Samad, T. and A. Annaswamy, eds. 2011. The impact of control technology: Overview, success stories, and research challenges. Report published by the IEEE Control Systems Society, February, 2011. Available at www.ieeecss.org/main/IoCT-report.

Saunders, P.T. ed. 1992. *Morphogenesis. Collected Works of A.M. Turing*. Amsterdam, the Netherlands: North-Holland.

Sauro, H.M., A.M. Uhrmacher, D. Harel, M. Hucka, M. Kwiatkowska, P. Mendes, C.A. Shaffer, L. Strömback, and J.J. Tyson. 2006. Challenges for modeling and simulation methods in systems biology. In: *Proceedings of the IEEE Winter Simulation Conference*, (Monterey, CA, December 3-6), pp. 1720–1730.

Scardovi, L., M. Arcak, and E.D. Sontag. 2010. Synchronization of interconnected systems with applications to biochemical networks: An input-output approach. *IEEE Transactions on Automatic Control* 55(6): 1367–1379.

van der Schaft, A.J. and J.M. Schumacher. 2000. *An Introduction to Hybrid Dynamical Systems*. London, U.K.: Springer-Verlag.

Schweitzer, F. ed. 1997. *Self-Organization of Complex Structures. From Individual to Collective Dynamics*. Amsterdam, the Netherlands: Gordon and Breach Publishers.

Scirè, A., I. Tuval, and V.M. Eguíluz. 2005. Dynamic modeling of the electric transportation network. *Europhysics Letters* 71(2): 318–324.

Sontag, E.D. 2004. Some new directions in control theory inspired by systems biology. *Systems Biology* 1(1): 9–18.

Spencer, S.L., M.J. Berryman, J.A. García, and D. Abbott. 2004. An ordinary differential equation model for the multistep transformation to cancer. *Journal of Theoretical Biology* 231: 515–524.

Strogatz, S.H. 1994. *Nonlinear Dynamics and Chaos. With Applications to Physics, Biology, Chemistry, and Engineering*. New York: Perseus Books Publishing.

Strogatz, S.H. 2000. From Kuramoto to Crawford: Exploring the onset of synchronization in populations of coupled oscillators. *Physica D* 143: 1–20.

Strogatz, S.H. 2001. Exploring complex networks. *Nature* 268: 268–276.

Strogatz, S.H. 2003. *Sync: The Emerging Science of Spontaneous Order*. London, U.K.: Penguin Books.

Sumpter, D.J.T. 2006. The principles of collective animal behaviour. *Philosophical Transactions of the Royal Society B* 361: 5–22.

Sun, X., J. Lei, M. Perc, J. Kurths, and G. Chen. 2011. Burst synchronization transitions in a neuronal network of subnetworks. *Chaos* 21: 016110.

Swan, G.W. 1990. Role of optimal control theory in cancer chemotherapy. *Mathematical Biosciences* 101: 237–284.

Tabareau, N., J.-J. Slotine, and Q.-C. Pham. 2010. How synchronization protects from noise. *PLoS Computational Biology* 6(1): e1000637.

Tabuada, P. 2009. *Verification and Control of Hybrid Systems: A Symbolic Approach*. New York: Springer-Verlag.

Tanaka, G., K. Tsumoto, S. Tsuji, and K. Aihara. 2008. Bifurcation analysis on a hybrid systems model of intermittent hormonal therapy for prostate cancer. *Physica D* 237: 2616–2627.

Tanner, H.G., A. Jadbabaie, and G.J. Pappas. 2007. Flocking in fixed and switching networks. *IEEE Transactions on Automatic Control* 52(5): 863–868.

Taylor, W., Z. Katsimitsoulia, and A. Poliakov. 2011. Simulation of cell movement and interaction. *Journal of Bioinformatics and Computational Biology* 9(1): 91–110.

Tomlin, C., I. Mitchell, A. Bayen, and M. Oishi. 2003. Computational techniques for the verification of hybrid systems. *Proceedings of the IEEE* 91(7): 986–1001.

Thompson, D'Arcy W. 1917. *On Growth and Form*. Cambridge, U.K.: Cambridge University Press.

Turing, A.M. 1937. On computable numbers, with an application to the entscheidung-sproblem. *Proceedings of the London Mathematical Society* 43: 230–265. Also available at www.scribd.com/doc/2937039/Alan-M-Turing-On-Computable-Numbers.

Turing, A.M. 1948. Intelligent machinery. Research report, National Physical Laboratory (NPL), Teddington, London, U.K.

Turing, A.M. 1950. Computing machinery and intelligence. *Mind* 49: 433–460.

Turing, A.M. 1952. The chemical basis of morphogenesis. *Philosophical Transactions of the Royal Society of London. Series B, Biological Sciences* 237(641): 37–72.

Uhrmacher, A. and D. Weyns, eds. 2009. *Multi-Agent Systems: Simulation and Applications (Computational Analysis, Synthesis, and Design of Dynamic Systems)*. Boca Raton, FL: Taylor & Francis Group.

Vicsek, T. 2008. Universal patterns of collective motion from minimal models of flocking. In: *2nd IEEE International Conference on Self-Adaptive and Self-Organizing Systems*, (Venice, Italy, October 20-24), pp. 3–11.

Vicsek, T., A. Czirók, E. Ben-Jacob, I. Cohen, and O. Shochet. 1995. Novel type of phase transition in a system of self-driven particles. *Physical Review Letters* 75(6): 1226–1229.

Venayagamoorthy, G.K. 2009. EFRI-COPN: Neuroscience and neural networks for engineering the future intelligent electric power grid. NSF-funded project, Division of Emerging Frontiers in Research and Innovation.

Venayagamoorthy, G.K. 2011. Dynamic, stochastic, computational and scalable technologies for smart grids. *IEEE Computational Intelligence Magazine* August: 22–35.

Villoslada, P., L. Steinman, and S.E. Baranzini. 2009. Systems biology and its application to the understanding of neurological diseases. *Annals of Neurology* 65: 124–139.

Wang, X.F. and G. Chen. 2003. Complex networks: Small-world, scale-free and beyond. *IEEE Circuits and Systems Magazine* First Quarter: 7–20.

Watts, D.J. 1999. *Small Worlds: The Dynamics of Networks between Order and Randomness*. Princeton studies in complexity. Princeton, NJ: Princeton University Press.

Watts, D.J. and S.H. Strogatz. 1998. Collective dynamics of "small-world" networks. *Nature* 93: 440–442.

Wellstead, P. 2010. Systems biology and the spirit of Tustin. *IEEE Control Systems Magazine* February: 57–71, 102.

Wellstead, P., R. Middleton, and O. Wolkenhauer. 2008. Feedback medicine: Control systems concepts in personalised, predictive medicine and combinatorial intervention. Technical report, Hamilton Institute, Ireland.

Willems, J.C. 1972. Dissipative dynamical systems. Part I: General theory. *Archive for Rational Mechanics and Analysis* 45(5): 321–351.

Winfree, A.T. 1967. Biological rhythms and the behaviour of populations of coupled oscillators. *Journal of Theoretical Biology* 16: 15–42.

Witsenhausen, H. 1966. A class of hybrid-state continuous-time dynamic system. *IEEE Transactions on Automatic Control* AC-11(2): 161–167.

Wu, C.W. 2007. *Synchronization in Complex Networks of Nonlinear Dynamical Systems.* Singapore, World Scientific.

Yazdani, A. and P. Jeffrey. 2011. Complex network analysis of water distribution systems. *Chaos* 21: 016111.

Ye, P., E. Entcheva, S.A. Smolka, and R. Grosu. 2008. Modelling excitable cells using cycle-linear hybrid automata. *IET Systems Biology* 2(1): 24–32.

Yu, H., J. Wang, Q. Liu, J. Wen, B. Deng, and X. Wei. 2011. Chaotic phase synchronization in a modular neuronal network of small-world subnetworks. *Chaos* 21: 043125.

13

Computing for Models of the World

Steve Johnson

CONTENTS

13.1 Introduction

Humanity has thousands of years of practice in trying to understand the world around us. Every culture comes to an understanding that certain things matter and others do not. We pay attention to the things that matter and build them into models of our external world. At various times and places, we have believed that things happen because of the actions of gods, demons, four (or five) elements, four humors, Qi, chemical or mechanical actions, rational decision making, invisible economic hands, historical forces, heredity, environment, predestination, or sheer chance.

Each of these models of the world emphasizes some aspects of our experience and ignores or suppresses other aspects. We spend billions of dollars crafting and dispensing pharmaceuticals to improve our health and, for the most part, ignore the positions of planets in the sky and the influence of prayer. Several hundred years ago, it was quite different.

We also tend to take the models we use to explain the outside world and apply them to understanding ourselves and others. In particular, we borrow ideas and metaphors that describe our outside world and use them to describe our inner state. If we have an idea, we think of a lightbulb turning on—clearly, this was not the way people thought of ideas 200 years ago. We use terms from engineering such as *stress* and *strain* to describe our inner state. Medicine is full of military terms—we fight disease and have a war on cancer. In the 1960s, people *tuned in*, a metaphor taken from AM radios. People have emotional *meltdowns*, a vivid radioactive image.

The advent of the *age of computation* has given us new vocabulary and shared experiences to describe our world, both inner and outer. Now you hear people wanting to *reboot* a conversation, wanting more input, or calling out *blue screen* when an argument confuses them. But computing has initiated subtle and powerful changes in how we look at the world and ourselves.

This chapter explores several of these changes and their implications. The goal is to make our unconscious use of language and metaphor more conscious, so we can better focus on our inner and outer reality—to look beneath the words and metaphors we use to describe it. We will spend some time discussing the scientific method, now under subtle attack by the forces of *big data*. We examine how our computing experiences may be changing our model of natural processes. And we note that social networking has the potential to radically change the very society it is networking.

13.2 Scientific Method

For much of the previous 200 years, a pervasive part of our culture has been our belief in the scientific method. The scientific method rests on some presuppositions, often held unconsciously. One is a belief that to predict something, either external or internal, we must understand how it works. And understanding it implies making up hypotheses of causality (if X happens, then Y will happen) and then testing these hypotheses by experimentation in as objective a fashion as possible. If the outcomes of these experiments are consistent with the hypothesis, the hypothesis is commonly accepted as a fact.

Perhaps the most powerful presupposition is a belief that the universe *can* be understood and that gaining control over the universe, and ourselves, can only be achieved by understanding these laws of causality and applying them to our outer and inner experiences.

The power of the scientific method in areas such as physics and engineering is undoubted, and by the mid-twentieth century, the word *scientific* was considered almost a synonym for *true*. But as the scientific method became an ever deeper part of our culture, it was increasingly applied in areas where these presuppositions are very questionable. The key to accepting a hypothesis such as *eating chocolate causes zits* lies not just in observing situations where we eat chocolate and get zits. We must also observe those situations where we eat chocolate and do not get zits, and where we get zits but have not eaten chocolate. These negative results may result from error or suggest other factors influencing our latent zit population—in the extreme case, they may suggest that the zits' arrival has nothing at all to do with eating chocolate.

The outcome of the scientific method is commonly taken to be *facts*. However, the scientific method in action is surprisingly contrary. To *prove* that chocolate causes zits, we first assume that zits appearing have nothing to do with chocolate (the *null hypothesis*) and then gather data and seek to *reject* this hypothesis. By this backdoor reasoning, if we cannot show that chocolate does not cause zits, then chocolate does cause zits. Ideally, we set up a *controlled experiment*, where we look for zits in two very similar populations, one of which eats chocolate and one of which does not. The key to designing a good experiment is, to the greatest degree possible, controlling *all* relevant factors in the experiment, not just the factors being tested.

In many cases, establishing a perfectly controlled experiment is difficult or impossible (e.g., experiments involving people are difficult to control well). Also, as suggested earlier, our choice of *relevant* factors to control is based on cultural norms that change over time. Experiments have often excluded factors that are later found to be important. Broader cultural factors also influence which experiments can obtain financial backing or be published in reputable journals. Cultural norms have, at times, made it difficult to study the motion of the planets, heredity, and the spread of AIDS, to name a few.

Applying the scientific method to complicated systems (such as treatments for disease) requires statistical methods to analyze the experimental data. The most common statistical methods report the probability that the results observed were due to chance. In other words, if we run an experiment involving zits and chocolate, and in reality they have nothing to do with each other, what is the probability that we would have seen the outcome that we saw? If this probability is very small, say less than 1%, then the null hypothesis is very unlikely, and we conclude that chocolate causes zits.

The rapid growth of computing power since the 1960s has challenged both the scientific method and its assumptions. As an example, consider a nutritional study that collects information on how much of 300 common foods 1000 people ate in the course of a week and also collects 50 pieces of medical information about these people (blood pressure, cholesterol, number of zits, etc.). Using this data, we can test nearly 45,000 hypotheses of the form "If someone eats tofu, they also eat broccoli," and the same number of hypotheses of the form "If someone eats Big Macs, they do not eat caviar." And we can also test another 30,000 hypotheses of the form "if someone eats sardines, they have low (or high) blood sugar." Applying the standard statistical tests, roughly 1200 of these hypotheses will appear to be correct at the 1% level *by chance alone*.

Before the advent of computers, it would have been a nearly impossible task to perform all these statistical analyses, but today they can be completed in seconds. Of course, in the pure scientific method, you really should make up the hypothesis before you collect the data, but the ability to examine all this data (often called *data mining*) is irresistible and sometimes even revealing. However, the purity of the scientific method is bent if not broken when this is done, because the assumptions that underlay these statistical techniques are violated.

A more subtle challenge to the assumptions of the scientific method arises from the huge amount of data that has become available over the past decade. As mentioned earlier, gaining control over our lives by understanding the fundamental laws of causality is a powerful presupposition of the scientific method. But computing has shown us another way to predict using data analysis instead of understanding.

Consider the problem of deciding what the weather will be tomorrow. In the past, weather forecasters spent decades painstakingly gathering information and developing models that included concepts like cold fronts, cloud cover, and temperature inversions. As the amount of available data increased, these models slowly became more precise, leading to dramatic improvements in the ability to predict tomorrow's weather.

But modern computers also allow us to do something quite different. Suppose we had an online database containing hourly readings of temperature, pressure, wind speed, and rainfall for the last 200 years. Then we might choose to predict tomorrow's weather by scanning this data to find a time in the past that looked a lot like today and see what happened on the next day. Note that we can use this data to predict tomorrow's weather without developing or validating any model involving cold fronts, etc., without requiring any understanding of the physics of the underlying processes, indeed without even having to believe that there are underlying processes at all. This method of forecasting does not depend on understanding the physics of the atmosphere—we could even apply the same method if we believed that the weather resulted from the actions of the (hopefully consistent) god Thor.

Because the focus is on getting results, we really have no hypotheses to test, no null hypotheses to reject. As mentioned earlier, when studying complex issues like ecology, human behavior, or weather, it is difficult or impossible (and sometimes unethical!) to perform controlled experiments. Using databases, we suddenly have the ability to predict behavior without *any* model or theory whatsoever (except perhaps a weak one such as "past performance predicts future performance").

There have been some spectacular successes with this technique, most notably in machine language translation, where decades of techniques that involved studying and analyzing languages, grammar patterns, parts of speech, and the structure of meaning were left in the dust by database techniques rather like those described earlier.

The deeper implications of this approach are startling. Taken to an extreme, there appears to be not much reason to study subjects such as linguistics (at least not much practical reason). Why spend time learning theories and techniques that clearly do not get to the true heart of understanding? Some teams working on machine translation have boasted that they have written translation programs without having anyone on the team who actually spoke the source language. Instead, the translation technology was based on a huge data base of existing translations, a computerized analog of the Rosetta Stone.

Databases, by their nature, are historical. Will having a huge database of language online slow down the evolution of language? Just as the invention of printing led to a standardization of spelling, will these huge linguistic databases lead to a standardization of thought? Will new words and idioms (such as LOL) be unable to enter the language because they are not already in the data base? Will new ideas fail to gain traction because there are no existing examples that would let it be translated? More generally, if the way we control our environment comes to rest on comparing the present against the past to predict the future, will the unpredictable and unexpected begin to feel upsetting and sinister?

13.3 Abstraction

Even over the course of my life, the need to remember things has diminished greatly. Most of us no longer need to remember multiplication tables, phone numbers, addresses, travel directions, and how to spell most words. We have small, portable devices that take care of this for us. Soon, with augmented reality, we may no longer need to remember names and faces, locations of stores and other services, business facts, rules and regulations, etc. This is a process that has been going on since the development of writing freed us from having to remember everything aurally, but it has accelerated greatly in the past couple of decades.

On the one hand, reducing the burden of memory frees us to *live in the moment*. But, on the other hand, it may be robbing us of something. We seem to have an inherent ability to abstract from specific experiences, to develop rules that distill these experiences. Much of our ability to learn language may be inherent in our neurology, where we learn to speak sentences that we have never heard before by abstracting from our previous linguistic experience.

In today's culture, these rules help us organize our experiences and guide our future behavior. Soon, *personal assistant* devices might instantly present a huge body of such knowledge to us, much of this information that we never directly experienced, and may do so even before we are even aware we need it. Will we still need an ability to abstract from our experiences? Will we, over time, lose this ability? It seems strangely paradoxical—we may gain instant access to a huge mass of human experiences, but thereby lose by atrophy the processes we use to organize our personal experiences.

Pushing this to the extreme is instructive. Perhaps children of the future will have personal assistants that learn their own personal language dialect and translate the thoughts of others into words they can understand. In fact, why would they need to learn others' languages at all, when each child could have their own, talking to others through their machines? Indeed, they might never have to experience an idea that they did not first think of themselves. Difficult, elusive concepts such as justice, empathy, fairness, and love might just be edited out of the experience of those people who did not happen to spontaneously discover them. Would these personal assistants challenge us to confront, tolerate, and accept ideas that differ from our own, or would they bowdlerize these ideas into something that we would find more comfortable?

Would this world be some hideous mating of *The Matrix* and 1984? Or would it be a world where we could be totally aware of each moment, our awareness informed and enriched by the wisdom of the ages?

13.4 Social Networks

Today, we are bombarded with information about the opinions of others. Facebook, Twitter, and Yelp tell us a lot about other people's opinions and actions almost instantly. We are engaged in a brave experiment to see what this ability does to our personal opinions and actions. If our favorite restaurant gets only two stars on Yelp, will we start to find the food and service less pleasing? If our friends like a certain movie, are we more likely to go see it? If they like it, are we more likely to like it, or at least say we do?

Of course, we have always been influenced by those around us. But it would have taken us weeks, or months, to collect the data we can access in seconds using social media. Will these media make it harder to be a nonconformist, even in small ways? The recent concern about cyber-bullying suggests that this might be so. Will we find ourselves not only prisoners of the past, but also prisoners of the present?

13.5 Atomism of Process

Another legacy of the computer revolution is a concept that might be called *atomism of process*, by analogy with the atomic theory of matter. The atomic theory of matter was based on the idea that you could divide matter into smaller and smaller pieces, but eventually you came to a piece of matter, an *atom*, that could not be divided further. The atomic theory of matter was instrumental in the development of chemistry and physics, although our current view is quite different than that of the ancient Greeks.

The atomism of process is based on similar reasoning—the notion that every process can be broken down into smaller and smaller processes that eventually end in atomic yes/no decisions based on data. This is the fundamental process behind writing computer programs. And indeed, programmers have been able to take this notion and apply it very successfully to many practical problems.

Note that the atomism of process shares one of the assumptions of the scientific method—that the universe is constructed using processes that are knowable and predictable and based in the physical world. In its purest form, sometimes called materialist reductionism, there is no place for God, soul, or anything else that cannot be based on physical data—what we see is all there is.

It is not too surprising that the atomism of process could be used to allow computers to play championship chess. It is somewhat more surprising that computers can do a rather good job of recognizing faces and speech, and we are now seeing computers driving cars and playing Jeopardy.

How far can we push this idea? Can *every* human process be broken down into atoms of process and replicated in a computer, given enough computing power? Can computers develop the ability to fall in love, write inspiring poetry and music, worship God, become a prophet of God, be jealous, or seek revenge? Are there no limits? If there are, what are they, and what, if anything, lies beyond them?

In the nineteenth century, people were amazed at the ability of a few individuals to do complicated compound interest computations in their heads. When I was in school, I was drilled in the ability to add two two-digit numbers in my head. Those of us who mastered this skill were considered *smart*. Today computers do these computations in nanoseconds.

It seems likely that computers will take over and dominate those activities that can yield to the atomism of process. In 50 or a 100 years, we may look back in amazement at the idea that humans actually drove cars and flew airplanes, when using computers is so much safer!

Is there a limit to this progression? Will we come to particularly value *human* qualities such as empathy, intuition, teamwork, and spirituality precisely because they are immune to being atomized and packaged into computers? Or will we come to think of ourselves as just slow messy error-prone flawed computers and feel the same kind of reverent awe for the best computers that earlier generations felt for Einstein or Gandhi? Wait and see …

Part IV

Citizen and Artisan Computing

14

AD-OPERA: Music-Inspired Self-Adaptive Systems

Paola Inverardi, Patrizio Pelliccione, Michelangelo Lupone, and Alessio Gabriele

CONTENTS

14.1 Introduction

Ubiquitous computing, as envisioned by Mark Weiser [37], promotes the view of systems as pervasive agglomerations of software, hardware, and people. We see applications enabled by ubiquitous computing replacing traditional systems that arose as the result of human activity. Ubiquitous applications are highly distributed, heterogeneous, mobile, and adaptive to both user necessities and operational environments. While recent research has produced useful results from programming down to physical layers, engineering approaches to effectively and efficiently support the development and execution of self-adaptive applications are still in their infancy. We envision a new computing paradigm that considers change as the norm rather than the exception. AD-OPERA will propose radically new approaches to SE, based on the investigation of, and inspiration derived from, a specific type of music that adapts itself in relation to the surrounding environment and to its history: *adaptive music* (AM).

We are referring to music that can be only computer generated. AM is able to adapt its behavior according to audience interaction and environment changes while still maintaining stylistic coherence. In other words, the style of the music will be maintained despite of adaptation: Style is an invariant of the music opera. AM aims at radically changing the traditional perception of a piece of music as a frozen work of art that eternally re-proposes its original conception. Therefore, AM promotes a perception of music as a *continuous present*, around an ever-ongoing dialog with the audience and with the environment. The audience and the environment play an active role in the composition process since the work of art is aware of them and adapts to their interactions and behavior.

Because of these characteristics of AM, we consider the study of AM as paradigmatic for visionary future systems. Frontier researches in AM will open new and unprecedented perspectives to the information technology world. We expect that this rethink of music, thanks to its intrinsic creativity, will generate a broad-mindedness that will lead to shape new theories and practices in ICT. Among the fine arts, music is characterized by a rigid structure that often offers mathematical interpretations. This makes it possible to establish a convergence between ICT and music that broadens the horizon of ICT by giving opportunity of technological developments and providing social benefits. From the music point of view, the structure of the sound becomes material for composition just as importantly as the pitches, the rhythm, and the dynamics are. The composer, free from limitations of both performer and acoustic instruments, puts in scene a completely new and fascinating sonorous dramaturgy that is loaded with fresh sensations. Similar to the revolution that architecture experienced with the use of new materials such as glass and ferroconcrete, the discovery and use of new materials, such as the structure of the sound, promote in music the creation of forms previously unimaginable. The music can then be perceived and experienced in a different manner through experimentation with new emotions and suggestions.

In AM, adaptation is performed according to a defined *musical score* in response to context variations; context variations could be captured, for example, by sensing human behavior or by sensing the perturbation in the environment produced by a water fountain. The musical score describes the temporal evolution of the system, and, more specifically, it contains the information that are used to reproduce the music opera. However, the score may easily be incomplete because the composer may have left some aspects purposely under-specified. Such aspects must be supplemented by adaptations at performance time, that is, when the score is interpreted. Furthermore, all the adaptations must be performed consistently with the composer's *style*. The style is the peculiar behavior of the elements that constitute the music; it is defined by the subjective rules and exceptions that the composer uses to manage the development in time of sound events and/or associated parameters. Style can be conceived as a distinctive and multidimensional quality of operas. Often, these different dimensions can be interrelated, and tradeoffs among different dimensions are to be expected. Examples of such dimensions are time and the quantitative characterization of pitch. Style imposes expressive coherence over time and typically imposes constraints on how goals in the score can be achieved. In other words, *the style is the key concept to define the boundaries of adaptiveness.*

We call *empathic systems* the specific class of self-adaptive ICT systems that share distinctive characteristics with AM and that will be the target of AD-OPERA. Self-adaptive systems have been extensively researched by the computing community. Empathic systems can have *goals* that change and evolve during the system's lifetime and that persist throughout the system's lifetime; these goals can be interrelated in various manners. The *causes* of change can be both internal and external and can be of different types. Moreover, the system behavior may *change* according to the (unplanned) context changes. The *consequences of adaptation* should be predictable since adaptations should preserve suitable invariants, that is, the style. Predictability covers qualitative and quantitative dimensions.

Empathic systems are heterogeneous in the sense that they integrate sensors, user interfaces, and devices that are able to perform complex computations. The context-awareness of empathic systems encompasses both environmental sensing (including user interaction) and system sensing (system resources and network), by means of a complex sensor network. This network must be able to interactively and selectively collect information. Empathic systems are characterized by a permanent and unavoidable user involvement: users serve the dual role of both suppliers and consumers of content and services. Empathic systems call for a radical change in the manner in which they are developed and executed. It is necessary to rethink the entire system life cycle process, from design, to validation, to execution. The traditional divide among static phases and dynamic ones becomes blurred, thus requiring new specification formalisms, new computing paradigms, and new validation techniques.

Current approaches assume that the environment in which applications operate is static (or changes slowly) and changes can be largely anticipated, so that the external context and the structure of the application, its components, and their relations and bindings can be defined at design time. Most of these assumptions are no longer true in the ubiquitous computing landscape. The usual closed-world assumption utilized when developing and validating a system cannot be adopted for AD-OPERA. Empathic systems are open systems whose definition becomes complete only at execution time, and they evolve by means of continuous adaptations throughout their lifetime. At design time, everything neither can be defined, nor should it be. Parts of the system specification can be given at a high level of abstraction, which describes only the essential *necessity*. They characterize the portion of the composition/system that the composer/designer leaves purposely *under-specified*. These parts will be concretized only at runtime as the effect of a continuous interaction of the system with its context, which includes the user. The completion/adaptation of the system must be performed by preserving the style/system invariant properties, thus implying a verification and validation step at execution time. This means that *a suitable engineering process must explicitly take into account complex verification and validation* (V&V) *steps at runtime,* when all the necessary pieces of information are available.

The chapter is organized as follows: Section 14.2 introduces adaptive music and Section 14.3 introduces adaptive systems. Section 14.4 discusses Ad-Opera challenges and describes the research methodology. Finally, Section 14.5 discusses concluding remarks.

14.2 Adaptive Music

The process of music composition has been subjected to fundamental transformations in recent decades, since the attention of composers has widened from the *mere* writing of a score to the formalization of the process, which leads to the definition of one among many possible scores. To clarify this concept, we can cite musical works that belong to the classical tradition. The first one is Mozart's *Musikalisches Würfelspiel* in which the parts are generated measure by measure by rolling dice to pick randomly from a list of multiple potential versions. The number of potential combinations is so large that any single instance of the waltz has probably never been heard and neither will be unless you write it down. The other example is Stockhausen's *Klavierstück XI* written in 1956. The piece is composed of 19 fragments spread over a single, large page. The performer may start playing any fragment and continue to any other, proceeding through the labyrinth until a fragment has been reached for the third time, when the performance ends. The compositional intention has not focused on one predetermined version of the score, but rather on the rules of a process, which lead to a definite outcome. Many other examples may be cited, as Morton Feldman's *Intermission 6* for one or two pianos of 1953, Earl Brown's first open-form piece *Twenty-five Pages*, or Bruno Maderna's *Serenata per un satellite* of 1969.

The affirmation of *computer music** has led to a substantial broadening of the idea of instrument and of the music generation. The general availability of powerful and automatic tools affects both how sound is generated, and the rules and processes on which composition relies. The user is strongly involved since his intervention may affect one or more parameters of the sound generated. The composer may also decide to use environmental figures as inputs to his work [12], since the plurality of sensor devices

---

* Refer to www.computermusic.org, www.mitpressjournals.org/cmj, or www.leonardo.info/lmj/

and connections available today allows him to consider the brightness of the room, or traffic flow elsewhere, as relevant to his invention. No matter what the interaction with the context, the key fact is that it is still the composer who is in charge of defining the behavior of the system over time, or, in other words, of defining the score. The score may possibly be non-linear, allowing for branches and diversions at certain places, but *the key point is that for an interactive opera, the entire plot is predetermined by the composer at composition time*. This leads us to discuss adaptive music.

In recent years, the term *adaptive music* has appeared, with a meaning drawn from the entertainment domain, interactive music. *Interactive music* [27] in the gaming field has the main concerns to produce a musical outcome that preserves continuity and relevance. Many implementations are built from a system that basically picks up musical fragments from a large database, arranges them according to the inputs received from the gaming engine, and finally synthesizes music. This approach becomes inadequate as soon as composition is employed by the composer to create music forms. We define adaptive music* [28] as a music that adapts itself in relation to the surrounding environment and past behavior and coherently generates its sounds.

A first prototype of an AM opera was realized in 2006 by Michelangelo Lupone for the work "Volumi Adattivi," in collaboration with visual artist Licia Galizia[†] and realized by Centro Ricerche Musicali—CRM of Rome. It is shown in Figure 14.1A and composed of different electronic devices made out of different materials. The audience can interact by moving and touching the mobile parts of the installation. After a while, if the audience is not interacting, the installation initiates a different process of evolution, focused on sensing the environment instead of the mobile parts. Thanks to this experience, the composer today can better define the composition rules, avoiding the risk of an acoustic degeneration of the artwork. Also remarkable in these attempts, and also beyond a simple interactivity, is that the work responses are always different, because of certain timing and environmental circumstances. It is absolutely essential in a research activity—melting together different expertises and technical domains—to achieve the implementation of a learning ability of the work, conditioning the reaction of the artwork to a previous history—also distant in time. This implies an auto-analysis ability.

Figure 14.1B shows an installation referred to as "Musica in Forma" that consists of two sculptural-musical works: "Trio plastico" and "In coro" (Lupone-Galizia, 2008; production CRM). The innovative conception of these works consists in the complete integration of the musical and the plastic form: the music score is in fact based on the timbres and pitches

---

* http://www.crm-music.it/it/risorse/documenti/doc_download/59-interactive-adaptive-evolutive-lupone
† http://www.youtube.com/watch?v=CTnhMC0EHSY

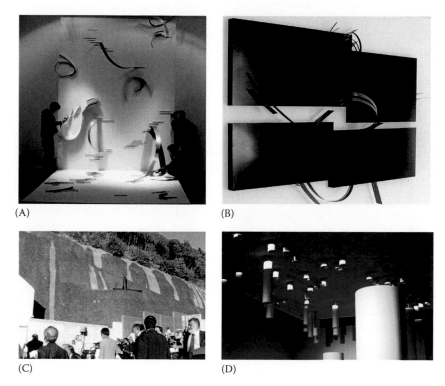

(A)   (B)

(C)   (D)

FIGURE 14.1
Examples of already realized adaptive works of art, which integrate music and plastic form ((A) "Volumi Adattivi," Lupone—Galizia, (B) "Musica in forma", Lupone-Galizia; (C) "Sorgenti Nascoste," permanent work of art, music, and sculpture, for the hydrogeological site in the Mountain Pizzuto—Solopaca, Italy, Lupone-Paladino; (D) "Spazio 1N," Gabriele).

generated by the forms when they are in vibration or are made to resonate with appropriate electronic devices, referred as Planephones® (CRM, 1998). Figure 14.1C shows an installation referred as "Sorgenti Nascoste" (Lupone-Paladino, 2007; production CRM), which is inspired by continuous modulations of water. The four seasons cause four major musical mutations by means of changes in timbre, pitch, and rhythm. Music evolution is influenced by the different climatic conditions and light. Finally, Figure 14.1D shows "Spazio 1N" (Gabriele, 2008 production Conservatory of Music L'Aquila), an installation that observes the audience by means of an optic and electronic surveillance system. Based on the observations, the reaction of the installation is a modulation of the sonorous space of the environment.

Another important goal of the research lies in the correlation among the data coming from every type of sensors. In the installations shown in Figure 14.1, the correlation of input data was based on a simple hierarchical principle of priority. The next step is a non-linear correlation among data flows to gain a behavior of the artwork approximating the ones of a living organism.

The examples described earlier are very different from the many implementations built from systems that simply retrieve musical fragments from a large database, arrange them according to the inputs received from a gaming engine, and finally synthesize music. The *interactive music* definition is not satisfactory for at least two reasons. The first is that this definition applies just to the gaming field, where the main concern is to have a musical outcome that preserves continuity and relevance.

Thus, the adaptive system of the definition is closer to a complex, automated interactive system—as Stockhausen's *Klavierstück XI*—as Mozart's *Musikalisches Wyrfelspiel*—than to a compositional system. Another reason is connected with control theory, where the term *adaptive* refers to the ability of a system to preserve an invariant in the face of changing conditions.

AM is synthesized by a computer through a dynamic process that constantly renovates itself. The legacy of forms and sounds already generated is not erased but archived and reprocessed to establish an immanent relationship with the listener, to focus on sense sharing and to stimulate the imagination. AM is capable of evolving as a living organism. The process involves the conception of an *open* musical form. The work is available for a dialogue with the user but possesses strong self-analysis processes and selective abilities to discriminate useless or prejudicial inputs. The term *adaptive* identifies a system that is *self-setting* in relation to both the actions of the user and the conditions of the surrounding environment. An adaptive work receives and/or detects the impulses arriving from outside and modifies its condition and its responses in an unpredictable or *partly* predictable manner. The user receives responses that take into account not only his current action, but also the sequence of his previous actions and of the entire environmental context. The system is to some extent able to *learn* and to *adapt itself* to the conditions around it. Typically, in these works, the user is free to choose the sequence of actions to realize, but the results are unrepeatable, in the sense that an identical sequence of actions at different times is not followed by an identical result.

Since the musical work is fully defined only at *runtime*, we speak of an interactive opera. The real challenge of AM is that the composer may leave parts of the composition *abstract*, by delegating their *concretizing* to the performer, to the listener, or to the environment. Adaptation of the music to audience interaction and to environment changes as well as the generation of sound materials is performed by necessarily preserving an expressive coherence. To achieve this, a composer defines his own *style*, that is, a set of constraints and rules regarding the sonorous matter, its temporal development, received input, and produced output. Examples of such dimensions are time, rhythm measure, and the quantitative characterization of pitch, etc.v

14.3 Adaptive Systems

In highly dynamic environments, it becomes impossible to manually manage tasks such as (re)configuration, maintenance, optimization, protection, or recovery. The central idea of autonomic computing [21] is self-management of software components. The initial four properties, collectively referred to as self-* properties, are self-configuration, self-healing, self-optimization, and self-protection [24]. In self-configuration, components install, set up, and integrate themselves, based solely on some high-level policies. In self-healing, internal, undesirable situations are automatically discovered and recovered from. In self-optimization, the system status is monitored and adjusted to increase performance. Finally, in self-protection, external threats are detected and mitigated [38]. The conceptual architecture for autonomic computing envisions an autonomic manager observing and controlling a managed element, thereby creating an autonomic element [1,22]. The autonomic manager consists of five key components, namely, monitoring, analysis, planning, and execution, all of which rely on a common base of knowledge (i.e., the MAPE-K cycle), forming a feedback loop. These five aspects are fundamental to any autonomous system, albeit some works assign different names to these steps [18,30]. Although different SE techniques have been studied to address constantly changing landscapes, these initiatives are mainly confined to specific research areas: software architectures (SAs) [11,16,20], middleware [26,34], component-based development [32], service-oriented systems [17], etc.

SAs are the most appropriate abstraction to use to reason about adaptability and to make adaptation effective and successful [20,25]. There is a wealth of architecture description languages and architectural notations that provide support for dynamic SA analysis [14]. The work in [20] introduces an SA-based self-adaptation framework, referred as Rainbow, which uses external mechanisms and an SA model to monitor and adapt a managed system. The work in [8] formalizes architectural style as a typed graph transformation system, while [15] uses algebraic graph transformations and [16] uses typed (hyper) graph grammars for representing and verifying self-adaptive systems. However, providing cost-effective techniques to address unplanned adaptation is still an open problem [22]. Moreover, architectural choices, and constraints and partial proofs performed at the architecture level, have to be reflected to the implementation so as to ensure safe adaptations, that is, adaptations that preserve desired properties, considered as invariants.

Adaptation in AD-OPERA must address efficient use of resources. Different approaches aiming to (over-)estimate resource consumption have been formulated. The MRG Project [2] proposes a framework for guaranteeing that programs do not violate resource bounds at runtime, work that was continued in the Mobius project [10]. In [36], a framework is presented that can estimate energy consumption of a distributed Java-based system. The worst-case execution-time problem has long been studied in the literature (e.g., [39]). In [4,29], there is a resource model and a static analysis approach to estimate Java resource consumption.

Adaptation in AD-OPERA must deal also with user preferences. This makes AD-OPERA strictly related to ongoing research in the field of services, where adaptation is necessary to ensure that users experience the best quality of service under changing conditions. In the field of adaptable services, different approaches have been formulated, often relying on user annotations. For instance, [31] proposes an approach that uses Java annotations to express metadata necessary to enable dynamic adaptation. Beyond this, [35] proposes a planning middleware- and architectural-based approach that supports dynamic service adaptation and re-configuration, relying on user-defined metadata. In [3,4], an approach is proposed for context-aware adaptive services supported by formal resource, service level specification models, and analysis. Adaptation happens at discovery time, so services are customized with respect to the context at binding time, but at runtime, they become frozen.

It is well known that there is a tradeoff between cost and predictability. The more dependable the system to develop, the more costly the development and the execution. So far the notion of cost has been driven by business criteria and social hazards severity. AD-OPERA deals with systems that are pieces of arts for which different notions of cost and possibly of dependability shall be devised. Accordingly, also the concept of process phases must be reconsidered. The latter is also demanded by the evolutionary nature of AM and by the fact that the audience is not a passive user but may become an active co-producer of the work of art.

AD-OPERA aims at providing a coordinated strategy that covers the entire life cycle within one large ambitious action. The future software process we envisage gives first-class importance to uncertainty and asks for explicit (formal) characterizations of software properties and assumptions for perpetual evolution. AD-OPERA will advance beyond the state of the art by providing an adaptation model able to support style-driven adaptation. This model must be suitable for efficient verification and validation that play an important role in regulating adaptability and evolution. This methodology promotes the use of verification and validation, which is typically confined to the critical systems domain, to every application domain.

The AD-OPERA framework is based on (i) a theoretical assume-guarantee framework that is presented in [22] to efficiently define conditions for adaptation at runtime while preserving a desired invariant, and (ii) a first-order logical modeling approach to describe evolvable computational systems that is presented in [9]. These formats are hierarchically composed of components that have associated supervisory processes that monitor and change the associated component as necessary. Supervisory processes monitor and adapt their dependent components, enabling evolutionary behavior to become integrated in system design.

14.4 Ad-Opera Challenges and Research Methodology

Ad-Opera's core challenge is to effectively deal with the interdisciplinarity that requires experts of ICT and artists to cooperate to significantly progress the state of the art in their respective domains (see Figure 14.2).

In the music domain, Ad-Opera will radically change the traditional perception of music by promoting dynamic self-expression. The result of Ad-Opera will enable artists/musicians to concentrate on the creative aspect of their work giving them resources to follow new manners of expression and the chance to think about advanced forms of fruition and interaction between users and the opera. AM will empower the relationship man–environment–artwork to improve the quality of life, stimulating processes of conscience, memory, comparison, and prediction. The active fruition of the musical work, characteristic of AM, will stimulate imagination, develop new manners of communication between individuals, and promote new avenues to explore how music can be brought to bear in a beneficial manner. The integration of music to dynamic, social, urban, environmental, and virtual contexts will welcome, accompany, and gratify human existence with high and civilized expressions.

The investigation into AM will provide radically new SE methodologies for adaptive applications in ubiquitous environments. Self-adaptive systems have been studied for decades within many different research communities in terms of both the basic self-adaptive mechanisms, as control loop, and the application areas and technologies, such as robotics and AI. The ubiquitous computing landscape has put forward the interest in self-adaptive systems. The challenge of self-adaptive systems is on defining cost-effective engineering techniques able to work in the presence of predictable degrees of dependability. In recent years, the SE community has recognized the relevance of such a challenge. The approaches proposed

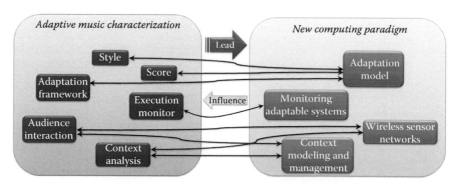

FIGURE 14.2
Transformative music-driven research.

so far follow quite a traditional path by attempting to extend/improve established SE methodologies, notations, verification and validation techniques, and supporting infrastructure to this challenging domain. All these attempts show their limits and partially work only in specific application domains.

We expect to change how an adaptive application is developed and utilized by promoting changes as the norm rather than exception but still providing support for system dependability. This process will contribute to the vision of an adaptive system as a well-behaved living organism that is capable of taking into account the unpredictable changes that occur in its context but still preserving the essential qualitative and quantitative properties expressed by the developer. AD-OPERA aims at providing support for a *cost*-effective principled development and management of empathic systems. It is important to notice that the notion of *cost* for empathic systems encompasses besides the usual business and technological dimensions, human-related concerns such as satisfaction, emotions, and creativity of the users. This aspect is increasingly important in an ever more user-centric world characterized by heterogeneous devices and ubiquitous networking, where systems are able to adapt to system changes and user requirements are a key paradigm in present and future integrated systems, such as collaborative systems, social networks, new media content based systems, games, and so forth. Therefore, AD-OPERA technologies will significantly extend the utilization and the creation of innovative empathic applications for urban and public contexts, for improving museum and archaeological sites, for multimedia applications, for games that enable user personalization according to mood, and many others.

As outlined in Figure 14.2, we believe that the study of a general framework and process that can be employed to compose and execute AM based on a formal notation to express the score and the style will create fresh synergies and major breakthroughs in different ICT research areas. Our research strategy is of a transformational nature based on an experimental approach that will put ICT experts in different ICT fields and musicians at work together. The basic common workbench will be the development of an AM piece of art.

As shown in Figure 14.3, the central cogwheel of the interdisciplinary research of AD-OPERA is the AM framework that moves the cogwheels of the ICT research. The resulting new field of empathic system engineering will produce empathic systems; examples of these systems domains are social networking, content provisioning, artistic applications, school and museum applications, etc.

The main contributions of AD-OPERA are a framework for AM; research on new forms of music composition; new SE models and methods for empathic systems, algorithms, and mechanisms for managing the adaptation; and the new discipline of empathic system engineering with its own concepts and practices. In the following we detail each of them.

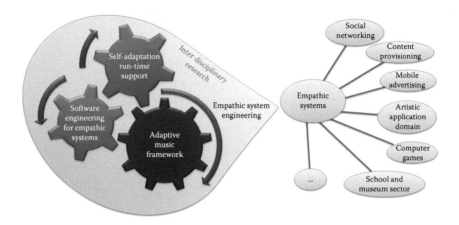

FIGURE 14.3
Overall strategy of the workplan.

14.4.1 Adaptive Music Framework

Investigating AM and realizing a framework for developing AM. The research on AM will move the cogwheels of the ICT research: inspiration derived from AM will be exploited in ICT. Both ICT researchers and musicians are called to collaborate within this research. There will be two intertwining threads of activities: on the experimental side and on the foundational one. The experimental side shall aim at the development of a *show case* of an AM work while the foundational one shall investigate and realize a framework for AM.

In order to define a framework for AM, the first step is to study and define the concepts of *score* and *style*, as well as formalisms to express them.

Score can be defined as the ordered set of symbols and/or graphic, text, and numeric elements, appropriate for a representation in time of sound events and/or their production and/or of parameters associated to them (e.g., the score of Water Walk by John Cage is a representation with graphic and text elements of the sound event production).

In AD-OPERA, the score must find another further meaning. It cannot just "represent in time the sound event, the sound production or variations of his parameters." In an adaptive opera, the sound is the impromptu result of many logical and computational processes and depends on environment status, the interaction with humans, and the analysis of current status in relation to previous sound conditions. An opera is non-deterministic or partially deterministic and capable of evolution; in other words, the work responds in a non-predictable manner, but the adaptation is coherent to behavior rules informing every constitutive process.

The *style* of a music work is the peculiar behavior of the elements constituting the music; it is defined by the subjective rules and exceptions that the

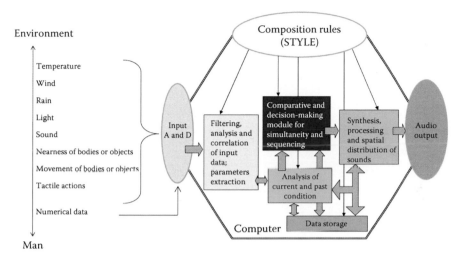

FIGURE 14.4
Adaptive music work (M. Lupone, design of music architecture of the permanent art sound installation "Gioco delle risonanze," at the Pompei ruins, 2010).

composer chooses to manage the development in time of sound events and/ or associated parameters.

Figure 14.4 shows an example of AM work design. Specifically, it is the design of music architecture of the permanent art sound installation "Gioco delle risonanze," at the Pompei ruins (2010).

The framework for AM has the aim to develop four computational areas (Figure 14.5 shows a preliminary architecture): *behavior, execution, environment,* and *control.* The behavior is determined according to the score, the style, and the other compositional elements specified for a determined opera (i.e., audio files, preset libraries, etc.). The *Opera unit* edits and manages the formal description of the compositional rules and style rules by means of a specific programming language for which a syntactic and semantic analysis is necessary. The *decisional unit* reflects and acts on the actual state of the system. The *decisional unit* must validate requests for adaptation, referring to the style invariants and actual system state.

The *execution unit* translates the score directives (compositional rules) into controls for the *synthesis unit,* according to the actual behavioral necessities for the opera (style rules). The *execution unit* organizes and schedules the audio events and control data to be processed. The *synthesis unit* provides for a programmable sound and music synthesis system. The *synthesis unit* can be made up from the scratch or using some third-part systems for sound management and elaboration (MaxMSP, Csound, etc.).

The *control and interface unit* coordinates the input and output devices of the Ad-Opera platform. It encodes audio, video, and control data for input and output, processes the data to constrain it into desirable bounds, and delivers

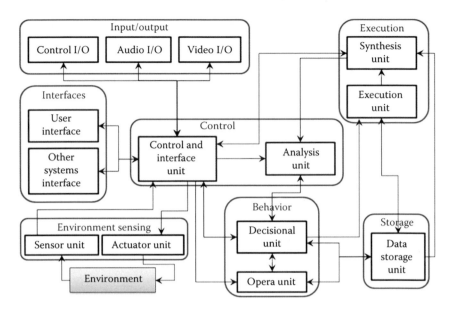

FIGURE 14.5
AM framework: preliminary architecture. (Figure taken from Gabriele, A., Adopera—definizione di un framework per la realizzazione di opere d'arte intermediali tempo-varianti (in italian), In *Master Thesis—Computer Science Department, University of L'Aquila*, 2009.)

data between the subsystems interfacing with sensors and actuators. The *analysis unit* provides libraries and procedures to examine the actual state, the input (control and audio data), the output (produced audio signal from the system), and extract data features.

The *sensors unit* provides data and communication between sensors and system. The sensors are essential to capture environmental features (light, wind, humidity, temperature, etc.) and interaction features (presence, proximity, position, contact, etc.). The *actuator unit* provides control and communication between the system and the actuators. The *environment* is the physical system (natural field and/or artificial infrastructure) that contains the opera and all the possible agents (system actors) that interact with the opera, for example, the audience. The environment behavior is, in general, unpredictable in the occurrence and sequence of the events.

14.4.2 New Software Engineering Models and Methods for Empathic Systems

The idea is to conceive new theories, methods, and tools for empathic systems—namely, context models, an adaptation model for verification & validation (V&V), continuous dynamic compliance checking, and a new life cycle process definition. Empathic systems perform adaptation as response to system or context changes. V&V regulates the adaptability of systems, since the adaptation in AD-OPERA is performed while preserving suitable invariants. The adaptation model shall encompass both bottom-up and top-down descriptions of under-specified system descriptions. The AD-OPERA adaptation model is based on assume-guarantee reasoning [23,33]. Each component comprising the system makes assumptions about its environment, and if these assumptions are satisfied, then the component will ensure other properties, which are referred to as the guarantees. By composing the set of assume/guarantee properties in an appropriate manner, it is possible to demonstrate the correctness of the entire system before constructing it. The assume-guarantee reasoning is based on four main dimensions: (i) the nature of the employed compositions; (ii) an efficient assumption generation technique; (iii) a decomposable property verification strategy; and (iv) the semantic coherence between the formalisms employed to write the system and the formalisms employed to specify the assumption/guarantee properties (e.g., the style). Adaptation is triggered by runtime monitoring, which supplies information regarding the degree of conformance of the execution to the stated invariants.

14.4.3 Algorithms and Mechanisms for Managing the Adaptation

The aim is to provide the runtime support for the modeling approaches: namely, design of wireless sensor networks (WSNs) for context sensing, context management, runtime monitoring and analysis, and synthesis of under-specified parts and corresponding runtime support to provide a framework for supporting the process life cycle of empathic systems.

Wireless Sensor Networks for Context Sensing: Despite the potential enormous benefits offered by the use of WSNs for empathic systems, there are some major challenges associated to the design of WSNs for this domain. The sensor network communication infrastructure should be designed so that it is able to efficiently support a reliable interaction with the environment and the users. Currently, there is no protocol for WSNs able to minimize energy consumption and satisfy the reliable and timely communications required by empathic systems. The management of the context will be possible by the development of new protocols and algorithms for WSNs. To sample the context accurately, these protocols and algorithms should overcome losses of communication over the wireless link and maintain a proper interaction with the system. We plan to include localization algorithms for WSNs able to track the context, such as the motion of one or more users that are not wearing any sensor device. The effects of the impairments coming from the wireless links behavior (e.g., path loss, shadowing, and interference) will be considered as part of the context and will be made available to the users.

Context Management: Empathic systems require awareness at the global level. To this end, context models must capture the overall environment—internal and external to the system. Most existing approaches to context modeling and management remain at a local level of awareness—lacking expressive concepts to describe the complex relations within the users, between system components, and the overall dynamic interaction between users and system. Context management no longer has to provide the best possible context data in terms of, for example, accuracy or level of details, but at a sufficient level of quality. Current context frameworks do not support the definition, control, and querying of context at multiple levels of granularity.

Runtime Monitoring and Analysis: Monitoring is carried out at different levels of abstraction. The properties predicated on the expected behavior are described very precisely when complete specifications are available at deploying time. In the case of under-specification, the oracle, that is, the mechanism to check the conformance of the execution with the invariant, is obtained by combining the invariant with information that becomes available at execution time and that may evolve unpredictably. The idea is to realize a flexible monitor, itself adaptive, which is able to address oracles that are deliberately incomplete and that depend so heavily on runtime and contextual information. The monitors will be evaluated in terms of intrusiveness, configurability, and impact on system performance parameters.

Synthesis of Under-specified Parts: Given a score with parts of the system under-specified, we wish to synthesize the complete specification. This activity corresponds to the interpretation activity in AM. The synthesis will use the user input and information obtained from the context. The synthesis of the concrete specification will follow the directives provided by the V&V techniques. The under-specification recalls the idea of abstract definition of a service or resource in service-oriented computing (SOC). The fundamental principle in SOC is the separation of a service's functional interface from its implementation. Selection of an abstract interface at design time is coupled with the selection of a service supporting the respective interface during runtime. Each service/resource implementing the same abstract interfaces provides the same functionality.

AD-OPERA promotes the late binding of the resource to a specific representation/interface. In AD-OPERA, we remove the assumption that the representation of the resource already exists, allowing missing pieces to be synthesized at runtime. This concept is inspired by composition in AM: the composer may leave parts of the composition undetermined, or *abstract*, by delegating to the performer or to the listener the possibility of *concretizing* the missing pieces. This task will be influenced by adaptation and sound generation algorithms. The synthesis algorithms proposed in this task must refine and evolve the abstract specification into a concrete one. The synthesis will be founded on a blend of evolving systems theory [9] and retrenchment theory [6,7], guided by an assume/guarantee perspective [22,33]. Retrenchment is a formalism mediated by extra predicates (compared with model-based refinement) that allow the relationship between an abstract operation and its concrete counterpart to be more flexible than that which is permitted by a refinement. It will capture the extent to which the developing performance is adhering (or not) to the desired style and will trigger adaptations, managed by the evolving systems theory where necessary, to either re-converge to the invariant demanded by the assume/guarantee framework or embark on a more drastic change, controlled by a higher level in the evolving systems hierarchy.

14.4.4 Empathic System Engineering

This new discipline targets systems that are beyond the AM and SE domains. This has then the double role of supporting and ensuring the cross-fertilization between AM and ICT and of leveraging the findings to the broader class of empathic systems. The new discipline of empathic system engineering with its own methodologies and practices beholds great promise for applications in domains such as social networking, content provisioning, mobile advertisement, computer games, school and museum sector, and artistic expression.

14.5 Final Remarks

AD-OPERA will propose radically new SE methodologies for adaptive applications in ubiquitous environments. We expect to change how an adaptive, context-aware application is developed and utilized by promoting changes as the norm rather than the exception but still providing support for system dependability. This will contribute to the vision of an adaptive system as a well-behaved living organism that it is capable of taking into account the unpredictable changes that occur in its context but still preserving the essential qualitative and quantitative properties expressed by the developer. AD-OPERA aims at providing support for a cost-effective principled development and management of adaptable systems. It is important to notice that the notion of cost for empathic systems encompasses besides the usual business and technological dimensions, the human-related concerns, such as satisfaction, emotions, and creativity of the users. This aspect is increasingly important in an ever more user-centric world characterized by heterogeneous devices and ubiquitous networking, where systems are able to adapt to system changes and user requirements are a key paradigm in present and future integrated systems, such as collaborative systems, social networks, new media content-based systems, games, and so forth. Therefore, AD-OPERA technologies will significantly extend the utilization and the creation of innovative AM-inspired applications for urban and public contexts, valorization of museum and archaeological sites, for multimedia applications, for games that enable user personalization according to mood and city customization by location-based facilities and sensor network, and many others.

AD-OPERA targets interdisciplinary research. It involves stakeholders of different nature who have very different backgrounds, competencies, and interests. The development, within the new discipline of empathic system engineering, of a common reference ground for communication and interaction defined in terms of a common vocabulary of concepts and a set of practices (both processes and tools) represents *per se* a contribution that we expect to export beyond the AD-OPERA-specific interdisciplinary context of artists, software engineers, and audience of the opera.

In the music domain, the result of AD-OPERA will enable artists/musicians to concentrate on the creative aspect of their work since they will be free from low-level technical details concerning the realization of the opera. The music domain can play an important role in ensuring that AD-OPERA results have a transformational effect on society. As with traditional works of music, listening in a participatory and conscious manner means following the logical and coherent flow of connection between the sounds. In this case, the sensorial stimulus spreads, since the sound is associated with tactile and visual experiences in a functional manner. Generally speaking, the work is enjoyed as a dynamic process of emotional learning in which the listener participates deliberately and for which he is partially responsible.

The synergy that can be established with the listener, with the environment and the architecture, with the materials and the sculptural forms, and with the passage of time and events, gives the public a correlated plurality of experiences, which sharpens responsiveness to the artistic and expressive gesture and stimulates cognition. On one side, in AM, the aesthetic experience takes the form of change, and consequently, culture becomes creative. On the other side, AM empowers the relationship man–environment–artwork to improve the quality of life, stimulating processes of conscience, memory, comparison, and prediction. The active enjoyment of the musical work, characteristic of AM, will stimulate imagination, develop new manners of communication between individuals, and promote new manners of exploring and exploiting music. The integration of music to dynamic, social, urban, environmental, and virtual contexts will welcome, accompany, and gratify human existence with civilized expressions.

Furthermore, a social consequence of AD-OPERA is its potential to break the psychological barrier that people may have while interacting with conventional ICT applications. Temporary music installations and permanent operas will put into practice this concept by promoting the interaction of a broader audience with the musical results of AD-OPERA.

References

1. An architectural blueprint for autonomic computing. http://www-03.ibm. com/autonomic/pdfs/AC%20Blueprint%20White%20Paper%20v7.pdf. A White paper, IBM corporation, third edition, June 2005. Accessed 24th May 2013.

2. D. Aspinall and K. MacKenzie. Mobile resource guarantees and policies. In *Proceedings of the Second international conference on Construction and Analysis of Safe, Secure, and Interoperable Smart Devices (CASSIS'05)*, G. Barthe, B. Grégoire, M. Huisman, and J.-L. Lanet (Eds.). Springer-Verlag, Berlin, Germany, pp. 16–36. 2005. DOI=10.1007/11741060_2 http://dx.doi.org/10.1007/11741060_2

3. M. Autili, P. Benedetto, and P. Inverardi. Context-aware adaptive services: The PLASTIC approach. In *Proceedings of the 12th International Conference on Funda-mental Approaches to Software Engineering: Held as Part of the Joint European Confe-rences on Theory and Practice of Software*, ETAPS 2009 (FASE '09), M. Chechik and M. Wirsing (Eds.). Springer-Verlag, Berlin, Germany, pp. 124–139. 2009. DOI=10.1007/978-3-642-00593-0_9 http://dx.doi.org/10.1007/978-3-642-00593-0_9

4. M. Autili, P. Di Benedetto, P. Inverardi, F. Mancinelli. A resource-oriented static analysis approach to adaptable java applications. In *CORCS'08*, Turku, Finland, 2008.

5. M. Autili, P. Di Benedetto, P. Inverardi, and D. A. Tamburri. Towards self-evolving context-aware services. *ECEASST Journal: Context-aware Adaptation Mechanisms for Pervasive and Ubiquitous Services (CAMPUS), DiSoTec'08*, Oslo, Norway, 11, 2008.

6. R. Banach, C. Jeske, and M. Poppleton. Composition mechanisms for retrench-ment. *Journal of Logic and Algebraic Programming*, 75(2):209–229, 2008.

7. R. Banach, M. Poppleton, C. Jeske, and S. Stepney. Engineering and theoretical underpinnings of retrenchment. *Science of Computer Programming*, 67:301–329, 2007.

8. L. Baresi, R. Heckel, S. Thöne, and D. Varró. Style based refinement of dynamic software architectures. In *Proceedings of the Fourth Working IEEE/IFIP Conference on Software Architecture (WICSA)*, pp. 155–164. Oslo, Norway, 2004.

9. II. Barringer, D. Gabbay, and D. Rydeheard. Modelling evolvable component systems. *Logic Journal of IGPL*, 17(6):631–696, 2009.

10. G. Barthe. MOBIUS—Securing the next generation of Java-based global computers. *ERCIM News*, 63:28–29, 2005.

11. N. Bencomo and G. S. Blair. Using architecture models to support the generation and operation of component-based adaptive systems. In *Software Engineering for Self-Adaptive Systems*, Vol. 5525 of *LNCS* pp.183–200, Springer, New York, 2009.

12. L. Bianchini. Designing a virtual theatrical listening space. In *Proceedings of ICMC International Music Conference*, Ann Arbor, MI: MPublishing, University of Michigan Library, 2000.

13. L. Bianchini and M. Lupone. Presentazione artistica (in italian). In *"ArteScienza"* ed. *FGTecnopolo*, Rome, Italy, 2011.

14. J. S. Bradbury, J. R. Cordy, J. Dingel, and M. Wermelinger. A survey of self-management in dynamic software architecture specifications. In *WOSS '04: Proceedings of the 1st ACM SIGSOFT Workshop on Self-Managed Systems*, pp. 28–33. ACM, New York, 2004.

15. A. Bucchiarone, H. Ehrig, C. Ermel, O. Runge, and P. Pelliccione. Formal analysis and verification of self-healing systems. In *Fundamental Approaches to Software Engineering (FASE 2010)*, Vol. 6013 of *LNCS*. Sringer-Verlag, Berlin, Germany/Heidelberg, NY, 2010.

16. A. Bucchiarone, P. Pelliccione, C. Vattani, and O. Runge. Self-Repairing systems modeling and verification using AGG. In *WICSA/ECSA*, pp. 181–190, 2009.

17. V. Cardellini, E. Casalicchio, V. Grassi, F. Lo Presti, and R. Mirandola. Qos-driven runtime adaptation of service oriented architectures. In *Proceedings of the Seventh Joint Meeting of the European Software Engineering Conference and the ACM SIGSOFT Symposium on the Fountations of Software Engineering (ESEC/FSE '09)*, pp. 131–140. ACM, New York, NY, 2009.

18. S. Dobson, S. Denazis, A. Fernández, D. Gaïti, E. Gelenbe, F. Massacci, P. Nixon, F. Saffre, N. Schmidt, and F. Zambonelli. A survey of autonomic communications. *ACM Transactions Autonomous Adaptation Systems*, 1(2):223–259, 2006.

19. A. Gabriele. Adopera—definizione di un framework per la realizzazione di opere d'arte intermediali tempo-varianti (in italian). In *Master Thesis—Computer Science Department, University of L'Aquila*, Italy, 2009.

20. D. Garlan, S.-W. Cheng, A.-C. Huang, B. Schmerl, and P. Steenkiste. Rainbow: Architecture-based self-adaptation with reusable infrastructure. *Computer*, 37(10):46–54, 2004.

21. P. Horn. *Autonomic Computing: IBM's Perspective on the State of Information Technology*, IBM Corporation, Riverton, NJ, October 15, 2001.

22. P. Inverardi, P. Pelliccione, and M. Tivoli. Towards an assume-guarantee theory for adaptable systems. In *SEAMS'09*, Washington, DC. IEEE Computer Society, 2009.

23. C. B. Jones. *Development Methods for Computer Programs Including a Notion of Interference*. PhD Thesis in Computer Science, Oxford University, TRnumber PRG25, 1981.

24. J. O. Kephart and D. M. Chess. The vision of autonomic computing. *Computer*, 36(1):41–50, 2003.

25. J. Kramer and J. Magee. Self-managed systems: An architectural challenge. In *FOSE '07: 2007 Future of Software Engineering*, pp. 259–268. IEEE Computer Society, Washington, DC, 2007.

26. H. Liu, M. Parashar, and S. Member. Accord: A programming framework for autonomic applications. *IEEE Transactions on Systems, Man and Cybernetics*, 36:341–352, 2005.

27. M. Lupone. Musica elettronica in acustica musicale e architettonica. In *Cingolani, Spagnolo, Torino, UTET*, 2005.

28. M. Lupone. Music and mutation. In *Hi-Art Review, Gangemi Editore*, 2008. ISBN: 978-88-492-1422-2.

29. F. Mancinelli and P. Inverardi. Quantitative resource-oriented analysis of Java (Adaptable) applications. In *Proceedings of the Sixth International Workshop on Software and Performance (WOSP '07)*. ACM, New York, pp. 15–25. 2007. DOI=10.1145/1216993.1216998 http://doi.acm.org/10.1145/1216993.1216998

30. M. Parashar and S. Hariri. Autonomic computing: An overview. In *Unconventional Programming Paradigms*, Vol. 3566 of *LNCS* pp. 257–269. Springer Verlag, New York, 2005.

31. N. Paspallis and G. A. Papadopoulos. An approach for developing adaptive, mobile applications with separation of concerns. In *COMPSAC*, pp. 299–306. IEEE Computer Society,Washington, DC, 2006.

32. C. Peper and D. Schneider. Component engineering for adaptive ad-hoc systems. In *SEAMS '08*, pp. 49–56. ACM, New York, 2008.

33. A. Pnueli. In transition from global to modular temporal reasoning about programs. *Logics and Models of Concurrent Systems*, pp. 123–144, Springer-Verlag, New York, 1985.

34. R. Rouvoy, P. Barone, Y. Ding, F. Eliassen, S. O. Hallsteinsen, J. Lorenzo, A. Mamelli, and U. Scholz. Music: Middleware support for self-adaptation in ubiquitous and service-oriented environments. In *Software Engineering for Self-Adaptive Systems*, Vol. 5525 of *LNCS* pp. 164–182, Springer-Verlay, Berlin, Heidelberg, 2009.

35. R. Rouvoy, F. Eliassen, J. Floch, S. O. Hallsteinsen, and E. Stav. Composing components and services using a planning-based adaptation middleware. In *Proceedings of Software Composition*, Vol. 4954 of *LNCS 4954*, pp. 52–67. Springer, Berlin, Heidelberg, NY, 2008.

36. C. Seo, S. Malek, and N. Medvidovic. An energy consumption framework for distributed java-based systems. In *ASE '07: Proceedings of the Twenty-second IEEE/ACM International Conference on Automated Software Engineering*, pp. 421–424. ACM, Atlanta, Georgia, 2007.

37. M. Weiser. The computer for the 21st century. *SIGMOBILE Mobile Computing Communications Review*, 3(3):3–11, 1999.

38. S. R. White, J. E. Hanson, I. Whalley, D. M. Chess, A. Segal, and J. O. Kephart. Autonomic computing: Architectural approach and prototype. *Integrated Computer-Aided Engineering*, 13(2):173–188, 2006.

39. R. Wilhelm, J. Engblom, A. Ermedahl, N. Holsti, S. Thesing, D. Whalley, G. Bernat, et al. The worst-case execution-time problem—overview of methods and survey of tools. *Transactions on Embedded Computing Systems*, 7(3):1–53, 2008.

15

Fuzzy Methods and Models for a Team-Building Process

Irena Bach-Dąbrowska

CONTENTS

15.1 Introduction

Nowadays, project management commonly adheres to an agile philosophy, which means projects are often lead by self-organizing, small teams characterized by transfer of decision powers and responsibilities from a single leader to all team members. In such conditions, task execution effectiveness depends mostly on interpersonal relationships within the team. Such an approach thus forces researchers to develop selection methods that allow assessing each team member (or prospective job applicant) with respect to the behavioral and psychological makeup of the entire team instead of applying methods that evaluate the hard skills of individual team members in isolation and how these skills match with the requirements of the vacancy.

This problem of team assessment has been observed by several authors (Harvey et al. 1998; Hlaoittinun et al. 2007; Alencar and Almeida 2010; Galinec 2010; Wojnar 2011; Bach-Dąbrowska 2012) who claim that the metrics to measure behavioral cohesion of the entire team with respect to the organizational structure of the team and the levels of responsibility and decision powers within the team structure is of significant importance. These authors postulated the necessity of providing methods and models combining assessment mechanisms of both hard and soft skills. They proposed models of such mechanisms based on multiple criteria with fuzzy logic as the underlying reasoning formalism.

The major goal of the proposed models is raising the efficiency of team cooperation and the execution effectiveness of adopted tasks by increasing the cohesion of the team in the sense of complementing competencies as well as interpersonal characteristics of the team as a whole. The use of the fuzzy-logic formalism is dictated by its primary advantage, an ability to define and measure imprecise information about behavioral and psychological features of a candidate, which is difficult to model in terms of conventional mathematics. Additionally, as in the selection process, information accuracy and quality depend on the information sources. Here, the fuzzy-logic formalism allows textual and linguistic data with varying (less or more) granularity in quantification.

In this chapter, selected methods and models are reviewed and grouped according to techniques combined with fuzzy logics that create alternative hybrids for selective and powerful decision support.

Section 15.2 includes the basic definitions related to teams, competencies, and the recruiting process. In Section 15.3, after a brief introduction to fuzzy-logic theory, multicriteria decision models implementing analytic hierarchy process (AHP), Myers–Briggs type indicator (MBTI), cognitive proximity (CP), and Saaty's fundamentals are discussed. For each presented model, the main assumptions of their use and proposed areas of application are defined. A summary section provides references to additional literature as an invitation for the reader to further explore topics related to developing, applying, and evaluating competencies in the team structuring process.

15.2 Human Resources Environment

This section provides a brief survey of basic definitions related to systems of human resources (HR) and explains the meaning of team, teamwork quality, HR abilities, and competencies.

In the most intuitive manner, a *team* can be defined as a group of independent individuals working together to complete specific tasks. This definition is quite simple and understandable for anyone from a popular science viewpoint. However, the term definition varies as the specific domain varies. A review of some basic classifications is presented in the succeeding text.

Firstly, there is a clear distinction between *work teams* as formal groups of people and *informal teams*. Work teams are understood as a group of employees who work semiautonomously on recurring tasks in situation where work content changes frequently and employees with limited skills and with a specific set of duties that are difficult to cope with (Business Dictionary, 2012). Informal teams are *other groups of people* who are assembled together in organizations. Work teams represent a specific type of *formal groups*, that is, their appointment requires formal time limits for their existence, planning their size, structure, and composition and ascertaining specific purposes of their implementation. Moreover, formal groups usually have the same group members through their existence. As to the *other group of people*, they are called *informal groups* that emerge naturally because of a specific area of activity or duties.

In Argote and McGrath's work (1993), groups are divided into *acting groups* and *standing groups*. In this classification, an acting group is a circle of people who perform interdependent activities, and a standing group is a circle of people who are labeled as a unit but who do not perform interdependent activities.

According to Sundstrom et al. (1990) definition, a team should share responsibility for specific outcomes of its actions. Moreover, according to West et al. (1998), an additional requirement is that the team ought to be identified as such by those within and outside the team.

Furthermore, if the period of cooperation is taken as a criterion, teams can be divided into *temporary teams* with members who work together only for the duration of a project and *permanent teams* that are together indefinitely. Following the successive extensions of a team definition, according to Stott and Walker (1995), it is necessary to identify the significance of relationships, the need for cooperation, and the degree of dependency among team members as some of defining features.

Finally, a relatively large number of authors (McClain and Romaine 2002; Davidson Frame 2003; Bolles and Fahrenkrog 2004; Schroder and Diekow 2006) emphasize that, whenever possible, project groups ought to be *heterogeneous* in order to ensure that diverse backgrounds, ideas, and creativity collide and combine.

For clarity of the presentation that follows, the term *team* is adopted as an alternative to each and every definition of a working group.

In most publications, a team-building process refers to a situation where teams are appointed for the specific purpose of handling multiple project tasks. Moreover, two main situations can be specified in a team-building process. One occurs when the team is established directly at the starting phase of the project, and the other occurs when the team must be extended with one or more new members. In each situation, the recruiting process must be performed. This then verifies whether a particular candidate is suitable to provide the content but also how precisely the personality fits into the team.

According to Harvey et al. (1998), teams can maximize organizational innovation when employees receive the right for an increased autonomy, participation, and ownership of executive decisions. Teams are provided with goals, or they develop goals together with their team leader. Then, they are free to decide how to achieve these goals. In addition to maximizing innovation, teams can provide a number of other advantages for the organization in which they operate:

- Teams make optimal use of HR as they are organized to gain access to an individual's knowledge and skills (IRS Employment Review, 1995).
- Teams enhance organizational learning because employees are able to experiment and create strategies that are best suited to their work (Wageman 1997).
- Teams can result in gains in an individual's productivity and efficiency and thereby create a synergy (Katzenbach and Smith 1993).
- Team work is associated with a greater variety of tasks and additional responsibility for team members that is likely to result in an increased level of job satisfaction, motivation, and employee commitment. This, in turn, may cause lower staff turnover and lower absenteeism and thereby reduce organizational costs and improve organization's base of knowledge (Kirkman and Shapiro 1997).

To achieve defined benefits from teamwork and to guarantee the *teamwork quality* understood as quality of the task-related and social interactions between team members, entrepreneurs must find a method allowing them to develop strong and successful team.

The analysis of literature shows that there are a number of elements that give teams their internal strength and ensure a successful of cooperation.

According to Schröder and Diekow (2006), each team member should identify herself or himself with the project and understand the requirements of its implementation much like everyone should be strongly motivated to achieve the team objective. The more comprehensible and clearer the goal is, the better is the congruence with the values, strategies, and opportunities for the associates.

Following the five characteristics of a strong team by Bednarz, in the team-building process, specific features should be examined (Bednarz 2012):

1. *Size.* Research has shown that most teams range from 2 to 25 individuals; however, the average number is less than 10 members. Smaller teams tend to be more effective as a larger amount of people has trouble with interacting constructively as a group. Large teams are also less likely to reach timely agreements on the arguable steps required to move the team forward. Large groups defined as teams often tend

to break themselves into smaller subteams that are responsible for secondary aspects of the project or the problem.

2. *Complementary skills.* In order to be effective, teams must be comprised of individuals who have the right mix of skills in three areas.

   a. *Technical or functional expertise*—team members should possess the appropriate technical or functional knowledge that is required by the team to accomplish its goals or objectives. Specific expertise is defined by the breadth and scope of the problem and the capabilities required to resolve it.

   b. *Problem solving and decision-making skills*—Teams must be able to identify any problems and opportunities that arise. They must be able to evaluate available options and make necessary trade-offs and decisions before proceeding. Therefore, it is important for the majority of, if not all, team members to possess advanced problem solving and decision-making skills that will allow them to move the team forward effectively.

   c. *Interpersonal skills*—teams depend on the effective communication and the constructive conflict resolution. These abilities depend on the interpersonal skills and emotional intelligence of individual members. They include respect, active listening, and empathy as well as the ability to handle criticism, remain objective, take risk, and build trust. As leaders appoint members to their teams, common sense should guide them in selecting individuals with the necessary complement of skills to achieve the team's purpose. While many teams are assembled based on more subjective personal criteria or according to a formal position description, these interpersonal areas should be considered first and foremost when selecting team members.

3. *Common purpose.* Teams develop direction, momentum, and commitment by working to shape a meaningful purpose. Effective teams invest extensive time and effort in exploring, defining, and agreeing on a purpose that belongs to them both collectively and individually. Once achieved, this gives teams an identity that reaches beyond the total number of individuals involved. Conflicts are both necessary and threatening to teams, but this identity keeps them constructive by providing a meaningful standard thanks to which a team can resolve clashes between individual and team interests.

4. *Common approach.* As teams require a common approach or an idea how they will work together to accomplish their purpose, individual members must agree on who will perform specific jobs, how schedules will be set and adhered to, what skills must be developed, how continuing membership is to be earned, and how the group will make and modify decisions, including when and how to change

its approach to getting the job done. The agreement on operational specifics and how they integrate individual skills and advance the team's performance lies at the heart of shaping a common team approach.

5. *Mutual accountability.* A team is not viable if it cannot hold itself accountable. Teams enjoying a common purpose and approach inevitably hold themselves individually and collectively more responsible for the team's performance. Teams develop further specific performance goals to provide clear yardsticks for accountability. Accountability provides a measure of the team's quality of purpose and approach. Groups lacking mutual accountability for performance have not shaped a common purpose and approach that can sustain them as a team.

Although all five characteristics presented earlier are valuable in a purpose of team building, the second feature, *complementary skills*, is crucial in a team structuring process and at the same time the most difficult to guarantee. The most practical methods applied in a team-building process are designed to assess the hard and soft skills of individual members of a group but not the team as a whole. Some of them are more common, popular, and typical than the others, and those are briefly characterized in the following. The most popular tools are

- *Formal documentation*: *application forms, resumes, cover letters, certificates*, or *rankings*. From an organization's perspective, the benefit of using an application form ensures that the same information is gained from all candidates. This helps achieve a level of consistency in the short-listing process. The format of CVs makes short-listing less consistent and certainly more difficult, while information is usually presented in a variety of ways, through different CV formats.

- *Interviews*: *structured, unstructured*. A structured interview is the most effective type of an interview. The interview process is formed through an identification of the key requirements of a job and drawing up a list of questions. A panel of interviewers works through each set of questions with each candidate and scores them on their answers. At the end of the interview process, the overall scores are taken into consideration and the best candidate is chosen.

- *Psychometric testing*: *abilities tests, psychological tests, personality profiling*. Those tests cover ability testing, aptitude testing, and personality profiling. The British Psychological Society (BPS) has two levels of qualifications. Level A covers ability and aptitude testing among which *general intelligence, verbal ability, numerical ability, spatial ability, clerical ability, diagrammatical ability, mechanical ability, sensory, and motor abilities* are distinguished. Level B covers personality profiling as it is commonly known that an individual's personality may affect their suitability for some positions.

- *Samples and simulation of work, group exercises.* Candidates are given a topic or a role-play exercise and are invited to discuss the topic or role play in a group. During the discussion/role play, observers who are looking for specific attributes award candidate with marks.

- *Assessment center/development center.* Assessment center is an external unit hired for conducting the selection process. Assessment methods are based on the principle of multiple testing procedures. This may include ability and aptitude testing, group exercises, in-tray exercises, presentations, and personality profiling.

- *Online screening and short listing.* Initial screening of applications is usually based on an assessment of a candidate's experience and qualifications against the job's requirements. Online systems that can filter applications automatically are now available. An online selection facility screens applications for a set of criteria through keyword searches. It may also provide a scoring mechanism.

- *References: written, recommendations confirmed by phone.* References are also used as a selection method. Occasionally, unsatisfactory references may affect a decision whether or not to choose a given individual

The review presented earlier is based on work of Juchnowicz et al. (2008) and GRB Career Matchmaking (2012).

Anderson and Cunningham-Snell (2000) provide a summary of research on how well methods predict future job performance. Chart 1 shows the following selection method scores as follows. Assessment centers are performing best (cf., score 0.68), whereas references provide the least value (cf., score 0.13). Work samples and ability tests prove to be reasonable methods with the score of 0.54. A perfect prediction is equal to 1.0 (Figure 15.1).

All presented methods as a single tool are more or less effective in the recruiting process; however, they work best when applied together. Unfortunately, only the wealthiest enterprises with the highest level of organizational development can afford the exploitation of specialized assessment centers. Most of other entrepreneurs use the structured/unstructured interviews and work samples (e.g., in the IT branch). Typical recruiting exams are primarily based on the knowledge, hard skills, and abilities (KSAs). They do not supply any relevant information about soft skills and other behavioral characteristics of a candidate. Thus, other characteristics, often crucial in ensuring the coherence to a team, are usually evaluated during the employment interview and probationary period. This phase is permitted when permanent employees are searched for, but it is not acceptable (because of time limits) when a project team is being appointed.

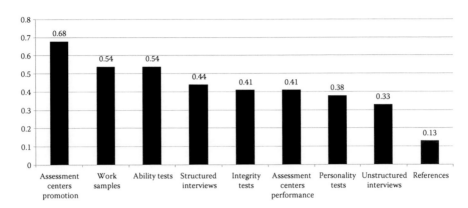

FIGURE 15.1
Ranking of methods according to their effectiveness in predicting future job performance.

Summarizing the presented overview, a number of main insights ought to be emphasized. Firstly, the determination of recruiting and selection methods should be matched based on the team structure and management style. Typical selection and recruiting methods are suitable for standard hierarchically structured teams but do not follow the needs of quickly modifying models of team management that transform from hierarchically to horizontally structured with a trend of middle-level management vanishing.

Secondly, the assessment of candidates ought to be strengthened by implementation of the metrics that allow measuring behavioral suitability for a given position and match with coworkers. It means that psychometric testing must be one of the crucial selection tools.

Finally, to guarantee project performance effectiveness, in case of establishing a project team, behavioral coherency of the entire team with respect to the organizational structure of the team and the levels of responsibility and decision powers inside of the team structure ought to be examined.

The next section of this chapter, after a brief introduction to fuzzy-logic theory, presents selected methods and models referring to the postulates earlier.

15.3 Fuzzy-Logic-Based Models of a Team Structuring

To submit a basis for a common vocabulary and increase simplicity and a better understanding of the concepts discussed in further paragraphs, a brief introduction to fuzzy-logic theory is first provided.

Fuzzy logic is a form of many-valued logic, based on imprecise data, implementing approximate rather than fixed and exact reasoning. In contrast to traditional logic theory, where binary sets have two-valued logic, true or false, fuzzy-logic variables may have a truth-value that ranges in degree between 0 and 1. Fuzzy logic has been extended to handle the concept of partial truth, where the truth-value may range between completely true and completely false. Furthermore, when linguistic variables are used, these degrees may be managed by specific functions (Garg and Gaur 2011).

The reasoning in fuzzy logic is similar to human reasoning. It allows for approximate values and inferences as well as incomplete or ambiguous data (i.e., fuzzy data) as opposed to only relying on crisp data (e.g., binary yes/no choices). Fuzzy logic is able to process incomplete data and provide approximate solutions to problems other methods find difficult to solve.

To apply the fuzzy-logic theoretical language, it is necessary to define its crucial elements. Authors, such as Zadeh and Hose (1975), Piegat (2003), Rutkowski (2005), and Alcalde and Burusco (2010), provide a detailed description of fuzzy logic. A subset of most important definitions necessary to understand and define of fuzzy-logic foundations is given in the following:

- *Linguistic variable* is a variable whose values are words or sentences instead of numbers, and that is characterized by a quadruple measure $[L, T(L), \Omega, M]$ where L is the name of the variable, $T(L)$ is a countable term set of labels or words (i.e., the linguistic values), Ω *is a universe of* discourse, and M is a semantic rule.

- *Linguistic value* is an enumeration of the values of a linguistic variable provided in words. For *example, if L*: speed, then linguistic values for L are high, medium, and low.

- *Fuzzy set* is any set that allows its members to have a different degree of membership, called membership function, in the interval $[0, 1]$. Fuzzy set A, defined in the numerical universal of discourse Ω, is a set of pairs:

$$A = \{(\mu_A^*(q), q)\}, \forall q \in \Omega \tag{15.1}$$

where
μ_A is a membership function of fuzzy set A
$\mu_A^*(q)$ is membership grade of the element q in a fuzzy set A, while
$\mu_A(q) \in [0,1]$

- *The membership function* ascertains each element q of a given variable a particular value from the range $[0,1]$:

$$\mu_A(q) : \Omega \to [0,1], \forall q \in \Omega \tag{15.2}$$

This value, called *grade of membership*, informs in what grade element q belongs to the fuzzy set A.

- *Singleton* is a set with exactly one element and one membership grade. For example, set $\{0\}$ is singleton and $\mu_A^*(0) = 1$.

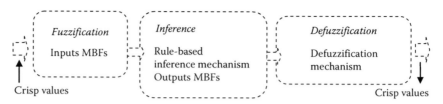

FIGURE 15.2
Fuzzy-logic system blocks.

The fuzzy-logic formalism and related inference mechanisms are implemented by expert systems based on three main modules (cf., Figure 15.2): *fuzzification*, *inference*, and *defuzzification*.

In the *fuzzification* module, crisp input values are translated into linguistic concepts that are represented by fuzzy sets. These concepts are called linguistic variables. Degrees of membership for all input values are assigned.

The decision-making process in fuzzy-logic systems is based on a set of rules. Linguistic rules are formed using operators that represent linguistic AND and OR. Finally, a computation of the applicability of the rules themselves, represented by a linguistic IF...THEN expression, is performed. The *inference* is a calculus consisting of a number of steps, such as aggregation, composition, and, if necessary, result aggregation. This last step is then called fuzzy *inference*.

In the *defuzzification* block, the result of the fuzzy *inference* is retranslated from a linguistic concept to a crisp output value.

The subsections presented in the succeeding text introduce methods and models dealing with the team-building task based on the fuzzy-logic formalism.

15.3.1 Employment Policy Using a Fuzzy Environment

This subsection provides a fuzzy model that represents a basic example of a fuzzy-logic implementation aiming at the selection process. It can be implemented for the purpose of team structuring to support choosing one person from a group of candidates. The selection criteria do not, however, cover the examination of behavioral profile of the candidate and her or his suitability to fit the team.

The presented model is based on decision making in the so-called *fuzzy environment*, consisting of *fuzzy objective*, *fuzzy constraint*, and *fuzzy decision*. It constitutes one of the basic fuzzy models for multiple-criteria decision making. The following description is a formalization of the aforementioned and is borrowed from Rutkowski (2005):

Let X_{op} designate a set of options (also called the elections or the alternatives), where

$X_{op} = \{x\}$. The *fuzzy objective* is defined as a fuzzy set G described in a set of options X_{op}. Fuzzy set G is described by a membership function that associates a certain value from the range [0,1] with each element x:

$$\mu_G(x): X_{op} \to [0,1], \forall x \in X_{op} \qquad (15.3)$$

This value, called *grade of membership*, informs about the extent to which the element x belongs to the fuzzy set G (i.e., fuzzy objective).

The *fuzzy constraint* is defined as a fuzzy set C described in a set of options X_{op}. Fuzzy set C is described by a membership function that associates a certain value from range [0,1] with each element x:

$$\mu_C(x): X_{op} \to [0,1], \forall x \in X_{op} \qquad (15.4)$$

This value, called *grade of membership*, informs about the extent to which the element x belongs to the fuzzy set C (fuzzy constraint).

Considering the process of decision making that is fulfilling both fuzzy objective G and fuzzy constraint C, a *fuzzy decision D* is defined. D is, in fact, a fuzzy set formed as a result of the fuzzy objective and fuzzy constraint intersection:

$$D = G \cap C \qquad (15.5)$$

where for each $x \in X_{op}$

$$\mu_D(x) = T(\mu_G(x), \mu_C(x)) \qquad (15.6)$$

Equation 15.4 allows for following an interpretation of the decision making in a fuzzy environment, that is, it allows "to attain a fuzzy objective G and fulfill the fuzzy constraint C." A precise form of notation in Equation 15.4 depends on the assumed *t*-norm.

The implementation of the model for an effective team-building process requires a definition of a set of criteria for candidate assessment (e.g., competency level, work experience) in terms of *fuzzy objectives* G and employment conditions (e.g., proposed wages, availability) in terms of *fuzzy constraints* C. Individual candidate x (where $x \in X_{op}$, $X_{op} = \{x_1, ..., x_n\}$, $k = 1, ..., n$) is assessed from the viewpoint of meeting all objectives G_n while satisfying constraints C_m that have been defined for a particular model. As the team structuring requires identification of more than one assessment criterion, the considerations presented earlier can easily be generalized for a multiconstraint and multiobjective environment.

For $n>1$ fuzzy objectives $G_1,...,G_n$ and $m>1$ fuzzy constraints $C_1, ..., C_m$ defined in a set of options X_{op}, fuzzy decision D is determined in the following way:

$$D = G_1 \cap ... \cap G_n \cap C_1 \cap ... \cap C_m \qquad (15.7)$$

where for each $x \in X_{op}$

$$\mu_D(x) = T\{\mu_{G1}(x),...,\mu_{Gn}(x), \mu_{C1}(x),..., \mu_{Cm}(x)\} \qquad (15.8)$$

Maximizing decision is an option $x^* \in X_{op}$, such that

$$\mu_D\left(x^*\right) = \max_{x \in X}\mu_D(x) \qquad (15.9)$$

The most suitable candidate for a vacancy is determined according to Equation 15.9.

15.3.2 Analytic Hierarchy Process and Fuzzy-Logic-Based Model

This subsection presents an approach to an optimal assignment of a project team based on the AHP and fuzzy-logic method proposed by Galinec (2010).

In his work, Galinec describes a method for project management and project team management based on multiple-criteria decisions. To this end, a set of situations is defined that are challenging from the team leader point of view. Then, the ability to predict possible events is discussed using means, such as identification of project-influencing factors and their complexity as an important management issue. The main factor ensuring the best work performance and an ultimate success is defined as human (i.e., staff) management. Thus, the team leader must be able to accelerate the accomplishment of the assigned tasks, and no other person should be assigned the same role at the same time. To avoid the subjective treatment that is innate to a human being, the leader is encouraged to develop fuzzy evaluation models. This practice allows making a more objective assessment by application of imprecise and uncertain values.

Following the model assumptions proposed by Galinec (2010), the developed evaluation system should serve as a means to constitute project teams based on their knowledge and skills. The system is used for a priori evaluation of a person's suitability for a particular job and a role in the team. One person can be a candidate for several roles in a certain team (be better than others and fit to be assigned to several jobs), whereas some jobs in the team may stay vacant because of a lack of appropriate candidates. To that address such lack of candidates, it is necessary to fuzzy evaluate an individual's ability for every job in each project using the set of conditions that are established by the current needs, a parallel development of the projects, and the specific demands of the jobs in future project teams.

Assigning the roles to the personnel and their integration in projects is mostly performed in each project phase (e.g., planning, analysis, design, construction), because in each of the development phases, it is necessary to intensify specific and varying knowledge and skills and then assign the personnel accordingly. As a rule, the same phases within different projects will not last for the same period of time. In fact, it is common that for all project phases, especially the later ones, end up behind schedule. This has a direct effect on the methods of assigning personnel according to the needs along the time axis.

15.3.2.1 Presentation of the Approach

In order to accomplish a goal, the business system in which the personnel configuration is operating must be open and adjustable, while the staff and management are required to be flexible. Such an approach includes a specially developed project team configuration process. One of the examples for such a process is application of fuzzy evaluation that accounts for the project-specific hypotheses. All of the earlier considerations point to the fact that such a complicated system can only be optimized by expert and competent staff who know both business processes and technology well and who met the people in person. Also, there is a necessity for a strong decision-making support system to minimize the subjective approach. Hence, except for the fuzzy model, the AHP method for estimating the weighted values of roles and projects is used as a complementary element.

AHP allows decision makers to model a complex problem in a hierarchical structure showing the relationships of the goal, objectives (criteria), subobjectives, and alternatives. Uncertainties and other influencing factors can also be included. AHP allows for the application of data, experience, insight, and intuition in a logical and thorough manner. AHP enables decision makers to derive ratio scale priorities or weights as opposed to arbitrarily assigning them. In doing so, AHP not only supports decision makers by enabling them to structure the complexity and exercise better judgment but also allows them to incorporate both objective and subjective considerations in the decision process. AHP is a complementary decision methodology because alternatives that are deficient with respect to one or more objectives can be compensated by their performance with respect to other objectives. AHP is composed of several previously existing but unassociated concepts and techniques such as hierarchical structuring of complexity, pairwise comparisons, redundant judgments, an eigenvector method for deriving weights, and consistency considerations.

15.3.2.2 Presumption

The m roles (tasks, jobs) $R_1 ... R_m$ of n individuals $M_1 ... M_n$ and of l projects $P_1 ... P_l$ are discussed.

15.3.2.3 Specifics

- The suitability of person M_n to perform certain roles R_m necessary for the execution of project P_l is evaluated. For each role, a grade V_{min} is set, and it represents the acceptability level of a person for an assignment. Mark a_{ij}, $i=1...m$, $j=1...n$, is used.
- Roles do not hold the same importance, that is, each role has its assigned weight, which sets its importance level. The mark used is p_i, $i=1...m$.
- Projects do not hold the same importance, that is, each project has its assigned weight that sets its priority level. The mark used is w_k, $k=1...l$.
- The assigning of the evaluated personnel on projects is done according to established priorities. The mark used is b_{ik}, $i=1...m$, $k=1...l$.

15.3.2.4 Optimization Criteria

What determines the assignment of the evaluated individuals to project jobs is their suitability to perform specific tasks a_{ij} and project weight wk. This is why project P_i has precedence constraints determined by its weight. For instance, if project P_i has precedence over project P_k, this means that assigning resources to P_k should not be started before P_i is fully assigned ($P_i < P_k$):

$$\max \sum_{k=1}^{l} a_{ij} w_k$$

15.3.2.5 Procedure

1. By using the fuzzy system, the level of assignment appropriateness of each person available for each job in the team is evaluated. If a person is not a good fit for a certain job, he or she is excluded from consideration for this job (the set of possible solutions is reduced).
2. AHP allows determining the weight of each task in the project and the weight of each project.
3. The most important position is filled by the person with the highest grade for this particular job (i.e., the best individuals are selected for the most demanding roles). If the person is able to assume several roles in the team, he or she is given the job for which they received the highest grade. In case several people have the same fuzzy grade for the same job, the choice is made by detailing the grade based on the defuzzification process.
4. Next, the acceptability threshold is set. The threshold is constituted by the lowest grade an individual can achieve in order to be assigned

to a job in a team. If no individual can be found for a particular job, new employees must be acquired. In exceptional cases, if a person's grade is below acceptable, but very close to it (i.e., if the result of defuzzification shows this tendency), the person is included in the candidates pool, or the option of reevaluation is considered.

5. For each project, starting with the one with the highest priority, each job is filled with one individual as long as there remains a job that is unfilled and a suitable individual is found.

6. Project appointment is analyzed, and if there is a disproportion in the quality of the assigned personnel, a reassignment is performed among the projects. Clearly, for this, a mutual agreement of the decision makers (e.g., project manager, members of the organizational unit for decision making and supervision) is necessary.

The approach presented herewith is a summary of a general model and method proposed by Galinec (2010).

15.3.3 Myers–Briggs Indicator-Based Methods

This section presents an implementation of a psychometrical testing method for soft skills evaluation as they often affect the suitability for a job posting. HR apply a wide range of specific psychological tests that recruiters can bring to bear in the hiring process.

As an example, the Thomas–Kilmann conflict mode instrument measures to what extent people display competing, collaborating, compromising, avoiding, and accommodating behavior in conflict situations. The task–people (T–P) leadership questionnaire examines to what level individuals focus on tasks versus how much attention is provided to people at work. The FIRO-B Awareness Scale examines people along three dimensions: inclusion ("Do you desire strongly to be included in group activities? Do you like to include others?"), control ("Do you prefer being in situations that are well under control? Do you feel a strong need to take control of situations?"), and affection ("Is it important to you to be liked? Do you express affection toward others?").

The most popular and adaptable method is, however, the MBTI (Cakrt, 1996; Myers Briggs et al., 1998), which allows determining which category out of sixteen possible psychological types a candidate belongs to. This typology is a simple, practical, and reliable tool that is used to make the requirements for specific functions the most precise. The method is also helpful in searching for complementary types of leaders and subordinates as well as in optimizing project teams. Reliability and accuracy of the MBTI method have been confirmed by reviews of millions of tests carried out for employees from large business companies such as Apple Computer, Exxon, AT&T, Citycorp, General Electric, Honeywell, McDonald's, and 3M. Only in 1986 in the United States the test was carried out on over one and a half million people (Greenberg and Baron 1993). Since then MBTI and other popular psychological tests such as Belbin Team Roles (Belbin 2010a,b) have become a standard in the recruiting and selection process.

The models presented in the following are based on a combination of fuzzy-logic theory and the MBTI method.

15.3.3.1 Team Members Selection Based on Individual and Group-Related Factors

This section presents a five-stage model proposed by Wojnar (2011). This model is based on the assumption that team skills and cooperation capabilities of a job candidate are equally if not more important factors than individual characteristics. It is based on a multiple-criteria decision model using two sets of criteria: individual and group related with an emphasis on MBTI and CP.

In the first stage of Wojnar's model, the existing team performance is analyzed to determine the team communication channels and knowledge transfers. This analysis is performed by a simple survey asking the team members a selected set of questions (e.g., "who they communicate with most often" or "who they ask for advice"). Such type of questions identifies the subject matter experts and the knowledge transfer within the team. The member who is a "knowledge source" for most of the team members should be considered as the main contact point for a new member in the process of induction into new tasks and responsibilities. The team members who attain a lower score should be considered for other roles. The results of the survey are represented by a digraph or by its matrix.

The second stage in the team analysis is based on the MBTI concept. It assumes that every person has a natural preference in perceiving the world and making decisions [6]. This preference is defined by four pairs of dichotomous attributes: extroversion versus introversion, sensing versus intuition, thinking versus feeling, and perceiving versus judging [5]. The combination of one attribute from each pair creates sixteen psychological types that a person can be described by. Each of the MBTI types is decomposed into single attributes, and those attributes are assigned a score using the cooperation capabilities perspective (Wojnar 2010). At this stage, Saaty's fundamental scale for pairwise comparison is implemented to enable the quantification of the relations between the attributes. This practice constitutes the first group-related criterion for the final model.

The second group-related criterion is a competence level of a candidate in relation to other team members. For that purpose, Walukiewicz's concept of CP is used. CP, also called technological proximity, defines the cognitive distance between actors working on a particular problem (Walukiewicz 2007). It consists of codified and tacit knowledge related to the problem that is being solved as well as to the problem-related experience, differences, and similarities of the actors. CP facilitates their creative cooperation and stimulates innovative processes involved in the act. It makes their communication easier and simplifies the learning process.

The goal of CP is to define and select an optimal group of actors working together from the perspective of creative problem solving or, in other words, from the perspective of knowledge absorption, productivity, and innovation.

These are two main group-related criteria that are used in the team member selection model. They are complemented by the analysis of individual criteria such as education, experience, academic results, age, language skills, and interpersonal capabilities. Individual attributes of the existing team members have a minor influence on the model as they are already incorporated in the current performance of the team.

The next stage in the selection model deals with an analysis of potential candidates, and it starts with the MBTI analysis. Then, it is followed by the analysis of an individual's attributes including her or his geographical location.

The model for a new team member selection based on group-related and individual attributes can be used, for example, in global telecommunication companies that have plans to outsource some of their operational tasks (e.g., IT support, marketing). After the outsourcing process, these tasks are jointly performed by internal resources and external consultants depending on the criticality level. A company that wants to outsource some of its activities is sending out a request for proposals to the outsourcing services. Then, the providers are proposing delivery of the services including HR allocated for the tasks being performed. Next, the requesting company can use such a model for selection of optimal external candidates to be working with an internal team in order to achieve the highest possible quality of the service to be provided.

15.3.3.2 Team Structuring Based on Three-Stage Selection

This section presents a general model for the team-building process proposed by Bach-Dąbrowska (2012). This model is based on the assumption that the selection process is divided into three stages where at each stage increasingly stronger constraints are introduced. They are defined based on the team roles and behavioral profile of the entire team. This model emphasizes the necessity of behavioral adjustment of every new team member to the most closely related coworkers.

In this model, the project is characterized by its complexity, specific/required character of management, and duration of its execution. The vacancies that are to be filled within a project team are characterized by the required competency level, experience, availability, and MBTI profile. The candidate for vacant job postings, in turn, is characterized by his or her competencies, work experience, psychological profile, and availability. Information about the project, vacancies, and candidates is formulated in a linguistic manner. Values assigned to the decision variables are defined in both precise (i.e., crisp) and imprecise (i.e., fuzzy) manners and can be expressed in the form of numbers as well as words.

In the model presented herewith, the team structuring process includes

1. Stage 1—definition of project requirements
   a. Determination of the expected competency level for each vacancy in a project
   b. Determination of the behavioral type for each vacancy in a project
   c. Project complexity
   d. Project character
2. Stage 2—*preliminary verification process* based on analysis of application forms and selection of candidates who fulfill
   a. Given set of basic criteria
   b. Given set of required competencies
   c. Adjustment to MBTI for selected vacancies in a project team
   d. Candidate's availability criteria for a given period of time
3. Stage 3—final selection process
   a. Determination of the final pool of candidates divided into a main and a reserve group, depending on the output variable ranges obtained in the previous stages
   b. Determination of a pool of candidates selected for alternative project teams, conforming to the psychological profile of the group

The method requires defining the so-called reference model for modeling and calculation purposes. The reference model describes a set of variables that are exploited during the verification process. Reference models are based on fuzzy-logic theory and are divided according to three stages:

- A project requirement reference model
- A preliminary verification reference model
- A final selection reference model

Each reference model presented in Tables 15.1 through 15.5 is defined based on the following assumptions:

1. A set of linguistic variables $V_i = \{V_1, \ldots, V_n\}$, $i \in N - \{0\}$ defines input and output criteria of candidate assessment in structuring the process of project team formation.
2. Linguistic variable V_i is characterized by a quadruple $[L_i, T_i(L), \Omega_i, M_i]$, where
   a. $L_i = \{L_1, \ldots, L_n\}$, $i \in N - \{0\}$ is a set of names of linguistic variables
   b. $T_i(L_i) = \{T_1(L_1), \ldots, T_n(L_n)\}$, $i \in N - \{0\}$ is a set of countable term set of labels or the linguistic values

TABLE 15.1

Project Requirement Reference Model

V_i	L_i	$T_i(L_i)$	t_{ij}	Ω_i	M_i	m_{ij}
V_1	Project complexity level	$T_1(L_1)$	t_{11}(low) t_{12}(medium) t_{13}(high)	*Expert knowledge*	M_1	*Expert knowledge*
V_2	Project character	$T_2(L_2)$	t_{21}(technical) t_{22}(technical/soft) t_{23}(soft)	*Expert knowledge*	M_2	*Expert knowledge*
V_3	Expected competency level for a vacancy	$T_3(L_3)$	t_{31}(medium) t_{32}(medium high) t_{33}(high)	*Expert knowledge*	M_3	*Expert knowledge*
V_4	Expected MBTI profile for a vacancy	$T_4(L_4)$	t_{41}(ISTJ) t_{42}(ISTP) t_{43}(ESTP) t_{44}(ESTJ) t_{45}(ISFJ) t_{46}(ISFP) t_{47}(ESFP) t_{48}(ESFJ) t_{49}(INFP) t_{410}(ENFP) t_{411}(ENFJ) t_{412}(INTJ) t_{413}(INTP) t_{414}(ENTP) t_{415}(ENTJ) t_{416} (INFJ)	$[1\div16]$ u: *points*	M_4	m_{41} [1:1] m_{42} [2:1] m_{43} [3:1] m_{44} [4:1] m_{45} [5:1] m_{46} [6:1] m_{47} [7:1] m_{48} [8:1] m_{49} [9:1] m_{410} [10:1] m_{411} [11:1] m_{412} [12:1] m_{413} [13:1] m_{414} [14:1] m_{415} [15:1] m_{416} [16:1]

c. $t_{ij} = \{t_{11}, t_{12}, ..., t_{nm}\}$, $i, j \in N - \{0\}$, $t_{ij} \subset T_i(L_i)$ is a set of the linguistic values of the linguistic variable

d. $\Omega_i = \{\Omega_1, ...,\Omega_n\}$, $i \in N - \{0\}$ is a set of a universes of discourse for linguistic variable V_i

e. $M_i = \{M_1, ..., M_n\}$, $i \in N - \{0\}$ is a set of semantic rules

f. $m_{ij} = \{m_{11}, m_{12}, ..., m_{nm}\}$, $i, j \in N - \{0\}$, $m_{ij} \subset M_i$ is a variability range for linguistic value t_{ij} with grade of membership equal to 0 or 1

15.3.3.2.1 Stage 1: Definition of Project Requirements

Project requirement reference model is of a multi-input/multi-output (MIMO) type, where project complexity and project character are the input variables, while the expected competency level for a vacancy and the expected MBTI profile for a vacancy are the output variables (Figure 15.3). The first stage allows for identifying the main requirements resulting from the project specification.

The MBTI profile includes 16 human behavioral types that are composed of the combinations of elements, such as *I* for *introversion*, *E* for *extraversion*, *T* for *thinking*, *F* for *feeling*, *S* for *sensing*, *N* for *intuition*, *J* for *judging*, and *P* for

TABLE 15.2

Preliminary Verification Reference Model

V_i	L_i	$T_i(L_i)$	t_{ij}	Ω_i	M_i	m_{ij}
V_5	Independent vacancy	$T_5(L_5)$	t_{51}(low) t_{52}(medium low) t_{53}(medium high) t_{54}(high)	*Expert knowledge*	M_5	*Expert knowledge*
V_6	Project team membership	$T_6(L_6)$	t_{61}(small) t_{62}(medium) t_{63}(high)	*Expert knowledge*	M_6	*Expert knowledge*
V_7	Candidate competency level	$T_7(L_7)$	t_{71}(low) t_{72}(medium) t_{73}(high)	*Expert knowledge*	M_7	*Expert knowledge*
V_8	MBTI profile	$T_8(L_8)$	t_{81}(ISTJ) t_{82}(ISTP) t_{83}(ESTP) t_{84}(ESTJ) t_{85}(ISFJ) t_{86}(ISFP) t_{87}(ESFP) t_{88}(ESFJ) t_{89}(INFP) t_{810}(ENFP) t_{811}(ENFJ) t_{812}(INTJ) t_{813}(INTP) t_{814}(ENTP) t_{815}(ENTJ) t_{816}(INFJ)	$[1 \div 16]$ *u: points*	M_8	m_{81} [1:1] m_{82} [2:1] m_{83} [3:1] m_{84} [4:1] m_{85} [5:1] m_{86} [6:1] m_{87} [7:1] m_{88} [8:1] m_{89} [9:1] m_{810} [10:1] m_{811} [11:1] m_{812} [12:1] m_{813} [13:1] m_{814} [14:1] m_{815} [15:1] m_{816} [16:1]
V_9	Availability	$T_9(L_9)$	t_{91}(consistent) t_{92}(inconsistent)	*Expert knowledge*	M_9	*Expert knowledge*
V_{10}	Vacancy character	$T_{10}(L_{10})$	t_{101}(independent) t_{102}(dependent)	*Expert knowledge*	M_{10}	*Expert knowledge*
V_{11}	Vacancy qualification	$T_{11}(L_{11})$	t_{111}(sufficient) t_{112}(insufficient)	*Expert knowledge*	M_{11}	*Expert knowledge*

TABLE 15.3

Input and Output Data for Second Stage of Verification Process

Step	Input Variables	Output Variables
1	V_5, V_6, V_{10}	V_{11}
2	V_7, V_{10}	V_{11}
3	V_8, V_{10}	V_{11}
4	V_9, V_{10}	V_{11}

TABLE 15.4

Final Selection Reference Model (Step 1)

V_i	L_i	$T_i(L_i)$	t_{ij}	Ω_i	M_i	m_{ij}
V_7	Competency level	$T_7(L_7)$	t_{72}(medium) t_{73}(high)	*Expert knowledge*	M_7	*Expert knowledge*
V_8	MBTI profile	$T_8(L_8)$	t_{81}(ISTJ)			m_{81} [1:1]
			t_{82}(ISTP)			m_{82} [2:1]
			t_{83}(ESTP)			m_{83} [3:1]
			t_{84}(ESTJ)			m_{84} [4:1]
			t_{85}(ISFJ)			m_{85} [5:1]
			t_{86}(ISFP)			m_{86} [6:1]
			t_{87}(ESFP)			m_{87} [7:1]
			t_{88}(ESFJ)	[1÷16]	M_8	m_{88} [8:1]
			t_{89}(INFP)	*u: points*		m_{89} [9:1]
			t_{810}(ENFP)			m_{810}[10:1]
			t_{811}(ENFJ)			m_{811}[11:1]
			t_{812}(INTJ)			m_{812}[12:1]
			t_{813}(INTP)			m_{813}[13:1]
			t_{814}(ENTP)			m_{814}[14:1]
			t_{815}(ENTJ)			m_{815}[15:1]
			t_{816}(INFJ)			m_{816}[16:1]
V_{11}	Posting adjustment	$T_{11}(L_{11})$	t_{111}(sufficient) t_{112}(insufficient)	*Expert knowledge*	M_{11}	*Expert knowledge*
V_{12}	Assignment	$T_{20}(L_{20})$	T_{201}(basic) T_{202}(reserve A) T_{203}(reserve B)	*Expert knowledge*	M_{20}	*Expert knowledge*

perceiving. They include the following combinations: ISTJ, ISTP, ESTP, ESTJ, ISFJ, ISFP, ESFP, ESFJ, INFP, INFJ, ENFP, ENFJ, INTJ, INTP, ENTP, and ENTJ.

As the models are defined at a general level, undefined values for universe of discourse for linguistic variables and for variability range of linguistic values occur. Those values must be defined by a domain expert (e.g., from management, psychology, and fuzzy system).

Reading key

Variable V_1

Name L_1: project complexity level

Set of linguistic values $T_1(L_1)$: {low, medium, high}

The universe of discourse of V_1, Ω_1: [0 ÷ 6]

Exemplary terms for linguistic values

m_{11}[0:1 1:1 2:0] LE (left external)

m_{12}[1:0 3:1 4:0] TA (triangular asymmetrical)

m_{13} [3:0 5:1 6:1] TRA (trapezoidal asymmetrical)

Graphical Representation (Figure 15.4).

TABLE 15.5

Final Selection Reference Model (Step 2: Team Adjustment)

V_i	L_i	$T_i(L_i)$	t_{ij}	Ω_i	M_i	m_{ij}
V_{13}	Posting A character	$T_{13}(L_{13})$	$t_{13,1}$(executive) $t_{13,2}$(subsidiary)	*Expert knowledge*	M_{13}	*Expert knowledge*
$V_{13'}$	Posting B character	$T_{13'}(L_{13'})$	$t_{13',1}$(executive) $t_{13',2}$(subsidiary)	*Expert knowledge*	M_{14}	*Expert knowledge*
V_8	MBTI profile candidate A	$T_8(L_8)$	$t_{8,1}$(ISTJ)			$m_{8,1}$ [1:1]
			$t_{8,2}$(ISTP)			$m_{8,2}$ [2:1]
			$t_{8,3}$(ESTP)			$m_{8,3}$ [3:1]
			$t_{8,4}$(ESTJ)			$m_{8,4}$ [4:1]
			$t_{8,5}$(ISFJ)			$m_{8,5}$ [5:1]
			$t_{8,6}$(ISFP)			$m_{8,6}$ [6:1]
			$t_{8,7}$(ESFP)			$m_{8,7}$ [7:1]
			$t_{8,8}$(ESFJ)	$[1 \div 16]$	M_8	$m_{8,8}$ [8:1]
			$t_{8,9}$(INFP)	*u: points*		$m_{8,9}$ [9:1]
			$t_{8,10}$(ENFP)			$m_{8,10}$ [10:1]
			$t_{8,11}$(ENFJ)			$m_{8,11}$ [11:1]
			$t_{8,12}$(INTJ)			$m_{8,12}$ [12:1]
			$t_{8,13}$(INTP)			$m_{8,13}$ [13:1]
			$t_{8,14}$(ENTP)			$m_{8,14}$ [14:1]
			$t_{8,15}$(ENTJ)			$m_{8,15}$ [15:1]
			$t_{8,16}$ (INFJ)			$m_{8,16}$ [16:1]
$V_{8'}$	MBTI profile candidate B	$T_8(L_8)$	$t_{8,1}$(ISTJ)			$m_{8,1}$ [1:1]
			$t_{8,2}$(ISTP)			$m_{8,2}$ [2:1]
			$t_{8,3}$(ESTP)			$m_{8,3}$ [3:1]
			$t_{8,4}$(ESTJ)			$m_{8,4}$ [4:1]
			$t_{8,5}$(ISFJ)			$m_{8,5}$ [5:1]
			$t_{8,6}$(ISFP)			$m_{8,6}$ [6:1]
			$t_{8,7}$(ESFP)			$m_{8,7}$ [7:1]
			$t_{8,8}$(ESFJ)	$[1 \div 16]$	$M_{8'}$	$m_{8,8}$ [8:1]
			$t_{8,9}$(INFP)	*u: points*		$m_{8,9}$ [9:1]
			$t_{8,10}$(ENFP)			$m_{8,10}$ [10:1]
			$t_{8,11}$(ENFJ)			$m_{8,11}$ [11:1]
			$t_{8,12}$(INTJ)			$m_{8,12}$ [12:1]
			$t_{8,13}$(INTP)			$m_{8,13}$ [13:1]
			$t_{8,14}$(ENTP)			$m_{8,14}$ [14:1]
			$t_{8,15}$(ENTJ)			$m_{8,15}$ [15:1]
			$t_{8,16}$ (INFJ)			$m_{8,16}$ [16:1]
V_{14}	Effectiveness of cooperation	$T_{14}(L_{14})$	t_{14}(low) t_{14}(medium) t_{14}(high)	*Expert knowledge*	M_{14}	*Expert knowledge*

15.3.3.2.2 *Stage 2: Preliminary Verification Process*

The preliminary selection reference model is of a multi-input/single-output (MISO) type. The implementation of this model is divided into four steps (Figure 15.5).

The first step of Stage 2 includes a verification of all applicants. At each subsequent step (2–4) of Stage 2, the number of candidates is reduced based

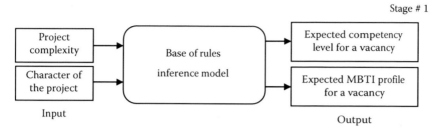

FIGURE 15.3
MIMO model.

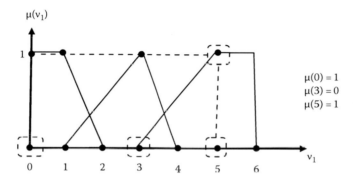

FIGURE 15.4
Graphical representation of fuzzy definition for variable V_1.

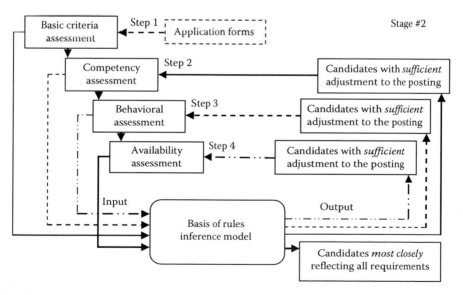

FIGURE 15.5
Four steps of the preliminary selection process.

on the previous stage result. Candidates with a *sufficient qualification for the vacancy* are graded using a predefined set of criteria. In the fourth step of verification, the output list of successful candidates is sorted based on their score. Only those applicants who achieved at least 0.5 grade of membership of all basic criteria are accepted for the selection process. At Stage 2, however, candidates are not yet categorized according to their achieved results, and also their psychometrical results are not yet "confronted with each other." That means the competencies of a selected candidate may suffice to fulfill the posting requirements, but there is no guarantee regarding their effective cooperation with other team members. Those two important elements are verified in Stage 3.

15.3.3.2.3 *Stage 3: Final Selection Process*

The final selection process is divided into two steps. The first step of Stage 3 allows for categorizing candidates selected in previous Stage 2 according to their achievements. As a result of this process, three lists of candidates are prepared:

- Basic list—includes candidates whose assignment to the vacant posting is not less than 0.8 grade of membership
- Reserve A list—includes candidates whose assignment to the vacant posting is between 0.6 and 0.8 grade of membership
- Reserve B list—includes candidates whose assignment to the vacant posting is between 0.5 and 0.6 grades of membership

In the second step of the final selection process, the comparison of selected candidates according to the vacancies and the relation between those vacancies and the MBTI profile of candidates is computed. As a result of the final step in the selection process, the alternative sets of team fulfilling given assessment criteria are generated.

To conduct the structuring process based the on presented reference models, it is necessary to implement those models in the fuzzy-logic system. A proposal of such a system is presented in Figure 15.6. The system structure includes seven rule bases. This solution comes about for two main reasons:

- The process of candidate verification is divided into stages and steps. This means that in each step, different criteria are checked to allow for a gradual selection of candidates. Hence, the output values from one stage are implemented as input values for the next stage.
- The division of rule bases allows minimizing the number of rules included in the inference process. For example, if in Stage 2 instead of four different rule blocks with a total number of rules equal to 66,

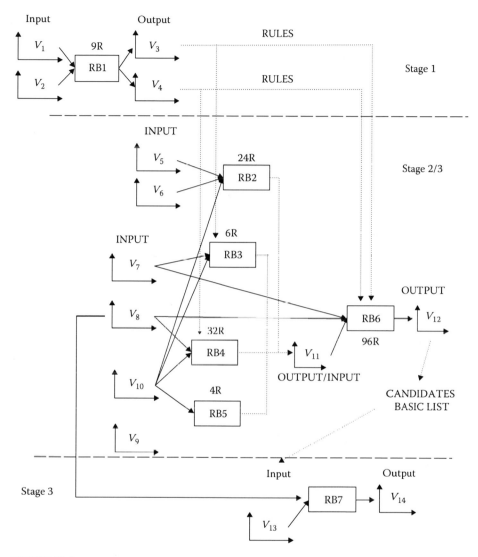

FIGURE 15.6
Fuzzy system structure.

only one rule block has been designed, the number of rules would increase to 1536. This, in turn, would cause difficulty in ensuring the consistency and correctness of the knowledge base as well as significantly increase the computation time.

In the proposed system, the output information from Stage 1 constitutes input project criteria defined in the rule base in Stage 2. The output variable from Stage 2 is applied as an input variable for Stage 3. As a result of the inference process in that stage, the basic list of best qualified candidates and two additional reserve lists are determined. In the next step (Stage 3), the candidates from the basic list are verified according to their psychological profile and effectiveness of cooperation in a given project team. This allows for forming the group of candidates for alternate project teams.

The general reference model proposed in this chapter is only the beginning of the research in the area of the multiple-criteria decision making for a team structuring process based on fuzzy-logic theory and psychological methods implemented in HR systems. The next step requires a verification and determination of membership functions for describing linguistic variables and their variability ranges. This includes studies and analyses based on the knowledge of experts from various fields (e.g., management, psychology, and decision support systems). The main goal of the research is to develop a consistent and general model of a selection process that combines an assessment of soft and hard skills and enables guaranteeing the behavioral cohesion of the team that ultimately translates into the effectiveness of decision making and project task execution.

15.4 Summary

This chapter contains an introduction to fuzzy-logic model-based approaches for a team-building process. Such approaches allow extending typical recruiting practice and selection processes to enable a wider and more precise assessment of a new team and/or existing team members, taking into account both their hard and soft skills. Moreover, as effectiveness of teams depends on the interpersonal skills and emotional intelligence of their individual members, the discussed models emphasize the value of behavioral criteria that are conducive to team cohesion.

Fuzzy logic, however, is not the only formalism supporting modeling and solving of multicriteria decision problems in a team structuring process. There are at least a few interesting proposals of deterministic models such as the clustering method presented in Hlaoittinun et al. (2007), the multicriteria group decision model presented in Hatush and Skitmore (1998), Topcu (2004), and Alencar and Almeida (2010), or a rule-based expert system implementing a method for team synergy examination presented in Mosterman (1990).

The presented material is hoped to encourage the reader to further study the topic by exploring any of the extensive list of references.

References

Alcalde, C. and Burusco, A. (2010). L-Fuzzy Concepts and linguistic variables in knowledge acquisition processes. In M. Kryszkiewicz and S. A. Obiedkov (Eds.), *Proceedings of the 7th International Conference on Concept Lattices and Their Applications* (pp. 38–49). Sevilla, Spain.

Alencar, L.H. and Almeida, A.T. (2010). A model for selecting project team members using multicriteria grupu decision making. *Pesquisa Operacional, 30*(1), 221–236.

Anderson, N. and Cunningham-Snell, N. (2000). Personnel selection. In N. Chmiel (Ed.), *Work and Organisational Psychology*. Oxford, U.K.: Blackwell Publishers Ltd.

Argote, L. and McGrath, J. D. (1993). Group processes in organisations: Continuity and change. In C. L. Cooper and I. T. Robertson (Eds.), *International Review on Industrial and Organizational Psychology* (Vol. 8). New York: John Willey & Sons.

Bach-Dąbrowska, I. (2012). Fuzzy model for structuring project teams. *Applied Computer Science, 8*(1), 3–24.

Bednarz, T. (2012). The five characteristics of strong team. Retrieved November 5, 2012, from http://www.pmhut.com/the-five-characteristics-of-strong-teams

Belbin, M. R. (2010a). *Management Teams: Why They Succeed or Fail*, 3rd edn. Oxford, U.K.: Butterworth Heinemann.

Belbin, M. R. (2010b). *Team Roles at Work*, 2nd edn. Oxford, U.K.: Butterworth Heinemann.

Bolles, D. and Fahrenkrog, S. (2004). *A Guide to the Project Management Body of Knowledge* (pp. 1–418). Seattle, WA: Project Management Institute.

Business Dictionary. (2012). No Title. Retrieved from http://www.businessdictionary.com/definition/work-team.html

Cakrt, M. (1996). *Typologie osobnosti pro menazery. Kdo jsem ja, kdo jste vy?* Management Press, NT Publishing Ltd, Praha.

Davidson Frame, J. (2003). *Managing Project in Organizations*, 3rd edn. (pp. 1–279). San Francisco, CA: Jossey-Bass.

Galinec, D. (2010). The framework for project teams optimal assignment. *The International Journal of Applied Strategic Management, 1*(3), 1–21.

Garg, S. and Gaur, V. K. (2011). Use of fuzzy logic and its implementation in software engineering. *International Journal of Research in Science and Technology, 1*(II) (pp. 1–8). Retrieved from http://www.ijrst.com

GRB Career Matchmaking. (2012). Selection method. *Recruitment Guide*. Retrieved from http://www.grb.uk.com/selection-methods, accessed date 20th October 2012.

Greenberg, J. and Baron, R. A. (1993). *Behavioral in Organisations: Understanding and Managing the Human Side of Word* (p. 216). Needham Heighs, MA: Allyn and Bacon.

Harvey, S., Millett, B., and Smith, D. (1998). Developing successful teams in organisations. *Australian Journal of Management & Organisational Behaviour, 1*(1), 1–8.

Hatush, Z. and Skitmore, M. (1998). Contractor selection using multicriteria utility theory: An additive model. *Building and Environment, 33*(2), 105–115.

Hlaoittinun, O., Bonjour, E., and Dulmet, M. (2007). A team building approach for competency development. *2007 IEEE International Conference on Industrial Engineering and Engineering Management* (pp. 1004–1008). IEEM. doi:10.1109/IEEM.2007.4419343

IRS Employment Review. (1995). *Key Issues in Effective Teamworking*, Vol. 592 (pp. 5–16). London, U.K.: Eclipse Groupe.

Juchnowicz, M., Rostkowski, T., and Sienkiewicz, Ł. (2008). *Narzędzia i praktyka zarządzania zasobami ludzkimi*. Warszawa, Poland: POLTEXT.

Katzenbach, J. R. and Smith, D. K. (1993). *The Wisdom of Teams*. New York: McKinsey & Company.

Kirkman, B. L. and Shapiro, D. L. (1997). The impact of cultural values of employee resistance to teams: Toward a model of Globalised Self-Managing Work Team effectiveness. *Academy of Management Review, 22*(3), 730–757.

McClain, G. and Romaine, D. (2002). *The Everything Managing People Group Book*, 2nd edn. (pp. 1–336). Avon, MA: Adams Media; F+W Publication Company.

Mosterman, P. J. (1990). NPI Workshop Team Synergy Subsystem, technical report #CIS-90-07.

Myers Briggs, I., McCaulley, M. H., Quenk, N. L., and Hammer, A. L. (1998). *MBTI Manual (A Guide to the Development and Use of the Myers Briggs Type Indicator)*. Palo Alto, CA: Consulting Psychologists Press.

Piegat, A. (2003). *Modelowanie i sterowanie rozmyte*. (L. Bolc, Ed.) (pp. 1–626). Warszawa, Poland: Akademicka Oficyna Wydawnicza EXIT.

Rutkowski, L. (2005). *Metody i techniki sztucznej inteligencji*. Warszawa, Poland: Wydawnictwo Naukowe PWN.

Schroder, J. P. and Diekow, S. (2006). *We Sie Projekte zum Erfolg fuhren*. Berlin, Germany: Cornelsen Verlag Scriptor GmbH&Co. KG.

Stott, K. and Walker, A. (1995). *Teams: Teamwork and Teambuilding*. New York: Prentice Hall.

Sundstrom, E., DeMeuse, K. P., and Futrell, D. (1990). Work teams: Applications and effectiveness. *American Psychologist, 45*(2), 120–133.

Topcu, Y. (2004). A decision model proposal for construction contractor selection in Turkey. *Building and Environment, 39*(4), 469–481.

Wageman, R. (1997). Critical success factors for creating Superb Self-managing Teams. *Organisational Dynamics, 26*(1), 49–60.

Walukiewicz, S. (2007). Four forms of capital and proximity. Working paper. *System Research Institute of Polish Academy of Science—Working Papers, 3*.

West, M. A., Borrill, C. S., and Unsworth, K. (1998). Team Effectiveness in Organisations. In C. L. Cooper and I. T. Robertson (Eds.), *International Review on Industrial and Organizational Psychology* (Vol. 13). Chichester, U.K.: Wiley & Sons.

Wojnar, J. (2010). MBTI as a measure of Emotive Proximity. In J. Hołubiec (Ed.), *Analiza Systemowa w Finansach i Zarządzaniu*. Warszawa, Poland: Instytut Badań Stosowanych PAN.

Wojnar, J. (2011). Multicriteria decision making model for the new team member selection based on individual and group-related factors. *Foundations of Management, 3*(2), 103–114. doi:10.2478/v10238-012-0045-4

Zadeh, L. A. and Hose, S. (1975). The concept of a linguistic variable and its application to approximate reasoning—II*. *Information Science, 8*, 301–357.

16

Opportunities and Challenges in Data Journalism

Susan McGregor and Alon Halevy

CONTENTS

16.1 Introduction: The Data Dilemma

On August 11, 2011, British Prime Minister David Cameron addressed an emergency parliamentary session called in response to the widespread U.K. riots of the preceding week. In his remarks, he stated that *Everyone watching these horrific actions will be struck by how they were organised via social media* and suggested that the U.K. government was exploring the possibility of limiting or banning access to social media during periods of public unrest [11].

Since the Arab Spring earlier that year, the role of social media as an organizing tool for demonstrations and protests had become accepted wisdom in many circles, providing a basis for Cameron's assertion that the government should have the right "to stop people communicating via these Websites and services when we know they are plotting violence, disorder and criminality" [11]. Less than 2 weeks later, however, a preliminary analysis and visualization of more than 2.5 million tweets by the U.K. newspaper *The Guardian* [3] indicated that riot-related traffic on the web service tended to spike *after* the violence began in a particular neighborhood, as shown in Figure 16.1. Over the following months, *The Guardian* partnered with London School of Economics and the University of Manchester, where more formal analyses confirmed that social media outlets such as Twitter and Facebook were not a factor in organizing the riots. In fact, they played a vital role in mobilizing cleanup efforts [2].

In 2011, more than 2.4 billion people were using the Internet worldwide [12], meaning that more than one-quarter of the world's population is either consuming or creating content on the web. Where mass communication was once the exclusive purview of governments and media outlets, today an organization of one can reach an audience of millions. In our information society, this is a huge disruption to the traditional balance of power in which governments and media organizations create the public narrative and individuals can only consume it. As the evolution of communication technologies has contributed to the collapse of both nations and news organizations, the journalism industry has struggled to make its own voices heard among the billions now echoing online. Yet as work like that of the *Guardian* demonstrates, there remains an elemental need for professional journalism executed with relevance, rigor, and accuracy. These are the qualities that define data journalism, both big and small, allowing it to be the much-needed signal in the noise that provides coherence and transparency, especially amid the cacophony that often constitutes conversations on the web.

Whether it is the protection of civil liberties in the United Kingdom supported by *The Guardian*'s work on the London riots, the prompting of privacy legislation in the United States triggered by the *Wall Street Journal*'s "What They Know" series, or the arrests and reform spurred by the *Los Angeles Times*' analysis of government payrolls—data journalism plays an increasingly important role in providing impetus and insight in our increasingly complex and interconnected world.

Over the course of this chapter, we will explore the issues that journalists face in using data as a source for their reporting, such as the time, skills, and tools required to process data, as well as specific technical issues related to discovering, cleaning, contextualizing, and publishing data-driven analyses. Along the way, well-discussed technologies like Google Fusion Tables and Google Refine have made this type of reporting faster and more accessible. Finally, we will highlight issues and opportunities still to be addressed in this promising and growing field.

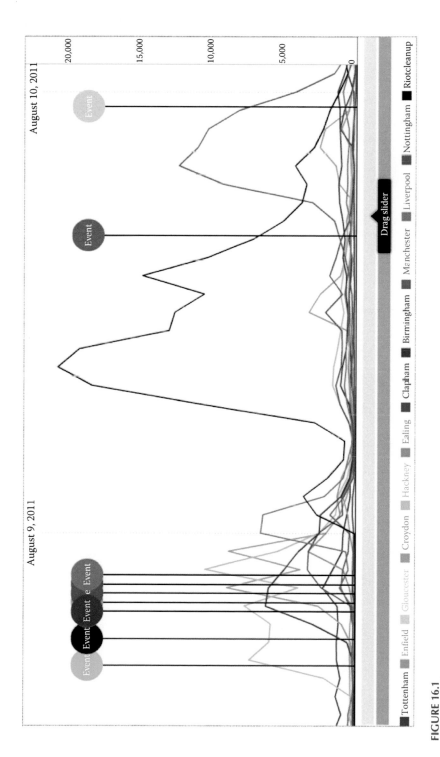

FIGURE 16.1

A screenshot of the interactive line chart published by the *Guardian* on its website on August 24, 2011. The large spike in the middle, shown in black, indicates tweets related to the riot cleanup.

16.2 Rise of Data Journalism: Contributing Factors

16.2.1 Need for Speed

Data-driven analysis and stories in the newsroom are nothing new. For decades, computer-assisted reporting has analyzed compilations of public records, private databases, and other specialized data sources to reveal newsworthy patterns and anomalies, or simply to identify leads for further investigation using more traditional reporting techniques.

Yet, while personal computers have been a newsroom mainstay for many years, most reporters typically have been ill-equipped to exploit the often powerful resources lying idle on their hard drives. The evolution of data analysis as a mainstream reporting technique was stymied by a range of obstacles: arcane and obscure data formats requiring highly specialized languages and skills to manipulate; limited technologies to support visualization and pattern-recognition, data literacy, and numeracy challenges among reporters; and outdated or difficult-to-obtain data sets. In an environment where news organizations are increasingly in competition with social media to break news, the expertise overhead and time lag involved in obtaining, cleaning, and analyzing data for many years made it impossible to use for deadline reporting, making the practice of data-driven journalism of secondary importance in most newsrooms, at best. As we will see later in this chapter, however, technologies such as Google Fusion Tables and Refine have begun to dismantle some of these obstacles, allowing real-time data journalism to become a core component of the twenty-first-century newsroom.

16.2.2 Event of the Network

Traditionally, big data has been defined solely by its size, and, specifically, whether a given data set required a supercomputer to process. The most relevant measure of big data today, however, is not the size of the file but the extent of the network [6]. Thus, a single YouTube video becomes big data by generating 100 million page views, as occurred with the Kony 2012 campaign [14]; likewise, so do the few hundred insurance companies identified by the Sarasota Herald Tribune through analysis of their networks of financial assets and obligations [13]. Because big data comprised of networks and relationships embodying phenomena that affect a large number of people, these types of data sets are especially germane to the journalistic enterprise. These data sets also provide a narrative- and context-rich complement to more abstract numerical data that, while offering its own opportunities for discovering news, can be much more challenging to use journalistically, as described in the next section.

16.2.3 Numbers Are Not Neutral

Even today, discussions of the journalistic ideal of objective reporting often hearken back to the heyday of broadcast news and the tenure of news anchors like Walter Cronkite and Edward R. Murrow. Cronkite's signature sign-off, *And thats the way it is,* embodied the concept of a single, encompassing viewpoint that laid claim to an unequivocal truth in news. With the rise of cable television and the *he said/she said* approach to balancing coverage, however, that reassuring sense of a single true narrative on important issues was lost, and with it, much of the apparent confidence of news audiences in the value of professional journalism. This erosion of the profession in the public view has only been exacerbated by the rise of the web and the access it affords to many thousands of interpretations of events and ideas, many of them persuasive and contradictory in seemingly equal measure.

At first glance, data journalism appears to offer a simple antidote to such problems. In a society long conditioned to view numbers as facts, employing the techniques of scientists and statisticians in analyzing the issues of the day seems to hold great promise for improving the scope and accuracy of journalistic conclusions. Yet much of the increasing pool of readily available, well-structured data accessible to journalists, especially on the web, provides little to no context about what it describes or how it was generated. Tempted by the availability and apparent truthiness of these typically numerical data sets, journalists may underreport the story behind the data and unwittingly misinterpret it or misapply their analyses. For example, climate scientists claim that in some cases climate projections have been made with data sets that do not have the appropriate fidelity for the particular projection.

In some cases, the absence of qualifying information around a particular data set may even be interpreted by journalists and audiences as an indication of the data's authoritative status, reinforcing the idea that data is an independent entity: privileged, objective, and free of human influence. The now-popular concept that data can and do speak for themselves, rather than the people who designed them [1], illustrates this phenomenon. Yet, presenting data without sufficient context not only risks its misapplication, it robs the resulting journalism of the essential narrative and relational qualities that make for the most powerful and affecting news stories.

16.3 Reality of Data Journalism: Facing Obstacles

16.3.1 Discovering Data on the Web

Preserving the context of data sets, however, is not just a philosophical problem, but a significant technical one. Many journalists rely on web searches to locate data sets and other structured information, but indexing such information on the web so that it can be discovered through search engines presents two primary challenges.

The first is that a great deal of the structured information that is available online resides in the deep web. Data on the deep web is the data we typically access through HTML forms (e.g., job search, airline ticket purchase, public records). The data itself is sitting in a back-end database and is served as an HTML page when a user enters a particular query or set of commands. Such content is often described as part of the *invisible web* because search engines cannot penetrate the forms and crawl the underlying content. Such data, therefore, may not be returned in relevant searches.

In addition to data on the deep web, many valuable data sources are locked up in organizational databases. In some cases, the data is intentionally less discoverable because it is proprietary or contains private information that should not be made available to the general public. In other cases, however, high-quality public data is locked away in these repositories for no truly good reason. Getting this data out involves several challenges. Some of the challenges of accessing this data are technical, such as determining how to extract the data from the back-end system in a format that anyone can access (e.g., CSV, XML).

Possibly more challenging is convincing data owners to release their data, whether it is private or nominally public. Data owners may fear that their data will be misused, will create liability issues, or they may simply be concerned about what extensive analysis by third parties might reveal. In some cases, data owners may release information to journalistic institutions, but only on the condition that the news organization's analysis alone—and not the underlying data—be published. While this can still provide substantive reporting value, data transparency is quickly becoming a part of the trust and loyalty relationship between news outlets and their audiences. As discussed earlier, however, publishing data directly requires additional work in the form of accurate and appropriate contexts and methodologies alongside, as well as taking precautions against dangerous invasions of privacy and individual security.

16.3.2 Making Sense of Discovered Data

Getting data out of the deep web or other repositories is only the first step. Once data is discovered, filtering out the good data and making sense of it are very challenging tasks. In particular, structured data on the so-called surface web poses almost as many technical challenges to discovery as that on the deep web. While more visible to humans, web pages that contain tables, HTML lists, or pages that have a repeated structure (such as the one shown in Figure 16.2) are in fact quite difficult for search engines to index properly for return in relevant search queries.

The main problem is that search engines are optimized for searching text documents, not structured data. For example, consider a query for zambia

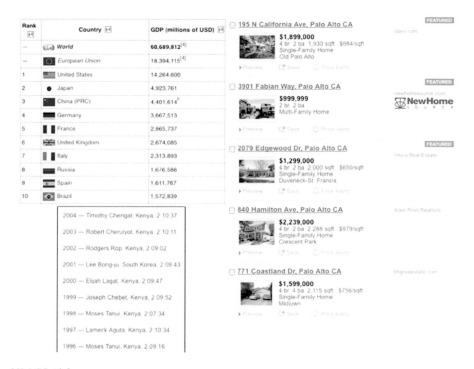

FIGURE 16.2

The different kinds of structured data on the web. The top left shows an HTML table of country GDPs, and the bottom left shows part of an HTML list of the winners of the Boston Marathon. On the right, we see a search result for real estate listings, where each result is formatted as a card.

population. A table that has this data will have a column header population, and close to the bottom of the table, it will have a row with the string zambia. However, the strings population and zambia appear very far from each other in the document, which a search engine will often take as a sign that those words are not closely related and therefore that the page is not relevant to the query. As we know, however, in the context of a table, the header of the column is as relevant to the last row as it is to the first.

Beyond this, it is hard for search engines to determine which pages on the web contain high-quality data. Less than 1% of the HTML tables on the web have good data in them, in part because so many of them are used exclusively for formatting purposes [4]. However, recent work [19] has shown that detecting semantic coherence of a column in a table is an effective signal for determining whether a table has high-quality relational content. For example, a table whose rows all contain the names of tree species is semantically coherent and therefore probably contains useful structured data.

While such insights are useful, the current reality is that we still know very little about the semantics of structured data on the web. Because the only schema information we have about tables are column names (at best), inferring the broader context of data is quite difficult. Any contextual information that may exist, including the relations represented by the data, is typically described in the surrounding text, unconnected to the table itself. For example, the fact that the table on the bottom left of Figure 16.3 lists the winners of the Boston Marathon is not in the table itself, but in the surrounding text. Moreover, the semantics of data-related queries can be very brittle—changing one token on the page (e.g., which year the data is from) can completely change the relevance of a given table to a particular query. Furthermore, some data sets may be available only as a set of smaller tables scattered on individual web pages, because their presentation is optimized for human consumption.

Consequently, for journalists, successfully locating structured data is only the beginning. Limited contextual data—wherever it may appear on a form or page—often makes determining the relevance of a data set to a particular journalistic question quite difficult. Moreover, while tools like ScraperWiki [18] are designed specifically to support and be accessible to journalists, knowledge of these tools and the skills to use them effectively are still beyond the expertise of most newsrooms, especially at smaller outlets. Where relational data sets are made available in their original form, it is often in proprietary formats that require costly software and/or expert skills to fully exploit, once again posing an obstacle to data access for increasingly time- and resource-strapped newsrooms.

16.3.3 Problem of Dirty Data

Assuming that the relevant data set can be obtained in a malleable format, it often needs to be *cleaned* before it can be used. Data often comes with errors, formatting irregularities, missing values, or using conventions that may not make sense outside of the context in which it was created. For example, if data is displayed on a map, then it is important that all the location names are recognized by the mapping system so it can provide the correct coordinates. Developing intuitive and easy-to-use data cleaning systems is a critical technical challenge.

For many years, the go-to tool for data cleaning has been Microsoft Excel, at least in part because of its ubiquity; as part of the Microsoft Office suite, it is a tool that has been available on most professional computer desktops for decades. For small-scale cleaning and analysis, Excel provides basic data sorting and editing features, including find and replace, and mathematical and statistical features such as functions for calculating sums, averages, and even standard deviations and z-indexes. More advanced users can also use Excel to create basic charts, graphs, macros, and pivot tables for data analysis.

Most journalists, however, have neither the time nor the training resources to become expert users of Microsoft Excel or virtually any piece of specialty software. While the spreadsheet interface is generally well understood even

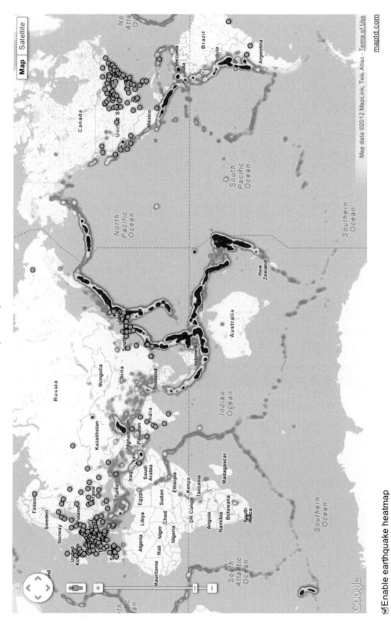

FIGURE 16.3

A map that combines two disparate data sources: (1) earthquake activity since 1973 and (2) location of nuclear plants. This map appears on http://maptd.com.

by novice users, the really powerful features of programs like Excel usually require expert instruction and many hours of practice to master. Moreover, a large part of understanding whether a given data set is relevant to a particular journalistic question rests on assessing its specific size, contents, and parameters, and tools like Excel require that much of this type of analysis still be done by inspection.

Fortunately, the recent introduction of Google Refine [9] has significantly streamlined the most fundamental tasks of data analysis. Rather than manually scrolling through an entire data set to see how many rows it contains, Google Refine indicates the current row count of a data set at the top of the screen as soon as it has been loaded. Refine also offers a powerful facet feature, which instantly identifies each unique data value present in a given data column and presents the value and the number of times it appears in a small summary pane at left. Because the default text-sort in the summary window is alphabetical, minor misspellings or letter-case differences will often appear adjacent to one another and can be quickly identified and mass edited directly in that same pane. In Excel, the Find and Replace feature will update only exact text matches, but one must identify all variations by direct inspection. All too often, small differences in the data values are not discovered until much later in the process, resulting in analysis and publishing errors.

Google Refine also provides functions for quickly viewing subsets of even complex data sets. For every column that is faceted, clicking on a value in the summary window will filter the data to show only rows containing that value. By selecting values in multiple summary windows, one can filter the data by several parameters with little more than the same number of clicks. As each filter is applied, the row count updates to indicate how many records match the current selections, and the filtered data set can be exported at any time without special configuration. Perhaps most importantly, all filters can be instantly removed and the full data set restored by simply closing out the summary windows.

16.3.4 Tools of Interrogation

At its most fundamental level, doing data journalism means asking journalistic questions of data. While most journalists have the skills to do document research, interview human beings, and draw meaningful interpretations from events, the need for special technical skills and other resources has prevented many journalists (and even whole newsrooms) from learning how to effectively interrogate data.

Although Excel has some applications as a tool for data interrogation, it is fairly static from the perspectives of data querying and management. For example, Excel requires that two related data sets be manually correlated into a single spreadsheet in order to be evaluated together. Because its analysis tools often operate on and change the underlying data, managing multiple views on even a single data set typically requires creating either multiple sheets or multiple workbooks. As files are correlated or updated, new copies must be transmitted to collaborators, usually via email. These files necessarily accumulate, and it can very quickly become difficult and time-consuming to identify the most current or appropriate version of the data set in question.

Those who are familiar with the principles of databases will quickly recognize that they are by far more appropriate for interrogating data than any spreadsheet program. Indeed, even the faceting feature of Google Refine discussed earlier essentially represents the simplest type of querying function supported by traditional database systems. Yet, while database systems are incredibly powerful, they are notorious for being difficult to use and requiring very skilled and dedicated personnel to use and manage them. Desktop database systems, such as Microsoft Access and SPSS, are costly, usually requiring a license or seat for each person that will use them, as well as specialized training and support. Web-based database systems require server space to be purchased, configured, and secured by specialists. Even if these resources are available, traditional database systems are not designed for relatively transient data management tasks done by people with limited technical or data management experience.

One illustration of the challenges involves the modeling of data. Database systems require that a model, or schema, of the data be created in advance. However, creating a model means that you need to know what the data is about and what questions you will ask of it. Even in cases where this can be discovered and the appropriate tables and indices set up, the query language used by most database systems, SQL, is rather involved.

Finally, most database systems do not have a straightforward method for associating high-quality metadata with individual tables or columns of a table. This is especially problematic in journalism, because the details and provenance of a data set are fundamental measures of its appropriateness for a given story. The source, recency, units, and other contextual information (e.g., Are these employment figures seasonally adjusted?) that do not comprise the data itself are among the most crucial elements for assessing its value. Database systems were built with the main goal of supporting high-throughput transactions and running complex SQL queries. Though this feature obviates the need to save multiple views of the same data set or email updated files, database tables' lack of contextual information can be profound. An emailed file has at least some associated source (the sender) and a timestamp (the date of the email). The text of the email itself is likely to contain some contextualizing information. Absent these cues about origin and context, even technically networked data may fail to live up to one of its foremost potentials: recombination and reuse.

16.4 Google Fusion Tables

Google Fusion Tables [8], launched in 2009, is a web-based tool that combines and extends aspects of spreadsheet, database, graphing, and mapping software to address the challenges described earlier and more. Google Fusion Tables enables easy import of data from CSV and spreadsheet files and will even guess the data types of each column (which can also be adjusted by the user). If there is any uncertainty, the system will also guess which row in the data represents the column names. On upload, the user is prompted to add meaningful metadata to the table, such as the source's name and web link, as well as a description of the data (with the useful prompt: *What would you like to remember about this data in a year?*). Once loaded, individual columns can be annotated with notes, such as format and unit information.

The default interface in Google Fusion Tables is a familiar spreadsheet format, but the functionality it offers is far more sophisticated. Like a database, Google Fusion Tables supports simple querying: users can filter their data by values and ranges of attributes and perform grouping and aggregation queries (e.g., combine multiple roes to show total population by state). Users can also join two data sets as long as they have read permissions to both, much like a simple JOIN query on a SQL database. Unlike SQL-driven systems, however, Google Fusion Tables does not require the user to *declare* a schema in advance or even be aware of the concept of schema. For example, one can join a table containing the coffee production of different countries (coming from the International Coffee Organization) with data about the coffee consumption per capita (coming from Wikipedia).

Google Fusion Tables also provides tools for instant data visualization, an important component of analyzing data for patterns and anomalies. Through the automatic schema-detection, the system tries to identify columns that can serve as keys for visualization. For example, Google Fusion Tables will try to recognize columns with locations and columns with time points. When it does recognize such columns, the visualization menu already enables map or time-line views of the data that can be further configured by the user. As such, Google Fusion Tables supports a very rapid flow from data ingestion to visualization in a variety of forms, including bar, line, and pie charts, as well as fully styled interactive maps. While some of these visualizations can be generated through advanced use of spreadsheet programs, the mapping feature deserves special consideration.

Much like data journalism and computer-assisted reporting, mapping in newsrooms has long been the purview of a small set of specially trained reporters and for many of the same reasons: creating maps required very expensive, complex software and substantial technical skill. Yet, in a data set where location is a relevant parameter, the data must be mapped in order to perform any meaningful analysis; there is no single mathematical construct that can act as a proxy for geography. Google Fusion Tables' location

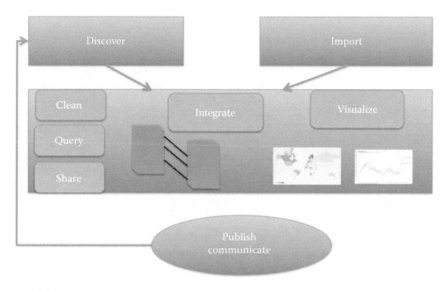

FIGURE 16.4
To fully realize the potential of structured data on the web, we must create an ecosystem around data that includes tools for discovering data on the web; importing data from existing repositories; tools for easily querying, cleaning, integrating, and visualizing data; and mechanisms for easily publishing data to the web.

data type, which can interpret many different forms of location information, including country and city names, latitude/longitude coordinates, and street addresses, allows a wide variety of geographic information to be mapped instantly. If it detects this type of information during the automatic schema-generating process, Google Fusion Tables will highlight the data in yellow and enable mapping in the visualization dropdown, marking the relevant locations. Even more robust maps can be created by joining data to a table containing the appropriate KML data. For example, see Figure 16.4, an example of data integration where two data sets are shown on the same map.

Finally, Google Fusion Tables uses sharing to help support collaboration and elaboration around data. When tables are initially imported into Google Fusion Tables, they are considered private—only the owner can see the data and make changes. However, as the owner works with the data, she may choose to share it with collaborators, who can be given permission to view, edit, or even annotate the data.

This is perhaps one of the most promising areas of future developments for Google Fusion Tables, because while data sets can, in isolation, provide many insights, the potential of combining data from multiple sources is explosive. Consider the map shown in Figure 16.2. A few days after the earthquake in Japan in March 2011, the creator of this map combined two data sets that were developed independently: data about earthquakes since 1973 and data about the location of nuclear plants. This visualization answered the question that was on the minds of billions of people around the world: Are they also in danger of experiencing the disaster that was unraveling in Japan?

Combining data from multiple sources, known as the field of data integration [7], faces many challenges. Of particular note is the challenge of *semantic heterogeneity*: When two data sets are developed independently, they will use different models. For example, the data sets may use different names to refer to the same columns, the same name to refer to columns with different data, and they may contain naming inconsistencies at the data level (e.g., one database may refer to IBM while the other will refer to International Business Machines). Such conflicts may exist even with individual data sets. Reconciling heterogeneity is a tremendous challenge, one that can be both time- and expertise-intensive. Yet, because data stored in Google Fusion Tables can be indexed, shared, accessed from anywhere, and meaningfully contextualized, it facilitates the kind of networked data augmentation that in many ways defines big data.

16.5 Presentation and Publishing

The same publishing capacity that gives journalists access to an unprecedented variety and quantity of structured data is exactly what allows them, in turn, to publish their work to the world. Publishing data to the web effectively, however, means reconciling many competing demands.

Journalists' primary professional function is to ask relevant, meaningful questions, answer those questions as thoroughly and fairly as possible, and then communicate what they have learned to their audience as clearly and completely as they can. In data journalism, however, an admirable desire for thoroughness can sometimes result in kitchen-sink publishing where as much information as possible is shoehorned into a single interface, with the interface itself dictating the content, rather than the reverse. Other times, misapprehensions of data's pristine status lead news outlets to publish stockpiles of data as is, with little to no supporting text or real context.

The best data journalism, however, arises through the methodical use of data and a strong consciousness of its contents and limitations. After the data has been carefully analyzed, it is often heavily edited—whether it is eventually presented as text, a visualization, in multimedia, or a combination of the earlier. Where possible, however, the complete underlying data set should also be made accessible as long as it is done responsibly, making sure that any data they publish is properly cleaned, verified, and, if appropriate, anonymized. It should be made available with its complete methodology in a standard, nonproprietary data format. If the data is already in a networked format, such as a public Google Fusion Table, it can simply be linked to.

This last approach has the added advantage of making the data crawlable by search engines, and so available to appear in the results of web search queries and therefore accessible to an even broader audience.

16.6 Remaining Challenges and Future Directions

Ideally, we would like to build an ecosystem around structured data that facilitates discovery, enhancement, and publication. Figure 16.2 shows the elements of such an ecosystem. The ecosystem will not consist of a *single* system from a single company, but rather a set of systems contributing in complementary (and sometimes redundant) ways.

Thus far, we have touched on some of the tools that are bringing journalists closer to that goal; we outline some components needed to realize its full potential in the following text.

16.6.1 Data Literacy

Data literacy needs to be made a priority in all professions and, with it, an understanding of the issues around collecting, accessing, and publishing data. Journalists must develop data literacy so that they can use it as a reporting tool and so that they can report knowledgeably on the issues that surround it. Judges must familiarize themselves with the material issues surrounding data formats so that they can rule appropriately in data-related cases. Especially in Freedom of Information Act cases, they should require that all metadata and other information necessary to evaluating public data sets be a requisite part of any settlement [15].

16.6.2 Safety and Privacy

While reporters can now file text, audio, photos—even video—from nearly anywhere, any use of current communication technologies leaves potentially dangerous data traces. Wireless and hard-wired service providers may share user information with government or other entities, endangering both sources and journalists. Even if the original requests for user data are later deemed illegal, there may be no legal liability for companies that comply [5]. Used maliciously, this information can create serious threats to civil and human rights.

16.6.3 Standards and Accessibility

The sheer number of expensive, complex tools that have been built to work with data is perhaps the clearest indication of its high economic value.

In the public sphere, however, data should be held to the highest standards of transparency and accountability. Any allegedly public data set should be released in a nonproprietary data format with all metadata intact and relational fields preserved.

Beyond this, the journalistic and humanitarian value of standardized data formats is difficult to overstate. For example, many government agencies release weather and other geographic data in KML because it is an accepted global standard. That it is also instantly consumable by Google Maps and Google Fusion Tables means that when there are floods in the Midwest, or a tsunami in the Pacific, lifesaving information about what areas are threatened can be published in a matter of minutes [14]. Developing tools and standards that eliminate the need for laborious data cleaning and correlation can improve the speed and accuracy of journalism, increase the transparency and accountability of government, and even save lives.

16.6.4 Cloud and the Crowd

Cloud-based data management services such as Google Fusion Tables go a long way to increase the usability of data, in part by making structured data shareable and accessible from anywhere. However, the cloud holds another important promise at the *logical level*. Specifically, if many high-quality data sets are put in the cloud, the cloud becomes a rich resource for data that can significantly enhance applications by enabling data reuse.

Imagine a scenario in which an analyst is looking at the latest trends in health data by county and considering additional locations to focus new efforts. She may be missing critical demographic context in her analysis, such as the population of each county or its average income—but this type of data is publicly available from reliable sources. If her data is in the cloud, she could potentially just ask for data about population. If the system could examine the locations in her database and find an appropriate data set that has population data for her locations, her analysis could be dramatically improved.

Adding this contextual data should be so easy that she should not even be aware she performed a database join. Importantly, the provenance of the additional columns should be made very clear, and she should even be able to choose from among any competing data sources. Some of these sources may be branded authorities; others may be crowd-sourced.

In this case, however, the crowd need not be a nameless set of individuals. Professional communities could collaborate to produce higher-quality data that they can share, such as scientists collecting and analyzing data about ecosystems [15], or coffee professionals assembling databases about farms and cafes worldwide [10]. The expertise of these communities might be confirmed by a trusted third party, much like Twitter's *verified user* system; alternatively, user rankings, recommendations, or other voting systems may be used to support source's claims to authority.

With such systems in place, a given community should be able to easily identify coverage gaps and places in their data, the quality needs improvement, as well as opportunities to resolve semantic heterogeneity. As these issues are identified, the community should be able to efficiently go about addressing these issues in a collaborative fashion. Because the community is built of motivated individuals (of many levels of expertise and cost), data collection could be done much more efficiently and effectively than it is today.

Looking at the industry today, there can be little doubt that big data will revolutionize journalism in the twenty-first century, both in concept and in practice. Its ascension also highlights, however, journalism's essential ongoing role in modern society. As Pulitzer Prize-winning data journalist Mo Tamman puts it:

Our job as journalists, more so now than ever, is to help people make sense of what's going on around them. There's so much noise out there it's deafening. Our role is to help them make sense in this deafening noise. It's what we've always done.

References

1. C. Anderson. The end of theory: The data deluge makes the scientific method obsolete. *Wired*, 16(7), June 2008.
2. J. Ball and P. Lewis. Twitter and the riots: how the news spread. *The Guardian*, 2011.
3. J. Burn-Murdoch, P. Lewis, J. Ball, C. Oliver, M. Robinson, and G. Blight. Twitter traffic during the riots. http://www.guardian.co.uk/uk/interactive/2011/aug/24/riots-twitter-traffic-interactive, 2011.
4. M. J. Cafarella, A. Halevy, Y. Zhang, D. Z. Wang, and E. Wu. WebTables: exploring the power of tables on the web. In *VLDB*, 2008.
5. Court of Appeals. 9th Circuit 2011 No.09-16676 (December 29, 2011). Hepting v. AT&T. https://www.eff.org/sites/default/files/filenode/20111229_9C_Hepting_Opinion.pdf, 2011.
6. b. danah and C. Kate. Six provocations for big data. In *A Decade in Internet Time: Symposium on the Dynamics of the Internet and Society, September 2011*, 2011.
7. A. Doan, A. Y. Halevy, and Z. G. Ives. *Principles of Data Integration*. Morgan Kaufmann, Waltham, MA, 2012.
8. H. Gonzalez, A. Halevy, C. Jensen, A. Langen, J. Madhavan, R. Shapley, W. Shen, and J. Goldberg-Kidon. Google fusion tables: Web-centered data management and collaboration. In *SIGMOD*, 2010.
9. Google refine. http://code.google.com/p/google-refine/, 2011.
10. A. Y. Halevy. *The Infinite Emotions of Coffee*. Macchiatone Communications, LLC, Mountain View, CA, 2011.
11. J. Halliday. David cameron considers banning suspected rioters from social media. *The Guardian*, 2011.

12. International Telecommunications Union. Internet users 06–11. http://www.itu. int/ITU-D/ict/statistics/material/excel/2011/Internet_users_01-11.xls, 2011.
13. P. S. John. Insurers risk of ruin. *Sarasota Herald-Tribune*, 2010.
14. G. Lotan. Kony 2012: See how invisible networks helped a campaign capture the worlds attention, 2012.
15. No. 10 Civ. 3488 (SAS) (February 7, 2011). National day laborer organizing network, et al. v. u.s. dept. of immig. & customs enforcement agency, et al. http://www.ediscoverycaselawupdate.com/National.pdf, 2011.
16. NOAA. National weather data in kml/kmz formats. http://www.srh.noaa. gov/gis/kml/, 2012.
17. PCAST Working Group. Sustaining environmental capital: Protecting society and the economy. http://www.whitehouse.gov/administration/eop/ostp/ pcast/docsreports, 2011.
18. Scraperwiki.
19. P. Venetis, A. Y. Halevy, J. Madhavan, M. Pasca, W. Shen, F. Wu, G. Miao, and C. Wu. Recovering semantics of tables on the web. *PVLDB*, 4(9):528–538, 2011.

17

Visualizing Emancipation: Mapping the End of Slavery in the American Civil War

Scott Nesbit

Humanists, no less than scientists, have begun thinking about increasingly large datasets in recent years. The National Endowment for the Humanities, in partnership with the National Science Foundation, the Institute for Museums and Library Services, and funding agencies in other countries, has issued calls for interdisciplinary teams to begin *Digging into Data*. At first glance, this seems a curious call to scholars accustomed to deep readings of literary and historical texts. Yet in recent years, humanists—often under the banner of digital history or digital humanities—have begun to find, produce, and organize large bodies of information, carefully managed agglomerations of the texts they have been accustomed to treat individually.

Many of these sources begin as simple, unstructured text. The work to structure these texts, making them profitable for future research, requires several transformations, breaking texts apart in order to put them together in new ways. As we work with these texts, we also have the opportunity to rethink how we go about our research, who might participate with us as we gather our evidence, and what kinds of scholarship we can produce for the public good.

My reflections on the processes and products of the digital humanities arise from "Visualizing Emancipation" (http://dsl.richmond.edu/emancipation), a project funded by a Digital Start-Up grant from the National Endowment for the Humanities, which I directed along with University of Richmond president Edward L. Ayers. "Visualizing Emancipation" is the first map of the most dramatic social transformation in American history, the freedom of four million slaves in the Civil War. In mapping this social transformation, it takes a new perspective on a significant scholarly question: where, when, and under what conditions did slavery fall apart? It brings together three kinds of evidence to answer this question, showing where slavery was protected by the U.S. government and where it was not during the Civil War; showing the approximate locations of U.S. troops during that war; and showing *emancipation events*, documented instances where the lives of enslaved men and women were changing, sometimes for good, sometimes for ill, during the war. By exposing the evidence on which it draws, it allows students and the public to access the primary sources of historical research to ask their own questions about emancipation and find out how their own locales and ancestors might have experienced the end of slavery and allows scholars to ask new questions about where, when, and how enslaved men and women escaped bondage, and what their lives may have looked like when they did so.

Building this project required access to large stores of qualitative information. Civil War historians, beneficiaries of large troves of digitized evidence, are fortunate in this respect. The *Official Records of the War of the Rebellion*, digitized by Cornell University Libraries, among other institutions, is a 160-volume series of documents—130 devoted to the army and an additional 30 volumes devoted to the navies—that cover the temporal and geographic expanse of the war. The Richmond *Daily Dispatch* is both a large and especially well-transcribed newspaper from the war that allows open-access courtesy of the University of Richmond Library. Researchers additionally have access to Ed Ayers' pioneering *Valley of the Shadow* project, a large archive of two communities, one northern and one southern, during the sectional conflict.

Our team began organizing information in these disparate sources by searching for evidence of changes in the institution of slavery, documents that indicated what we began to call *emancipation events*. Ayers and I decided early on to partner with students in order to find this evidence, for two reasons. First, we wanted undergraduate students at the University of Richmond to participate in the careful weighing of evidence and close readings of documents that lie at the heart of historical research. Second, we asked them, in ways that we could not ask a machine, to describe the kinds of evidence they were collecting, to begin placing labels on the events they found in conversation with each other and with the project directors.

Finding and encoding emancipation events required much more nuance than we could achieve using algorithms alone. Students became involved in recursive, careful weighing of evidence and refinement of our hypotheses about what emancipation looked like in the Civil War. While we knew that finding evidence of men and women becoming free would be a complicated task, we did not anticipate the difficulty we had in judging who was becoming free and who was not during the war. We quickly decided that we would look for a much more general set of events; we asked students to look for any document in which slavery was changing, or any evidence of African Americans acting (outside their normal course of duty as members of the U.S. Colored Troops). While giving this broad directive, we asked students to describe what they found. After a few months of describing these emancipation events without a controlled vocabulary, we began refining the ways that we discussed emancipation events, combining some categories with large overlap, eliminating others that seemed too vague.

Together with our student researchers, we pared our list to nine emancipation event types that described much of what we found in the *Official Records* and other sources.

In order to make geographic sense of the places where African Americans were becoming free, we also gathered data on where U.S. regiments moved over the course of the war. Frederick Dyer, a bugle boy in the Union Army, compiled and published such a list in 1908, the product of years of painstaking research. Nearly 100 years later, scholars at the Tufts University Perseus Digital Library scanned Dyer's Compendium of the War of the Rebellion and encoded it according to standards developed by the Text Encoding Initiative. In encoding the text, the Perseus library group developed software for automated geotagging of dates and place names, yielding robust annotations on Dyer's spare list of movements.

Transforming this text into data that we could plot on a map required that we extract elements from the Dyer text, including the names of regiments, the places they went, and the locations, and link them in new syntactical relations. We joined together in our database dates and places that were clearly related in the original text, turning the phrase "Movements on Dalton May 5–9" from a section on the 1st Alabama Regiment Cavalry into a data point indicating the Cavalry unit was in Dalton, Georgia, on the fifth through ninth days of May 1864.

In the process of mapping what would have taken an army of researchers years to accomplish, our automated process introduced some errors into our dataset. Some places mentioned in Dyer's text were too ambiguous or vague for the software to hazard a guess as to which place on the earth was mentioned, and sometimes places familiar in the Civil War era have disappeared. The geotagging algorithms on which the dataset relied originally tagged *Ft. Sumter* as a place on the Pacific coast of Oregon. Occasionally, our map tagged events occurring in 1864 as occurring on the same date in 1863. Our editing process is ongoing, identifying mistakenly assigned places or dates and rectifying the errors. In the meantime, our resulting dataset includes a list of approximately 40,000 data points describing the movements of 3,500 regiments over the course of the war.

Together, these data are enabling scholars and the public to come to fresh understandings of the geographic relation between war and the end of slavery. Enslaved men and women were more likely to find freedom in some places than others. A cursory look at the mapped evidence from the Official Records, tagged as *fugitive slave* escapes, show how freedom and Union arms pushed into the Confederacy by water and rail (see Figure 17.1). Enslaved men and women living along the Atlantic seaboard had the greatest and earliest opportunities to find freedom. Enslaved men and women living along the South's major rivers had a greater chance, too, especially those on the plantations of the Mississippi delta, along the Tennessee River in northern Alabama, and living and working near the rivers cutting across Virginia's Tidewater region.

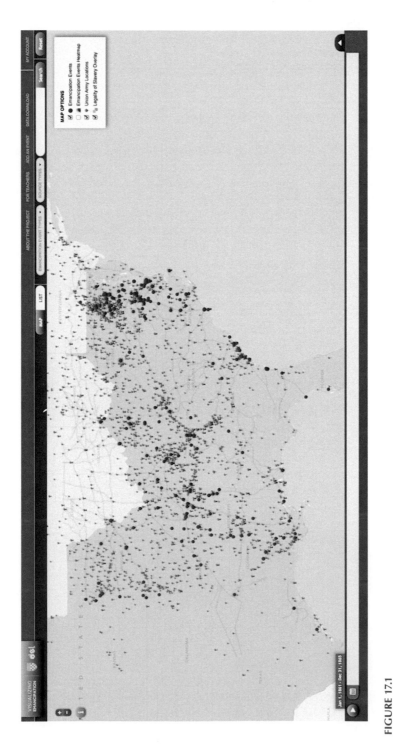

FIGURE 17.1
Visualizing emancipation. Fugitive slave escapes documented in the *Official Records of the War of the Rebellion* and Union regiment locations in the American Civil War.

Those living or working along the South's 10,000 miles of railroads were also more likely to find freedom. Confederate civilians along the line between Corinth, Mississippi, and Decatur, Alabama, complained to their government at the close of 1862 that in the past year their enslaved workers "had been carried off in very large numbers, declared free, and refused the liberty of returning to their owners." Union officers had "pressed all the negroes in this country" around the Nashville–Decatur line by the end of 1863. Before he followed the rails through Georgia, Gen. William T. Sherman moved his troops along the Jackson–Meridian line in Mississippi with a train of refugee families extending as far as the column itself, or in Sherman's turn of phrase, "10 miles of negroes." Some ran to U.S. lines of their own accord, others were dragged without their assent. Once under Union protection, men and women found themselves in a legal state of freedom, yet with immediate constraints no less coercive than those they experienced under slavery, as they were put immediately to work cooking, digging, farming, or marching to war.*

Emancipation consisted of much more than the rush of enslaved people to Union lines. "Visualizing Emancipation" breaks the actions and experiences of enslaved men and women into what we have called emancipation event types, each carrying a pattern distinct from but related to the others. We were particularly interested in the experiences that marked the end of slavery. Both armies conscripted men and women into service, pulling them away from their homes and the forms of slavery they had known before. People of color took part in irregular fighting, raiding plantations while not enlisted in any military unit. They passed intelligence to the U.S. Army and served as guides to troops. African Americans also suffered abuse, were rushed away from oncoming Union soldiers so that their owners might protect their human property, and were at times re-enslaved once they had escaped slaveowners' control.

The patterns made by a few of these kinds of events suggest how emancipation begins to look differently once mapped in time and space, and broken apart by the different experiences black southerners encountered. Our research into the *Official Records of the War of the Rebellion* shows that African Americans were victims of war-related abuse more frequently once black men began fighting for the United States. Scholars have written poignantly of the abuse of men enlisted in the U.S. Colored Troops, particularly Confederates' cold-blooded murder of surrendered black soldiers at Fort Pillow. Our analysis shows that attacks against nonuniformed black southerners also rose after 1862.

---

* This interpretive section borrows from the introduction to "Visualizing Emancipation," http://dsl.richmond.edu/emancipation/introduction/. W. G. Thomas, *The Iron Way: Railroads, the Civil War, and the Making of Modern America* (New Haven: Yale University Press, 2011) 27; Civilians, "To the Hon. Secretary of War of the Confederate States of America," Florence, AL, January 6, 1863, *Official Records of the War of the Rebellion* (hereafter *OR*) I.20.ii, 442–443; G.M. Dodge to U. S. Grant, Pulaski, TN, December 9, 1863, *OR* I.31.iii, 366; W. T. Sherman to H. W. Halleck, Meridian, MS, February 29, 1864, *OR* I.32.ii, 498–499.

Occasionally, this abuse came at the hands of undisciplined U.S. soldiers, such as those commanded by William Dwight, who raped the enslaved women they found at New Iberia, Louisiana, 4 months after the Emancipation Proclamation went into effect. More often abuse came at the hands of Confederates, who killed unarmed men and women at Goodrich's Landing, Mississippi, on Hutchinson's Island, South Carolina, at Helena, Arkansas, and a large number of other places dispersed throughout the South. Irregular violence against African Americans composed a greater part of the conflict in the west than the east. Bushwackers attacked black men and women to control Missouri. Women and children who worked in U.S.-owned plantations along the Mississippi were at constant risk of attack by small, marauding units of Confederates.

Freedom was more secure in the eastern theater of war, particularly in Virginia and North Carolina, than those in the West, but more dangerous to achieve. In Virginia, escaping slavery was a dangerous proposition because of the highly mobile and numerous Confederate units operating throughout the state. The likelihood that an enslaved man or woman would be caught while attempting to get to Union lines was great, even if they were accompanying a U.S. unit. Confederate troops were eager to attack smaller commands that had moved in advance of the main body of U.S. troops. They captured hundreds of escaped slaves after halting Brig. Gens. James Wilson's and August Kautz's raid along the Danville Railroad in June 1864.*

Yet once behind Union lines in a refugee camp, fugitives from slavery were relatively safe from attacking Confederates, even if they found there neither comfort nor health. Few raiding parties penetrated Union lines to seize black southerners living around Fortress Monroe in Virginia or to New Bern, North Carolina. The tens of thousands of African Americans who left their farms in the tidewater regions of Virginia and North Carolina were secure in their freedom after the Emancipation Proclamation. Refugee camps and U.S.-owned plantations along the Mississippi River did not share the natural geographic advantages of the Atlantic seaboard. These farms and villages were often lightly guarded and suffered frequent raids, some of which re-enslaved hundreds of men and women. The view of emancipation brought to light through our project is above all one variegated by space.

As important as we believe the gains in scholarly understanding might be, we are likewise enthusiastic about the ways in which "Visualizing Emancipation" offers new models of interaction between humanities scholars, students, and the public. Historians have always tended to think of their work as public-facing. We often write in accessible, relatively jargon-free language in part because we hold out hope for members of the public to engage with our texts. Yet until recently, historians have not been as quick to think about ways of integrating and acknowledging the intellectual

---

* "Reports from Petersburg," Richmond *Daily Dispatch*, July 1, 1864.

work and research of ordinary citizens or undergraduate students into their own scholarship.

Visualizing Emancipation and its model of undergraduate research bring valuable lessons for undergraduate education in the humanities. Humanists have often labored under the assumption that undergraduates are not able to do the kinds of careful work required for effective scholarship in the humanities. Our experience with this project leads us to believe that, with proper guidance, undergraduates can be effective researchers in large-scale humanities projects. This guidance consists principally in limiting the scope of the undergraduate research assignment, breaking a larger project into assignments that, while manageable, still demanded intellectual labor and creativity. Our goal was to help undergraduates do work that was useful to the field of history *and* intellectually stimulating.*

We created three distinct opportunities for students to make controlled, interpretive decisions that did not rely on large bodies of contextual knowledge or a deep reading in secondary literature. Students engaged in small-scale intellectual labor, appropriate for those who had not yet read deeply in the historical literature of the American Civil War and emancipation. First, they helped determine, with guidance, what *counted* as an emancipation event. This interpretive act required potentially controversial decisions about how slavery was changing in wartime. They grappled with questions about whether to include incidents in which people of color remained slaves but negotiated some greater autonomy in the slave regime, as well as instances where their limited autonomy under the antebellum regime seemed even more tightly constrained. In each case, undergraduates were making interpretive choices that affected the outcome of the project.

As they decided whether to include events in our database, students also made decisions about how to encode this information. At first, we gave them little guidance, instead of asking students to describe what they found in their documents. Students quickly came up with relatively long, idiosyncratic lists. Through a series of conversations, we pared these lists down to 10 or so categories of action that reflected, we hoped, the evidence in our broad survey of documents rather than a preordained list of actions we had expected to see. Undergraduate interpretation—labeling work that would have been unrealistic to accomplish with computation alone—helped produce the event types at the core of our research results.

When this small-scale research was complete, our undergraduates found that they had developed significant bodies of knowledge about particular places and times of the war and could use these to formulate their own

---

* For other examples of productive undergraduate research in the humanities and digital humanities, see Christopher Blackwell and T. R. Martin, "Technology, Collaboration, and Undergraduate Research," Digital Humanities Quarterly 3.1 (2009) http://www.digitalhumanities.org/dhq/vol/3/1/000024/000024.html. The following account draws on Scott Nesbit and E. L. Ayers, "Landscapes of the American Past: Visualizing Emancipation, White Paper Submitted to the National Endowment for the Humanities," June 2012.

research agendas. Some students had become quite knowledgeable about the geography of eastern North Carolina, while others developed a detailed understanding of the passes and hills of Appalachia. One student, a history major, pivoted from her work on emancipation events in the Tidewater to a summer research fellowship investigating how men and women living on a plantation near Petersburg, Virginia, acquired food during the transition from slavery to freedom. "I discovered my love for research" while working on these projects, she told the University of Richmond student newspaper.*

By asking students to describe in a few words the actions they found within the documents, we enabled them to practice historical interpretation on a very modest scale. By recursively moving from the texts they studied to their determinations of emancipation event types, they did historical work manageable for many undergraduate students while opening up opportunities for later, more robust interpretive work.

Since our initial release of Visualizing Emancipation, members of the public have built on these emancipation events, adding accounts they found in newspapers and manuscripts held in archival collections across the country. They have documented African Americans building fortifications in Kentucky, moving between Union and Confederate lines in South Carolina, and fleeing plantations in Virginia. The invitation to members of the public to contribute their own expertise and research is not uncommon today— Wikipedia is built on millions of hours of publicly contributed labor. Yet it represents a remarkable democratization of history, as citizens both identify sources for inclusion in an open-access historical map of emancipation and interpret the events collected in that map, events that in large and small ways pulled slavery apart. As Americans commemorate the 150th anniversary of their Civil War and emancipation, scholars, students, and members of the public can use computationally aided work to come to an increasingly sophisticated understanding of the geography of slavery's end.

---

* University of Richmond undergraduate student and Digital Scholarship Lab intern Megan Molnar, quoted in "President Ayers Creates Digital Scholarship Lab," *The Collegian*, University of Richmond, April 21, 2011.

18

Synergistic Sharing of Data and Tools to Enable Team Science

Ross Mounce and Daniel Janies

CONTENTS

18.1 Introduction

Our goals are to compare and contrast examples of workflow efforts in the context of team science that have been heavily funded by governments (i.e., infectious disease genomics) with efforts that have been supported more by enthusiastic individual scientists (i.e., paleontology). We present lessons learned from these examples in terms of interoperability, agility, data sharing, metrics for scientific impact, and openness. In conclusion, we make suggestions for progress in synergistic sharing to enable team science.

18.1.1 Examples

18.1.1.1 Infectious Disease Genomics

Rapid sharing of genetic information from infectious organisms is crucial to global public health and vaccine design. There are several data-sharing initiatives in infectious disease genomics. Many efforts focus primarily on influenza data whereas others are general resources for molecular biology.

Most of these efforts are funded by governments, and we provide several examples in the following text.

The German Federal Ministry Food, Agriculture, and Consumer Protection (BMELV, http://www.bmelv.de/EN/Homepage/homepage_node.html) is the current supporter of the Global Initiative for Sharing of All Influenza Data (GISAID, http://gisaid.org). GISAID previously had the support of the U.S. Centers for Disease Control (CDC, http://cdc.gov). The CDC is part of the U.S. Department of Health and Human Services (HHS, http://hhs.gov). The HHS currently supports the Influenza Research Database (IRD, http://fludb.org) via contracts administered by the National institute of Allergy and Infectious Diseases (NIAID, http://www.niaid.nih.gov) and the National Center for Biotechnology Information (NCBI or GENBANK, http://ncbi.nlm.nih.gov). The Swiss institute of Bioinformatics (SIB, http://www.isb-sib.ch) supports the OpenFlu database (OpenFluDB, http://openflu.vital-it.ch).

Other efforts, termed Bioinformatics Resource Centers (BRC), were funded in 2004 by NIAID with over $100M USD to focus on molecular data for priority A–C pathogens (reviewed in Greene et al., 2007). As of 2009, many of the BRC's sites lost their NIAID funding and went dark (e.g., ApiDB, http://www.apidb.org; BioHealthBase, http://www.biohealthbase.org; ERIC http://ericbrc.org; NMPDR http://www.nmpdr.org; Pathema; http://pathema.tigr.org). Their resources and data have been consolidated into efforts for eukaryotes (http://eupathdb.org/eupathdb), bacteria (http://patricbrc.vbi.vt.edu), and viruses (http://www.viprbrc.org). Including and beyond infectious disease genomics, the NCBI collaborates with the DNA Data Bank of Japan (DBJ, http://www.ddbj.nig.ac.jp) and the European Molecular Biology Laboratory (EMBL, http://www.embl.org) as the International Nucleotide Sequence Database Collaboration (INSDC, http://insdc.org).

Most of the efforts listed earlier support biomolecular data submission, storage, query, and retrieval but not workflows. There are some initiatives to develop these resources as workflow platforms. For example, IRD supports geographic metadata in latitude and longitude forms and phylogenetic analyses. A user can calculate a distance-based tree in GENBANK in which Basic Local Alignment Search Tool (BLAST; Altschul et al., 1990) is used to align database sequences to the query sequence, rather than a multiple or progressive sequence alignment preferred by phylogeneticists (http://www.ncbi.nlm.nih.gov/Web/Newsltr/V15N1/blastlab.html; Philips et al., 2000).

However, the application coverage and strength of underlying computational resources among these efforts are variable. Most of these efforts are built on the model of interaction such as service provider to user. These models tend not to evolve as users create content and tools. In order to foster evolution of some of these resources into ecosystems for team science, we have developed and provided modular web services for high-performance computing as applied to genetic surveillance of infectious diseases and other biogeographic problems. Supramap (http://supramap.org [Janies et al., 2011]) and Geogenes (geogenes.org [Janies et al., 2012]) are

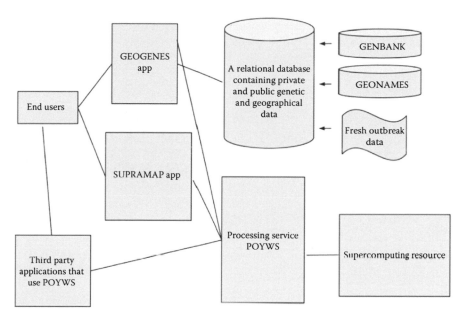

FIGURE 18.1
Diagram of the interactions of the users, applications, data, services, and computing resources put together to enable a workflow for genetic surveillance of infectious diseases.

two different clients to demonstrate the modularity of the POY web service (poyws.org [Janies et al., 2012]) (Figure 18.1). Our aim is to make the service available to the community so that users with new ideas and data can create novel end-to-end workflows in any field that can take advantage of progressive sequence alignment, phylogenetic tree search, and/or geographic projection of trees and mutations. To accomplish this, we have published an application programming interface (API; http://poyws.osc.edu/PoyService. asmx). Similar efforts support a wide variety of alignment and tree search tools such as the CIPRES Science Gateway (http://www.phylo.org/index. php/portal).

18.1.1.2 Paleontology

In paleontology, computational team science approaches are relatively young compared to other research fields. Traditionally, paleontology was a descriptive science, but increasingly it is becoming highly analytical (Adrain, 2001). For example, paleontologists use large-scale meta-analyses to synthesize published data. Attempts at synthesis represent a significant computational challenge because fossilized evidence of organisms can be found from sub-fossil assemblages just a few thousand years old, to inferential evidence from at least 3Ga ago (10^9 years ago; Brasier et al., 2006; Wacey et al., 2011). Moreover, it is commonly estimated that 99% of species biodiversity that has ever existed is now extinct and thus may only be found in fossil form—if preserved at all. Current estimates of modern diversity posit that there are roughly between 1 and 100 million species alive today (Costello et al., 2011; Mora et al., 2011). If this is just 1% of all the biota there has ever been, then clearly computational approaches are necessary to compile and relate information about known fossil diversity.

Despite this challenge, paleontology, aside from petroleum applications, is not well funded. This landscape has led to the evolution of different models of community interaction than seen in infectious disease genomics. The major database that paleontologists use to collate information (PaleoDB; http://paleodb.org) relies on data entry and curation from a small team of voluntary, trusted contributors. Using this model, this effort is unlikely to be able to accurately encompass, curate, and further develop the database.

Another strategy in paleontology and other natural sciences is crowd-sourcing, for example, the Open Dinosaur Project. The problem that this attempts to solve is deceptively simple: to gather all published, standardized measurements of dinosaur bones. Yet in practice, because much of the literature is old, difficult to find, and difficult for computers to read in an automated way, it is left to human volunteers to compile these data. At least 50 serious contributors have helped this project, and a publication is in preparation.

In similar fields, social networks have advanced scientific progress. An excellent example is a Smithsonian field research team in Guyana using social media for rapid fish identification (Sidlauskas et al., 2011). The problem the team faced was, due to local customs laws, they had to identify all the fish they had collected within a week or leave them behind to rot. By uploading pictures of their samples to Facebook (http://tinyurl.com/Sidlauskas) for everyone to see, they got 90% of the thousands of samples identified in 24 h. They could not have achieved this by themselves. This ingenuity is an example of small groups of researchers on a small budget responding to problems as they present themselves.

18.1.1.2.1 Lessons Learned

Despite differences in funding level, or whether the initiative originates from government or from working scientists, computation is enabling aggregation of large datasets. What is needed going forward are shared data, code, and computational resources that provide focal points for researchers to come together into agile teams that develop productive workflows. Workflows that capture parameters of a computational pipeline are also important for experiments to be repeatable and extendable by researchers external to the team (Stodden, 2010; Peng, 2011). Thus, new experiments can be well controlled with only certain variables being adjusted at a time (e.g., new data added or analytical parameter being shifted).

Moreover, teamwork provides a means to achieve consilience as envisioned by Wilson (1998). Modern problems are complex, and science succeeds if several types of people work together—those making observations, those conducting analyses, those making the analyses efficient, and those communicating the results. Achieving this goal will require standards for interoperability and a culture reset toward sharing of data, working in teams, and providing new vehicles and metrics of impact and credit for scientists working toward these aims.

18.1.1.3 Interoperability

For example, in modern evolutionary meta-analyses that attempt to 'Link Big' (Sidlauskas et al., 2010), it is important that one shares data with standardized annotations and formats. As a prerequisite, this requires research communities to establish data formats. Also, as is noted by Parr et al. (2012) and others, relevant data is often scattered across tens of thousands of articles or databases in many hundreds of different formats. This makes it difficult but not impossible to automatically extract the raw observations. We see the promise in the work of Knobloch and colleagues (e.g., Parundekar et al., 2010; Goel et al., 2011), who can identify images and link information from across the web rather than siloing the data.

18.1.1.4 Agility

Approaches by various governments to these problems have often been to make large awards to contractors to build centralized databases. We see two major limitations to these large efforts: (1) data is getting too large to centralize, and (2) there remains resistance to sharing. In the case of genomics, the cost of sequencing has tumbled 100,000 fold in the past decade and continues to fall. Contractors that must detail every hour of effort for several years henceforth do not have the flexibility or foresight to respond to increasing large datasets and changing observational and analytical technologies. In contrast, agile teams need not be permanent but can form and reform as problems require and communities evolve. Some limitations are not technological but sociological. Evidence of data withholding and/or weak data availability abounds (e.g., Campbell et al., 2002; Wicherts, et al., 2006; reviewed in Janies et al., 2010). Making data hard to use is counterproductive, because when data is openly available, it is more likely to indicate a sound piece of repeatable research (Wicherts et al., 2011), it is more likely to be cited (Piwowar et al., 2007), and can be combined with new data to address the problem going forward.

18.1.1.5 Data Sharing

Thus, one would think that it would be mandatory to make all underlying data openly available. This is not however the case. Despite prominent calls for public phylogenetic data sharing (Sanderson et al., 1993), a lack of mandates and other factors has resulted in less than 4% of phylogenetic tree data being publicly available online (Stoltzfus et al., 2012). Mandates apply to only a very restricted set of data types such as DNA sequences and genome-wide association studies that generally must be submitted to a repository such as NCBI's GenBank before publication of analyses using such data. In some cases, data is not shared due to constraints such as patient privacy laws and national priorities, which are context dependent (treated in Janies et al., 2010).

Scientists are under pressure to publish original research, yet there is little or no reward provided for other activities such as data sharing. Some recent initiatives may however bring change to the sociological aspects of data sharing (or non-sharing): the invention of the *data paper* as explained by Chavan and Penev (2011) in which data can be published (e.g., Narwade et al., 2011) and cited (Edmunds et al., 2012) for academic credit, with suitable explanatory metadata, without any particular analysis or traditional paper published with the data. Organizations such as DataCite (http://datacite.org), a global not-for-profit organization, encourage and incentivize the early publication of data as a separate *first-class research object* (Reilly et al., 2011) so that further use of such valuable primary data is not delayed until after the associated first publication comes out.

18.1.1.6 Metrics for Scientific Impact

The broadening of ways in which academic impact is measured, away from the Journal Impact Factor to Article Level Metrics and other altmetrics (http://altmetrics.org), using web tools will further increase the importance of data, code, and other *nonpaper* contributions to research and evaluation thereof because usage of these products by others will be increasingly countable. At the moment, academics are evaluated for tenure by weak proxies for good work such as number of papers, as first or senior author, whether they are published in well-cited journals, and the amount of grant funding obtained. The altmetrics ethos seeks to expand the ways in which the success of academia is quantified. Components of the altmetrics suite such as Wikipedia cites, Citeulike and Mendeley bookmarking statistics, and public discussions are tangible metrics for scientific impact (Priem et al., 2012). Some measures like Facebook *Likes*, Tweets, and Google +1's are more likely to be measures of popularity, controversy, and/or outreach but can be important to reach a wide audience including the general public, policy makers, and journalists.

18.1.1.7 Openness

There is wide acknowledgment of barriers to research that can arise from works that are not openly licensed (as outlined in Carroll (2011) and Hagedorn et al. (2011)). The Creative Commons Attribution License (http://creativecommons.org/licenses/by/3.0/) as used by some Open

Access journals explicitly allows free reuse of any such materials that are published with it and thus allows repeatable scientific research. BioMedCentral has recently proposed that scientific publishing should be *open by default* using the CC-BY license for publications and the CC0 waiver for data where possible (Hrynaszkiewicz and Cockerill, 2012) to enable maximum return on investment of research. Other less permissive licenses are more commonly used in subscription-access scientific publishing. Many licenses require prospective users of data to first contact the copyright holder(s) to obtain permission before any materials can be reused. This permissions-based system does not scale well and causes particular problems for those seeking to reuse previously published materials on large scale (e.g., text-mining applications). With more than 50 million published academic articles in existence (Jinha, 2010) and articles appearing at an ever-increasing rate, well into millions per year (Kell, 2009), text mining tools and applications are necessary to make sense and summarize academic works.

In the realm of infectious disease research, the GISAID project encourages data sharing within the influenza research community by requiring users of its resource to agree to acknowledge data providers. This is a laudable effort; however, closely circumscribed projects and database policies are not sufficient to spread the wider goals of code and data sharing. Thus in 2009, the Panton Principles (http://pantonprinciples.org) were drafted. These are a set of principles devised on the basis of the observation that "Science is based on building on, reusing and openly criticizing the published body of scientific knowledge. For science to effectively function, and for society to reap the full benefits from scientific endeavors, it is crucial that science data be made open" (Open as per http://opendefinition.org)

These data principles also compliment the principles of the Science Code Manifesto (http://sciencecodemanifesto.org), which argue for source code availability, clear copyright and ownership statements, code citation (akin to data citation), appropriate credit for all software developers, and curation of code for the lifetime of its plausible usage. Most importantly, with well-curated and shared workflows, teams can work together, and other workers can repeat experiments while controlling variables.

We take a personal interest in these principles. In paleontology, one of us (R.M.) has tried to bring about change at a grassroots level with respect to data-sharing and data publication standards. In their open letter to the paleontological research community (Mounce et al., 2011; discussed in Callaway, 2011), fellow researchers were invited to publicly affirm their support for increased utilization of the Internet to communicate data in more interoperable forms along with traditional research publications. This approach was deliberately chosen because what is necessary is *sociological* change. Technical barriers to the effective transmission of nearly all types of data and code utilized in paleontology are almost nonexistent. Databases and other necessary infrastructure arguably already exist in usable forms (albeit with room for improvement). Therefore, the main impediment to change in paleontological knowledge synthesis is a widespread use of databases and infrastructure as an integrated part of the normal publication process in the way GenBank serves molecular biosciences.

Open data publication in itself will not completely remove all barriers to knowledge synthesis and reuse—it needs to be provided in recognized formats so that it can be easily integrated. Acceptable data formats must be agreed upon with a community-led approach to standardization (Brazma, 2001). Molecular biologists have been quick to create data standards (e.g., Gene Ontology terms; Ashburner et al., 2000), and some of these have been successful as demonstrated by their wide adoption. In paleontology, data standards such as Virtual Anatomy eXtensible Markup Language (Sutton et al., 2012) for 3D virtual specimen imaging data have been proposed such that data from published works can be more interoperable.

18.2 Conclusions and Suggestions

We do not believe that mandates will be sufficient to compel researchers to make this change toward increased sharing of code, data, and workflows. Some clear data-sharing mandates fall well short of 100% compliance. For example, Prayle et al. (2012) document 22% compliance for clinical trial data. But in research disciplines where data sharing is encouraged such as molecular biology via GenBank submission of sequence data commensurate with publication or prepublication of influenza data in GISAID, significant progress in data sharing can be made. Piwowar (2011) documents an increase from 5% to 30%–35% for microarray datasets. Thus, if change is to occur, it needs the support of the research community. One change that we would advocate from government would be to fund many communities of scientists who are committed to compete and collaborate in agile and open team science approaches (e.g., Open Chemistry Research; Woelfle et al., 2011; and the Open Tree of Life http://opentreeoflife.org).

References

Adrain, J. M. 2001. Systematic paleontology. *Journal of Paleontology* 75(6):1055–1057.

Altschul, S., W. Gish, W. Miller, E. Myers, D. Lipman. 1990. Basic local alignment search tool. *Journal of Molecular Biology* 3:403–410.

Ashburner, M., C. A. Ball, J. A. Blake, D. Botstein, H. Butler, J. M. Cherry, A. P. Davis, K. et al. 2000. Gene ontology: Tool for the unification of biology. The gene ontology consortium. *Nature Genetics* 25:25–29.

Brasier, M., N. McLoughlin, O. Green, and D. Wacey. 2006. A fresh look at the fossil evidence for early Archaean cellular life. *Philosophical Transactions of the Royal Society B: Biological Sciences* 361(1470):887–902.

Brazma, A. 2001. On the importance of standardisation in life sciences. *Bioinformatics* 17:113–114.

Callaway, E. 2011. Fossil data enter the web period. *Nature* 472:150.

Campbell, E. G., B. R. Clarridge, M. Gokhale, L. Birenbaum, S. Hilgartner, N. A. Holtzman, and D. Blumenthal. 2002. Data withholding in academic genetics: evidence from a national survey. *JAMA: The Journal of the American Medical Association* 287(4):473–480.

Carroll, M. W. 2011. Why full open access matters. *PLoS Biol* 9(11):e1001210.

Chavan, V. and L. Penev. 2011. The data paper: A mechanism to incentivize data publishing in biodiversity science. *BMC Bioinformatics* 12:S2.

Costello, M. J., S. Wilson, and B. Houlding. 2011. Predicting total global species richness using rates of species description and estimates of taxonomic effort. *Systematic Biology* 61(August):871–883.

Edmunds, S., T. Pollard, B. Hole, and A. Basford. 2012. Adventures in data citation: Sorghum genome data exemplifies the new gold standard. *BMC Research Notes* 5:223.

Goel, A., M. Michelson, and C. A. Knoblock. 2011. Harvesting maps on the web. *International Journal on Document Analysis and Recognition* 14(4):349–372.

Greene, J. M., F. Collins, E. J. Lefkowitz, D. Roos, R. H. Scheuermann, B. Sobral, R. Stevens, O. White, and V. Di Francesco. 2007. National institute of allergy and infectious diseases bioinformatics resource centers: New assets for pathogen informatics. *Infection and Immunity* 75(7):3212–3219.

Hagedorn, G., D. Mietchen, R. Morris, D. Agosti, L. Penev, W. Berendsohn, and D. Hobern. 2011. Creative commons licenses and the non-commercial condition: Implications for the re-use of biodiversity information. *ZooKeys* 150:127–149.

Hrynaszkiewicz, I. and M. Cockerill. 2012. Open by default: A proposed copyright license and waiver agreement for open access research and data in peer-reviewed journals. *BMC Research Notes* 5:494.

Janies, D., L. Pomeroy, J. Aaronson, S. Handelman, J. Hardman, K. Kawalec, T. Bitterman, W.C. Wheeler. 2012. Analysis and visualization of H7 influenza using genomic, evolutionary and geographic information in a modular web service. *Cladistics* 28:483–488. doi: 10.1111/j.1096-0031.2012.00401.x

Janies, D. A., T. Treseder, B. Alexandrov, F. Habib, J. J. Chen, R. Ferreira, U. Çatalyürek, A. Varón, and W. C. Wheeler. 2011. The supramap project: Linking pathogen genomes with geography to fight emergent infectious diseases. *Cladistics* 27(1):61–66.

Janies, D. A., I. O. Voronkin, M. Das, J. Hardman, T. W. Treseder, and J. Studer. 2010. Genome informatics of influenza a: From data sharing to shared analytical capabilities. *Animal Health Research Reviews/Conference of Research Workers in Animal Diseases* 11(1):73–79.

Jinha, A. E. 2010. Article 50 million: An estimate of the number of scholarly articles in existence. *Learned Publishing* 23:258–263.

Kell, D. 2009. Iron behaving badly: Inappropriate iron chelation as a major contributor to the aetiology of vascular and other progressive inflammatory and degenerative diseases. *BMC Medical Genomics* 2:2.

Mora, C., D. P. Tittensor, S. Adl, A. G. B. Simpson, and B. Worm. 2011. How many species are there on earth and in the ocean? *PLoS Biol* 9(8):e1001127.

Mounce, R., A. Butler, K. Davis, A. Dunhill, R. Garwood, J. Lamsdell, D. Legg, G. et al. 2011. An open letter in support of palaeontological digital data archiving. http://supportpalaeodataarchiving.co.uk. Accessed on 25 May 2013.

Narwade, S., M. Kalra, R. Jagdish, D. Varier, S. Satpute, N. Khan, G. Talukdar, V. B. et al. 2011. Literature based species occurrence data of birds of northeast india. *ZooKeys* 150(150):407–417.

Parr, C. S., R. Guralnick, N. Cellinese, and R. D. M. Page. 2012. Evolutionary informatics: Unifying knowledge about the diversity of life. *Trends in Ecology & Evolution* 27(2):94–103.

Parundekar, R., C. A. Knoblock, and J. L. Ambite. 2010. Linking and building ontologies of linked data. In *Proceedings of the 9th International Semantic Web Conference (ISWC 2010)*, Shanghai, China.

Peng, R. D. 2011. Reproducible research in computational science. *Science* 334(6060):1226–1227.

Phillips, A., D. Janies, and W. Wheeler. 2000. Multiple sequence alignment in phylogenetic analysis. *Molecular Phylogenetics and Evolution* 16:317–330.

Piwowar, H. A. 2011. Who shares? who doesn't? factors associated with openly archiving raw research data. *PLoS ONE* 6(7):e18657.

Piwowar, H. A., R. S. Day, and D. B. Fridsma. 2007. Sharing detailed research data is associated with increased citation rate. *PLoS ONE* 2(3):e308.

Prayle, A. P., M. N. Hurley, and A. R. Smyth. 2012. Compliance with mandatory reporting of clinical trial results on ClinicalTrials.gov: Cross sectional study. *BMJ* 344(January):d7373.

Priem, J., H. A. Piwowar, and B. M. Hemminger. 2012. Altmetrics in the wild: Using social media to explore scholarly impact. http://arxiv.org/abs/1203.4745.

Reilly, S., W. Schallier, S. Schrimpf, E. Smit, and M. Wilkinson. 2011. Report on integration of data and publications. Technical report for the Opportunities for Data Exchange (ODE) project. http://www.alliancepermanentaccess.org/wp-content/uploads/downloads/2011/11/ODE-ReportOnIntegrationOfDataAnd Publications-1_1.pdf

Sanderson, M. J., B. Baldwin, G. Bharathan, C. Campbell, D. Ferguson, J. M. Porter, C. V. Dohlen, M. F. Wojciechowski, and M. J. Donoghue. 1993. The growth of phylogenetic information and the need for a phylogenetic database. *Systematic Biology* 42:562–568.

Sidlauskas, B., C. Bernard, D. Bloom, W. Bronaugh, M. Clementson, and R. P. Vari. 2011. Ichthyologists hooked on facebook. *Science* 332(6029):537.

Sidlauskas, B., G. Ganapathy, E. Hazkani-Covo, K. P. Jenkins, H. Lapp, L. W. McCall, S. Price, R. Scherle, P. A. Spaeth, and D. M. Kidd. 2010. Linking big: The continuing promise of evolutionary synthesis. *Evolution* 64(4):871–880.

Stodden, V. 2010. Reproducible research in computational science: Problems and solutions for data and code sharing. *ICML Workshop Machine Learning Open Source Software*, Haifa, Israel.

Stoltzfus, A., B. O'Meara, J. Whitacre, R. Mounce, E. Gillespie, S. Kumar, D. Rosauer, and R. Vos. 2012. Sharing and re-use of phylogenetic trees (and associated data) to facilitate synthesis. *BMC Research Notes* 5:574.

Sutton, M. D., R. J. Garwood, David J. Siveter, and Derek J. Siveter. 2012. SPIERS and VAXML; A software toolkit for tomographic visualisation, and a format for virtual specimen interchange. *Palaeontologia Electronica* 15(2):4T, 14p; http:// spiers-software.org/links.htm.

Wacey, D., M. R. Kilburn, M. Saunders, J. Cliff, and M. D. Brasier. 2011. Microfossils of sulphur-metabolizing cells in 3.4-billion-year-old rocks of western australia. *Nature Geoscience* 4(10):698–702.

Wicherts, J. M., M. Bakker, and D. Molenaar. 2011. Willingness to share research data is related to the strength of the evidence and the quality of reporting of statistical results. *PLoS ONE* 6(11):e26828.

Wicherts, J. M., D. Borsboom, J. Kats, and D. Molenaar. 2006. The poor availability of psychological research data for reanalysis. *American Psychologist* 61(7):726–728.

Wilson, E. O. 1998. *Consilience: The Unity of Knowledge*. Vintage, New York.

Woelfle, M., J.-P. Seerden, J. de Gooijer, K. Pouwer, P. Olliaro, and M. H. Todd. 2011. Resolution of praziquantel. *PLoS Neglected Tropical Diseases* 5:e1260.

Part V

Visionary Pondering and Outlook on Computation for Humanity

19

Evolution of the Techno-Human

Jim Pinto

CONTENTS

Evolution took millions of years to progress to the development of primates and then Homo sapiens.[1] It took humans thousands of years to get to the industrial revolution, after which technological developments continued with an exponential rate of change—performance doubling at an accelerating rate.[2,3]

Toward the end of the twentieth century, significant changes in the human landscape have occurred in just decades. In the next century, technology acceleration that is centered on cheap and abundant computing power will continue with major evolutionary consequences.[4]

19.1 Man with Tools *Superior*

Evolution is survival of the fittest—Homo sapiens with tools survived, while the equivalent, sans tools, did not.[4,5] The development of tools continually extended human powers through the agricultural and industrial revolutions until today, when we have developed vast computational capability combined with instant worldwide communications and connectivity. Evolution moves exponentially. The developments of the last century have dwarfed progress of the past millennium and have themselves been exceeded by those of the past decade.

History makes it evident that man with tools inevitably survived and conquered man without.[6] Even spiritually advanced humans (Indians a few centuries ago, Aztecs, and American Indians within the past two) were quickly subjugated by technological prowess.

That was societal change—what about individual and personal? Are the tools of today evolving a new form of techno-human?[7] More and more people are becoming walking conglomerations of various electronic devices. Many people never travel without a multipurpose wireless-connected gadget that is a conglomeration of several functions—telephone, digital camera, Internet-connected TV, search engine, news, GPS, and many more features and functions. Armed with these tools, most people can do lots of things that would have been considered magic only a few years ago![8,9]

Increasingly in the future, people will be constantly connected to the world through intelligent processors and communication.[10] These devices will be everywhere—in our clothes, our beds, our homes, all over, and even inside our bodies. Because we will interact with the world and each other primarily through electronic devices and communications, the distinction between reality and nonreality will blur.[11]

Three technology laws can be identified that are bringing about an advance in civilization that has significant and far-reaching consequences that are unfolding with awesome speed.[12] Moore's law (exponential increase in processing power), Gilder's law (exponential increase in communications bandwidth), and Metcalfe's law (exponential increase in connected intelligence) are together bringing irreversible change to the human species.[13]

19.2 Robots and Intelligence Augmentation

Robotics is a significant application of computers for humanity. All robots have three elements: brain (the computer intelligencer), bone (the connecting mechanisms and linkages), and brawn (the motor power).

Rodney Brooks is a leading *roboticist* and computer science professor and director of Massachusetts Institute of Technology (MIT)'s Artificial Intelligence Lab. His book, *Flesh and Machines: How Robots Will Change Us*,[14] promises that machines will have emotions, desires, fears, love, and pride. Brooks's robots are still primitive, but he believes that intelligent machines will come into their own as the boundaries between what is now fantasy and what is fact are breached.

With intelligent robots come a host of ethical problems, as we wrestle with the implications of Asimov's laws of robotics (in order of priority: a robot must not injure a human being, must obey a human, and must protect itself) and with the very real possibility that we have created a new kind of slave. There is no way of getting around this future, and Brooks is convinced that the human species will change. Says Rodney Brooks: "With all these trends we will become a merger between flesh and machines."

Brooks believes that robots in the future will probably be nothing like such all-knowing brain machines as HAL from the movie 2001: *A Space Odyssey*, nor will they resemble the sleek cyborgs of other Hollywood nightmares. Rather, they will be simple, ubiquitous, curious little machines that will have more in common with humans than one might think. Brooks and his fellow researchers suggest that the focus of much artificial intelligence (AI) and robot research has been to develop superhuman devices that operate at the highest intellectual levels. It is much better, he says, to make a lot of simple, cheap robots that can perform only a few tasks but do them well.

Brooks' book begins with a brief but comprehensive overview of the field of research into AI and robotics, then dives quickly into his and his fellow enthusiasts' work as they engineer one strange, insect-looking (and weirdly human-acting) metallic creature after another. Brooks points the way toward a future where humans work in tandem with his fast, cheap creations not a science fiction utopia, but a future where people have more and better tools to work with.

Brooks and other robotics gurus clearly believe that AI is on the brink of a major breakthrough.[15,16] They predict personal robotic servants, computers that can program themselves, and robots that cost a fraction of a human worker—all within the next decades. The many derivatives of AI research are becoming increasingly vital to our economy and civilization.[17]

Virtually every industry extensively uses intelligent algorithms. The trend now is that the *narrowness* of the intelligence of AI systems is becoming less narrow, with many applications beginning to combine multiple methodologies. *Strong AI* represents the culmination of these ongoing and accelerating trends.

19.3 Synthetic Intelligence

Many leading futurists believe that synthetic intelligence will advance beyond human capabilities within the next few decades.[18] Will it be possible to create a self-aware and reasoning entity that has the capacity for love and all the other (good and bad) qualities we ascribe to human beings? If that intelligence can love, can it also hate with equal intensity?

If we create something that loves us, do we have a responsibility to love it in return? Will synthetic intelligence develop spirituality? At some *critical intelligence* level, will a robot have a soul? Must we believe that a human creation of superior intelligence contradicts the laws of the human creator? Or is it reasonable to think that any creation is simply an evolutionary extension? Will future techno-humans (robots or part-humans) perhaps be equivalent to another species or race? These questions are all generated by thinking about the applications of computers to humanity.[19]

Steven Spielberg's movie, *Artificial Intelligence*, is about a new age possibility—synthetic love. Of course, *Artificial Intelligence* is only a movie— but it generates intense thinking on all kinds of questions relating to robots and synthetic intelligence, as a precursor of what a lot of futurists think will inevitably happen in this new century.

In his book, *The Age of Spiritual Machines—When Computers Exceed Human Intelligence*, Ray Kurzweil argues that synthetic intelligence will inevitably exceed human capabilities within the next few decades.[20] According to Kurzweil, as technology continues its accelerating progress, computers will attain human intelligence levels within about three to four decades and advance to the combined intelligence of all humankind by the end of the century. Some fringe thinkers are already discussing transhuman and posthuman societies.

Several important questions arise regarding the applications of computers in this arena. At what stage will a machine have an independent legal identity to protect its life, liberty, and pursuit of happiness? As the development of artificial body parts advances to the replacement of entire human segments, perhaps even the brain, at what stages will the human identity cease and the machine identity commence?

If humans can download their entire consciousness to a machine and their physical body shows inferior characteristics (aches and pains), at what stage will they choose to survive in synthetic form and discard the organic original? And, when the organic body is shut down, what are the social, moral, legal, and theological implications? Will synthetic beings maintain legal status?

Ray Kurzweil's follow-on book, *The Singularity is Near: When Humans Transcend Biology*, tracks the future of consciousness and intelligence.[21] Humankind, says Kurzweil, is at the threshold of an epoch, the *Singularity*, that will see the merging of human biology with the staggering achievements of genetics, nanotechnology, and robotics, to create a species of unrecognizably high intelligence, durability, comprehension, memory, and so on.

Kurzweil argues that there are virtually no constraints on human capacity. He envisions the transformation of the species, eliminating old human limits.

Kurzweil is one of the world's most respected thinkers and entrepreneurs, and he backs up his techno-optimism with charts and great detail. His extrapolations seem plausible; yet the thesis he postulates in *The Singularity is Near* is so extreme that many are skeptical. He leaves many important questions relating to politics, economics, and morality unanswered.

The New York Times Book Review describes the book as the *Manhattan Project model of pure science without ethical constraints.* Kurzweil's vision requires technology, which we continue to build. But it also seems to demand leaps of faith.

The website longbets.org[22] is an arena for competitive, accountable predictions, a forum for focused discussion and debate about prediction. With a bet of $10,000, Ray Kurzweil bets Mitchell Kapor, the founder of Lotus Development, that a computer or *machine intelligence* will pass the Turing test (be able to successfully impersonate a human) by 2029.

Today, there are many who postulate transhuman and posthuman futures. They see biologically based humans as physically vulnerable, with flawed intelligence. They imagine transhumans who will advance beyond human frailty, with a combination of technological prowess supplementing regular human capabilities. The concepts of posthumans extrapolate many steps further, to where intelligence will advance far beyond conventional bio-chemical bodies.[23,24]

19.3.1 Technological Progress: Threat or Benefit?

In his article "Why the Future Doesn't Need Us" in the April 2000 issue of *Wired*, Bill Joy, cofounder of and chief scientist at Sun Microsystems, argues that our current pace of technological progress poses a very real threat to the future of the human race.[25] He proposes a new ethical standard to guide innovation—and he even recommends that scientists halt potentially dangerous research. His apocalyptic vision of the future has provoked a debate about innovation throughout the high-tech industry.

One gets stuck in this metaphor—synthetic versus human intelligence—until the recognition dawns that the combination has the potential for tremendous good. That combination is already extrapolating rapidly today, with a vast and independent intelligence developing in and around us connecting the gadgets we utilize, through the amorphous and omnipresent Internet.

The web intelligence already exists, survives, and grows, independently and inexorably. Today, when connected to the web, anyone can search out valuable knowledge with a click of a mouse, trade stocks effectively without being a stockbroker, provide at least a preliminary diagnosis of an illness without being a doctor, and perform many similar functions previously considered the domain of experts. Sure, there is potential for bad and evil, but human society always seems to come up with solutions, to develop the antibodies that eliminate a virus and stimulate the cure.

The synergy that can develop between natural and synthetic intelligence generates optimism. In his March 2000 book *The Symbiotic Man—A New Understanding of the Organization of Life and a Vision of the Future*, Joel de Rosnay describes the developing *symbiosis* between man and machine.[26] He insists, "We are witnessing the origin of a new life-form on Earth—a still embryonic macro-organism made up of the totality of human beings and machines, living creatures, networks and nations—trying to live in symbiosis with the planetary ecosystem."

Indeed, this is the ultimate extrapolation of the use of computing power to extend the powers and capabilities of humanity.[27]

19.4 Biotech Revolution

Biotech is impacted in a major way by the advances in computing power to the extent that it is one of the technology shifts that will soon have a disruptive impact on humanity at large.[28]

In the new century, genetic engineering (modifying a living organism's gene structure) is making regular, major advances. J. Craig Venter, the genome-sequencing pioneer and a principal player in the successful push to sequence the human genome, states unequivocally that the acceleration of computing power has had a major impact on genetic engineering and the biotech revolution.

Europe has now approved genetically modified (GM) corn, designed to thwart a parasite that attacks traditional corn, thus bringing greater yield per acre. We may soon find that GM food is safer, more abundant, perhaps better tasting, and less costly (better yields).

Undoubtedly genetic engineering will eventually start to be offered for humans—for resistance to viruses such as cold and flu, or perhaps for cancer resistance. There are broad issues that genetic engineering and cloning will raise, beyond *technical* feasibility, which is certain.

19.4.1 Human Cloning

In the 1930s, Aldous Huxley's Brave New World forecast a society based on clones, with a population optimized for productivity and happiness—humans bred for work, skills, administration, leadership, etc. In our century, that prospect looms.[23]

After the discovery of DNA, the Human Genome Project was launched to construct detailed genetic and physical maps of the human genome.[29] This determines the complete nucleotide sequence of human DNA and localizes the estimated 100,000 human genes, to help determine their individual effects and functions. This 20 year international research program heralded the start of amazing new progress in the biotech arena.

It seems inevitable that sequencing of the human genome will bring about a new level of knowledge of our biological makeup. As we begin to understand the subtle yet significant differences, the consequences are awesome. Technology will enhance biology through the virtual elimination of disease and significant increase in longevity.

With new biotech tools, the process of aging is itself under intense investigation. In the absence of disease, why do humans age? Someone will inevitably discover that tiny ingredient of human DNA that is depleted when cells split in the growth process and that, when replenished, will eliminate, and even reverse, the aging process. So, what are the implications of human life expectancy of 250 or 500 years?

The possibility of human cloning, raised when Scottish scientists at Roslin Institute created the much-celebrated sheep *Dolly*, aroused worldwide interest and concern because of its scientific and ethical implications.[30]

Cloning technology is advancing rapidly to where humans will be cloned—sooner or later. Once this becomes technically possible, it will happen increasingly in countries that have no ban on cloning. Clones may be developed to *harvest* their organs.

With so many unknowns concerning reproductive cloning, the attempt to clone humans at this time is considered potentially dangerous and ethically irresponsible.

The first human clone will probably be born in a country, where work on human cloning is proceeding without ethical constraints. Wherever the child appears, its birth will undoubtedly electrify the world; people will want to know if it remains healthy and how it develops as it grows to adulthood. One imagines that other cloned children will follow and become commonplace. This will then become a new commodity in the growing emporium of human reproduction.

The entire subject of genetic engineering and cloning is directly connected to the impact of advances in computing and demonstrates yet again the dramatic influence of computers on humanity.

19.5 Intelligence and Consciousness in the New Age

The intrinsic definitions of intelligence and consciousness are changing rapidly. Today, with personal computer tools and connection to the vast knowledge base of the Internet, traditional tests of knowledge and intelligence are virtually obsolete.

As you read any book or article, you do not really know whether the original author can spell properly or write grammatically correct language because word-processing programs do automatic corrections. Similarly, knowledge and experience on virtually any subject are supplemented, even replaced, by Internet access.

Just as an inexpensive calculator endows a middle-school junior with the math capability of a savant, an Internet connection represents an extension of human capabilities that provides vast power and knowledge to the user.

Beyond individual enhancement, consciousness seems to be taking a step forward with artificial intelligence (AI). Technology products are taking on new functions as companions and friends. Visual recognition systems and virtual assistants allow communications with voice and gestures. Many people have emotional attachment to devices, even a sort of spiritual connection.

As these technologies advance, interaction is becoming increasingly intuitive. Devices are gaining a semantic understanding of speech beyond simple voice commands, allowing for more natural conversations. Emotion sensing capabilities analyze facial expressions and tone of voice, making interactions richer and more effective. A new category of "living" objects is emerging—intelligent devices that interact in ways similar to living beings.

Medical technology now verges on incorporating computers and Internet communications directly into human anatomy to enhance intelligence, improve health and longevity, and extend lifespan. Human enhancement technology (HET) combines human genetic engineering with nanotechnology, biotechnology, information technology, and cognitive science to improve human performance.

Another aspect of ongoing evolution is the collective intelligence generated by widespread connectivity. Large number of people converge upon the same knowledge, and the individuals achieve enhanced intellectual performance. Crowd-sourced technology projects produce vastly better results than individuals. More and more major news sources are getting better information from user-created content such as blogs, social media, and web-based forums.

Consciousness is diffusing across integrated human-technology networks. It is not just humans networking with other humans, but integrated techno-human intelligence and consciousness.

19.6 Uncertainties of the New Century

Today, we are poised on a narrow ledge of uncertainty in a global village that considers the bounty of the wealthy world greedy, isolationist, and selfish. The wealthy few are surrounded by hordes of helpless and hungry hangers-on and an angry element that seems bent on destruction.

The new century brought into focus the stark reality of our vulnerability in a supposedly free world. The most powerful country was suddenly nervous and jittery about terrorists that precipitated decade-long wars that have cost literally more than a trillion U.S. dollars. Further, sinking stock markets and broken budgets have ruined retirement for many. These, and a hundred horrible happenings, bombard the news every day—constant cacophony and vociferous videos with a half-hour refresh time.

The success of capitalist freedoms was suddenly tainted as stories of excesses by corporate executives and failures of financial institutions became commonplace. The heroes of just yesterday, the same ones who had mingled with presidents and politicians, those very corporate captains fresh from cover stories in financial glossies, were suddenly doing *perp walks* for the public. The rebellion of the 99% against the perceived excesses of the 1% signals widespread dissatisfaction.

The majority of the people on this planet have always been poor. But the situation now is different for three basic reasons:

1. The numbers of poor are increasing exponentially—overloading the carrying capacity of underlying social support systems.

2. The poor now have TV access—to view how the wealthy live.

3. The poor have access to weapons that can produce large-scale destruction.

Most experts agree that population in developed nations will begin a serious decline during the next few decades while continuing to surge ahead in underdeveloped nations. In a democratic world, the more populous nations should, by the definition of democracy, be in control. In any event, their very numbers will inevitably lead to conflicts and splintering. And in a world made ever smaller by the instant access of media and information, comparatively luxurious lifestyles stick out like a sore thumb!

As the world enters an age that professes to eschew race, color, and creed, it is indeed disappointing that ethnic and religious conflagrations continue. In some parts of the world, the local education and environment generate beliefs and values that are diametrically opposed to freedom and democracy and, if unchecked, will fester and spread like some awful, virulent disease that will try to destroy everything in its path.

Anyone who contemplates the future should be concerned about these fundamental trends. The collision of cultures during the next decade will very likely make the world a very difficult place to live. Further, it seems that there is no significant way to resolve these problems before they become much bigger and resolution becomes much harder, if not impossible.

19.7 Humanity's Hard Problems

The latter part of the twentieth century brought science and technology that dramatically and irreversibly changed society and civilization.[31]

Ocean liners that bridged continents and railroads that conquered the old West gave way to inexpensive and widely available jet travel that reduced the world to a global village. Electric power brought automation, which quickly relegated manual labor to menial jobs at the low end. Agricultural advances quickly produced a surplus, and farms today only employ a small percentage of the population in most advanced countries. The age of electricity and radio rapidly switched to TV, electronics, computers, and the Internet—electronic consciousness. The development of computing power—ever faster, smaller, and cheaper—brought startling new developments in widely different segments of human endeavor.

At the turn of the century, satellite communications connected the world for simultaneous viewing, and some six billion souls on this planet went through the countdown together, excitedly and anxiously, as time ushered in a new year, a new century, and a new millennium. For a brief shining moment, it seemed that humankind was close-knit and indomitable, looking positively for a powerful extrapolation of past success.

But, in year one of the new century, on a date equivalent to an emergency telephone number that will remain forever etched in human memory, the dream of a capitalistic Camelot had a sudden and rude awakening. More than anything else, September 11, 2001, represented a benchmark, a transition to a new century where the old solutions are no longer applicable.

New, hard problems arose, which defied the hard technology solutions of the past. Warships, cruise missiles, and fighter jets now look like obsolete, lumbering things. Indeed, the low-tech box cutters that were used in the hijacking served to demonstrate the exposed vulnerabilities of an open democracy. The subsequent widespread cutbacks and the efforts to provide relief through government aid demonstrate the fragility and inadequacy of capitalistic enterprise with a short-term profit motive.

The world must seek solutions to the hard problems that have been exposed. Both capitalism and democracy will need to adapt to the realities of the new age. It is clear that the problems are hard and cannot be solved by the old hard solutions that might have been effective in the past.

19.8 Computers Bring Soft Solutions

New, soft solutions are needed. The coming century is a century of soft things. It is my contention that computers are at the center of the solutions that are emerging.[27]

The global village exacerbates the gap between rich and poor; vast populations view our excesses while they subsist. Let us educate the less fortunate so that, rather than see the seedy side, they are exposed to the freedoms and warmth of our culture. The use of computers and the Internet enables fast, widespread, inexpensive education for the masses.

As we enter an age that professes to eschew race, color, and creed, it is indeed disappointing that ethnic and religious conflagrations continue.

A new, enlightened, global community must find soft ways to eliminate hatred and prejudice.

Today, the power of social networks is enabling outright rebellion of the masses and causing dictatorships to crumble. This is the magic of small, handheld computers connected via the Internet, another stark example of how the computer is having a major effect on humanity.

The removal of communication restraints in the new age redefines the neighborhood, workplace, and playground. The Internet now shrinks the world to a global village and vastly expands participation.

In India, even though only relatively few people actually own computers, the cost of a seat in an Internet café is about \$0.50/h, and they are always crowded. I have often wondered what these people were doing endlessly, hour after hour, day after day. And then it dawned on me—they were working, playing, socializing, and romancing in the global village.

In the past, sailors and other migrants would physically travel to the places where their work was valuable. In today's knowledge-based society, knowledge is easily accessed with an Internet connection. So a graphic artist or a software developer can generate a regular and significant income while still being physically remote.

This had led to significant business in new financial markets. For example, Infosys and Wipro, two software companies currently trading on NASDAQ (Infosys) and the New York stock exchange (Wipro) with market caps in the billions, are located in Bangalore, India.

Socializing on the Internet is a new age phenomenon stimulated by free e-mail, live-keyboard chat, and webcam video.[32] I know several people who communicate frequently with e-friends around the world, exchanging family photos, news, views, and even *local* gossip, and all the time they hardly know their neighbors next door.

In the past, people socialized in their neighborhood, playing chess or cards with neighbors and friends. Today, people can play chess, bridge, or backgammon on the web. There always seems to be a steady stream of players available, matched by experience and skill level. Players can keyboard chat with their partner, or simply turn off their comments if preferred.

Romance on the Internet is already fairly common. Of course, any new cultural shift has its downside, and everyone has an unpleasant story to relate. After exchanging messages and photos for over a year, one romance ended when the e-mail pal traveled halfway across the world to find out that the girl of his dreams was really a middle-aged, pot-bellied plumber.

On the good side, I know several couples that met happily through e-mail. One couple exchanged e-mails for a couple of years, finding love and understanding from a far away friend to fill the gap of personal loneliness; they are now happily married to the soul mate they met in cyberspace.

To work, socialize, or play in the virtual society of a global village, it is almost too easy to avoid actual physical movement, and one wonders how humans will evolve in this new environment.[34] Since there is no *virtual food*, the chronically connected often forget to eat. If the trend persists, perhaps humans will deteriorate physically to slender wisps like those alien space beings we see in the movies.

19.9 Future Workplace

In the postindustrial economy, the way jobs are distributed is changing completely. Factory-based and central-office jobs are steadily declining. The old, punch-the-time-clock type of factory work is a relic of a bygone age. And nobody needs a central office to read and compose documents, send e-mails, or browse the Internet.

Crowdsourcing is the act of outsourcing tasks, traditionally performed by employees or local contractors, to an undefined group of people (a crowd), through an open call. This gathers those who are most fit to perform tasks and contribute with the most relevant and fresh ideas, not only to do the work but enhance the result.

As the amount of digital work increases and the amount of physical work decreases, our notions of employment and work change profoundly. Digital work does not require roads and factories; it requires an Internet connection with equipment that people have access to in their homes. The need for industrial-era offices, supervisors, and employment arrangements diminishes.

Companies are able to access, in real time, the perfect person for a given job—the one who will do the best job, enjoy it the most, or do it the fastest. This changes the landscape and the definition of *work*. This is the path to lower unemployment. Manufacturing and office automation has led to some jobs becoming obsolete. But new kinds of automation can actually drop unemployment rates. They reduce the amount of time it takes unemployed people to find jobs.

Technology has the power to create a much more efficient market, connecting job seekers to employers instantly. Today, most jobs are found via the Internet. For the past decade or so, companies have been looking offshore for cheap labor. Today, it does not matter where work originates—as long as the suppliers are connected via the Internet. This is the future of work.[35]

19.9.1 Twenty-First-Century Prognostications

The first decade of the new century and millennium has come—and gone. On the surface, the world continues to turn—but there are significant changes and differences everywhere. Some of these changes have already been occurring inexorably over the past decades—the advancing of Alvin

Toffler's *Future Shock*.[36] But the world, reflected through its omnipresent media, seems to require a single event to jar its consciousness.

On September 11, 2001, the new century woke up to just such an event, one that is being compared with Pearl Harbor. While Pearl Harbor brought America into World War II, the New York twin towers disaster has taken its fighters, bombers, and troops into other major conflagrations that have lasted more than a decade.

This hard reality brings the recognition that a new society is emerging— new demographics, institutions, ideologies, and problems. Things will be quite different from the society of the late twentieth century and different from what most people expect. Much of it will be unprecedented. Most of it is already here or is rapidly emerging.

In spite of the decade long show of force by the United States and its allies in Iraq and Afghanistan, the new society will be dominated by brain, not brawn. The world will be borderless for business, with upward mobility for everyone. In a borderless world, democracy will attain a new meaning.

In all developed countries, the dominant factor in the next society will be rapid growth of the older population and rapid shrinking of the younger generation. The replacement birthrate is 2.2 births per woman of reproductive age; many European countries are already as low as 1.5, and Germany and Italy at about 1.3. Half the population in Germany (the world's #3 economy) will soon be over 65. At the current rate, Italy, now about 60 million, will be 20 million by the latter part of this century.

India is the world's largest democracy, with a population now exceeding 1.1 billion. Its birthrate, which is not controlled like China's, exceeds 3.3, and this means India will have the world's largest population midway through the century.

The American population will continue to grow, primarily because of immigration and the higher birthrate of immigrants. Politically, immigration will become an important—and highly divisive—issue in all rich countries. Unlike Japan and most other developed countries, America has learned to accommodate and thrive with immigrants.

The twentieth century saw the decline of agriculture—farming employment in the United States declined from about 30% to less than 2% today, and protective farming subsidies are common. Similarly, manufacturing, which employed 35%–40% early in the twentieth century, is now about 15% and will decline to about 10% in most developed countries. With new technology, more production will be performed with fewer workers, with more and more labor-orientated production going offshore. Soon, manufacturing too will see more protectionism and subsidies.

Current corporate business structures will change. As presently defined, the job of the CEO is undoable. The illusion that paying the leaders more and more attains better performance is antiquated. Like the President of the United States, the CEO will emerge primarily as a figurehead and cheerleader. Corporate leaders will provide only mission and vision, while each corporate segment will have different local objectives. In the future, there will be not just one kind of corporation but a range of models to choose from.

The decoding of the human genome is a significant new inflection point with revolutionary and startling consequences for humans and life as we know it. Technology will enhance biology through the virtual elimination of disease and significant increase in longevity. Human cloning has already arrived, and though politicians and governments decry the advances, they cannot stop it. Thousands of scientists all over the world are racing to see who will develop the first human *Dolly*. And when that development becomes a practical reality—which it will within decades—the lottery of conventional biological birthing will seem unnecessarily risky and even quaint.

Underlying all this societal change, technology advances will continue to drive the economy and, with it, business and social environments. The three technology laws will still rule—computer power (Moore's law) and bandwidth (Gilder's law) will continue to double every year. Metcalfe's law—effectiveness increasing exponentially with connectivity—will continue to drive growth of the Internet. Throughout the world, everything and everyone will become electronically linked, causing major changes in the ways business will be done and breeding more closeness in the global village.

19.10 Conclusion

Get ready for change! Stimulated by the use of computing power, the new century is bringing with it enormous changes in all areas of human consciousness. Significant philosophical, ethical, moral, legal, sociological, and spiritual questions must be answered, as we move forward into the new century and millennium.

References

1. J. Diamond. The 3rd Chimpanzee. *The Evolution and Future of the Human Animal*. Harper Perennial Edition, New York, 1992, ISBN-10: 0060845503.
2. R. Leaker. *Origins: The Emergence and Evolution of Our Species and Its Possible Future*. Penguin Books, New York, 1991, ISBN-10: 0140153365.
3. E. Mayr. *What Evolution Is*. Basic Books, New York, 2002, ISBN-10: 0465044263.
4. P. Watson. *The Great Divide: Nature and Human Nature in the Old World and the New*. Harper, 2011, ISBN-10: 0061672459.

5. History.com Staff (April 28, 2011), Did Homo Erectus craft complex tools and weapons? http://www.history.com/news/did-homo-erectus-craft-complex-tools-and-weapons/

6. Dennis O'Neil, Evolution of modern humans: A survey of the biological and cultural evolution of archaic and modern homo sapiens. Behavioral Sciences Department, Palomar College, San Marcos, CA, http://anthro.palomar.edu/homo2/

7. Annalee Newitz, American Journalist, How did humans really evolve? http://io9.com/5774949/how-did-humans-really-evolve/

8. C. T. Rubin. Artificial intelligence and human nature. http://www.thenewatlantis.com/publications/artificial-intelligence-and-human-nature

9. Academy of Evolutionary Metaphysics, Advancing technology. http://www.evolutionary-metaphysics.net/advancing_technology.html

10. B. Allenby, D. Sarewitz. *The Techno-Human Condition*. MIT Press, Cambridge, MA, 2011, ISBN-10: 0262015692.

11. P. H. Kahn Jr. *Technological Nature: Adaptation and the Future of Human Life*. MIT Press, Cambridge, MA, 2011, ISBN-10: 0262113228.

12. M. Ma. The evolution of innovation. http://www.psychologytoday.com/blog/the-tao-innovation/201204/the-evolution-innovation

13. J. Pinto. Intelligence in a new age. http://www.jimpinto.com/writings/newageintelligence.html

14. R. Brooks. *Flesh and Machines: How Robots Will Change Us*. Pantheon Books, New York, 2002.

15. C. Q. Choi. *Human Evolution: The Origin of Tool Use*. http://www.livescience.com/7968-human-evolution-origin-tool.html

16. Mapping out future of intelligent robots, *Science Daily* (July 29, 2008). http://www.sciencedaily.com/releases/2008/07/080729075109.htm

17. MIT Press—Series. Intelligent robotics and autonomous agents. http://mitpress.mit.edu/catalog/browse/browse.asp?btype=6&serid=51

18. J. Bach. *Principles of Synthetic Intelligence PSI: An Architecture of Motivated Cognition*. Oxford University Press, New York, 2009, ISBN-10: 0195370678.

19. J. R. Anderson. *How Can the Human Mind Occur in the Physical Universe?* (Oxford Series) on Cognitive models and architectures, ISBN-10: 0195398955.

20. R. Kurzweil. *The Age of Spiritual Machines*. Orion Business, 1999, ISBN-10: 0752820788

21. R. Kurzweil. *The Singularity Is Near: When Humans Transcend Biology*. Penguin Books, 2005, ASIN: BOOOQCSA7C

22. Web site. The arena for accountable predictions. http://www.Longbets.org

23. T. Peters. Trans-humanism and the post-human future: Will technological progress get us there? http://www.metanexus.net/essay/h-transhumanism-and-posthuman-future-will-technological-progress-get-us-there

24. Amanda C.R. Clark, Trans-humanism and Post-humanism. http://www.ignatiusinsight.com/features2010/aclark_tranhumanism_mar2010.asp

25. Bill Joy, Why the future doesn't need us. http://www.wired.com/wired/archive/8.04/joy.html

26. Joel de Rosnay. *The Symbiotic Man—A New Understanding of the Organization of Life and a Vision of the Future*. McGraw-Hill, New York.

27. Science Daily (March 26, 2009), Synthetic biology: The next biotech revolution is brewing. http://www.sciencedaily.com/releases/2009/03/090325091809.htm

28. D. Miller. Brave new world and the threat of technological growth. http://www.studentpulse.com/articles/509/brave-new-world-and-the-threat-of-technological-growth
29. Human Genome Project Information, About the human genome project. http://www.ornl.gov/sci/techresources/Human_Genome/project/about.shtml; http://genomics.energy.gov/
30. Human Cloning Foundation, Website of the Human Cloning Foundation. http://www.humancloning.org/
31. Human condition: Wikipedia, the free encyclopedia. http://en.wikipedia.org/wiki/Human_condition
32. eBiz MBA, The eBiz Knowledgebase. Top 15 most popular social networking sites. http://www.ebizmba.com/articles/social-networking-websites
33. J. Pinto. Soft solutions for hard problems. http://www.jimpinto.com/writings/softsolutions.html
34. J. Pinto. Work, play and romance in the new age neighborhood. http://www.jimpinto.com/writings/newneighborhood.html
35. Time Magazine Lists: The future of work. http://www.time.com/time/specials/packages/article/0,28804,1898024_1898023_1898169,00.html
36. A. Toffler. *Future Shock*. Random House, 1970, ASIN: BOO8E69TRW

20

Computation in Human Context: Personal Dataspace as Informational Construct

Thomas Boettcher

CONTENTS

We humans, social creatures with solitary challenges, create informational mechanisms just as we make tools to harness our contextual situation, our surrounding environment. Increasingly, our ability to survive and thrive is predicated by the success of our informational constructs, the basis from which we compute our relationship to self, social structure, and environment.

We define ourselves by our computational capabilities and the behavioral models derived from their use. We therefore express our humanity through our surrounding information architectures and the computational constructs that drive them, our powerful mechanism to survive and thrive.

20.1 What Is Our Computational Construct and How Is It Composed?

Our computational construct includes both our *onboard* abilities to sense, perceive, and process (those nested in our brains and nervous system) and our enhanced, externalized computational tools. In combination, we get a working mechanism for how each human engages the daily tasks of living. We start with what we already can do by ourselves, we enhance that capability with learning, and then we surround ourselves with processing capability in the form of tools.

From what do we compose these tools? Probably the most plastic, flexible, and expandable device is the microprocessor, a mass-produced and programmable wonder of modern technology. From the cell phone in our hand to the MRI, we undergo at the request of our physician the applications of microprocessing that are endless. But what is the result of the use of microprocessing-based tools? Data. All of our computational mechanisms generate data, the data that describes us and guides our decisions and predilections. We define the new human through the enhanced capabilities of our computation upon great volumes of data. But how do we harness all these disparate sources of data, from cell phone to MRI and even the conversations we conduct from human to human?

20.2 Foreshadowing a Computational Need

How does a human-devised computational construct help us live every day? How do we improve our lives with computation? The human story of advanced cognition, tool using, and social construct dates back some 15,000 years. It is within the past few hundred years that algorithmic representation of natural phenomenon through the scientific method helps humans harness their environment. That viewpoint is quantitative, and quantitative rationale requires qualitative perspective. We cannot fully harness such perspective without a broad view of the interrelationship among all our sources of data. We need to find the appropriate architecture, a place of residence, a home, for our computational construct. Let us take a look at some alternatives.

20.3 Elemental Computational Construct: DNA

Perhaps the greatest of all computational constructs is the code of life, DNA. It uses simple and repeatable base pairs to represent plans for the assemblage of vital proteins from which life constructs itself. From the computational construct comes the organic construct. What kind of a computational construct is the organic cell growth cycle guided by DNA code? DNA computation produces proteins, and the epigenetic layer guides those production engines to adapt to environmental stimuli. This is the essential working mechanism for sensing our environment. Sensing provides data that is then computationally enacted upon as the organism protects itself or changes for advantage.

In a simplistic sense, we are organic models harnessed to enact our DNA guidelines, dynamic manifestations that must move and live in an ever-changing environment. We live for a while, we carry our genes around, we reproduce in best manner possible, and we pass on our genetic computational code. Claude Shannon, the renowned information theorist for research juggernaut Bell Labs, created seminal work in the transmission of all signaling systems that might apply to DNA as well. He describes that, "practically, we are not interested in exact transmission when we have a continuous source, but only in transmission to within a certain tolerance. The question is, can we assign a definite rate to a continuous source when we require only a certain fidelity of recovery, measured in a suitable way" [1]. Perhaps humans can be observed as merely computational constructs with limited organic fidelity.

This begs the question: What is the computation at the operational, behavioral layer—the hunting, the moving, and the daily surviving as opposed to the synthesis layer?

20.3.1 Computation to Survive and Thrive

Humans have computed for as long as there have been humans. How fast is that mastodon? Can five of us bring him down? Alone or in groups, we humans analyze, synthesize, and design systems as we define ourselves. Why? Because we need. Because we want. Because we strive to survive. Judging distance, interpreting noise, discerning gaze vectors, and reckoning strength levels required computational capability at the very dawn of human behavior, and success in such areas conferred advantage upon the best early brains. It probably did not hurt to brandish a sharp spear or convincing club as well. From our earliest days, the tools we made fit our need yet simultaneously defined our brains because of the computational models that accompanied those tools.

20.3.2 Computation in Modern Motion: Rather Routine

Imagine this scene: You are sitting on an airplane, flying through a relentless snowstorm. Nothing can be seen around this complex aluminum tube hurtling through the air with 150 people on board at 550 mile/h. Two jet engines provide the thrust, but computation leads the way for our fearless pilot. Instrumentation enabled by sensing and interpreted by computation makes this phenomenon truly a miracle but also quite a routine occurrence. This is the amazing aspect of computation: it makes the amazing seem rather routine. We can make our lives highly predictable through computation. We can make chaotic environments behave as dynamic systems. Moving systems with thousands of parts can be thrust into multivariate environments such as storms and dense fog, yet still be expected to behave as predictable systems.

Perhaps, hundreds of years from now, the concept of a plane will seem quaint, naïve, even vulgar in its consumption of fuel and raw, rough efficacy. We do know that computation will persist in some manner, whether on board our brain or outside our bodies in a manner almost unpredictable to us today. For today, we must embrace the task of understanding just how our computational construct must be designed as well as how it must serve us.

New Situation …

Currently, we enjoy an environment rich with resources and possibilities that we access through computation. The stepwise process of expanding beyond our onboard computational capabilities might have started with pictorial symbols, or perhaps simple uniform objects that enabled counting. Externalized devices such as pencil and paper expanded the abilities provided by vocalization and standardized alphabet, yet our brains had to have the cycles and capability to interpret these methods. Now, however, we find that externalized computation devices of all sorts can supplement our innate ability to compute, to think, even to reason.

What is our primary tool today? Human context is defined by a plethora of inexpensive devices coupled with surrounding support systems.

When the device and the system fade to commodity, the information of self becomes the operand, and our ability to store, retrieve, share, restrict, and analyze becomes the focal functionality. In short, the computation becomes the context, and our capability to leverage our information of self becomes self-defining. We are our most comprehensive daily creation, and each of us uniquely designs self through that which we think and do each day. We approach the point now where we can change almost anything about our self-definition, even the information of our own genetic code, demonstrating the premise that every single bit of our own information has value and creates its own entity: the infoindividual.

20.4 New Computational Context: The Infoindividual

We constantly strive to define ourselves; our wants, our needs, our virtues, and our vices all feed into our definition of self. We express that self, that individualism, with the "ultimate private act and public assertion of self, the revelation of the infoindividual. That which we control personally is also that which we can leverage publically as a new and separate societal institution. The individual can finally assume the role of societal institution based solely on the information of self" [2]. The infoindividual exists separately from the individual and grows in proportion to the individual in thought and action. The infoindividual shadows the individual. Every transaction, every conversation, and all our choices and responses fill an ongoing, comprehensive pool of data that sloshes without form unless harnessed. "Every form has its own meaning. Every man creates his meaning and form and goal" [3].

Acting individually and autonomously, we should control the information that describes our intrinsic nature and charts our behavioral course, for daily survival depends upon our awareness of self. We must protect our definition of self and build the foundation upon which thriving behavior may flourish. In protecting and asserting the value of our infoindividual, we assert a new point of view, a new way of looking at our universe. Our new viewpoint asserts that "our information of self must exist as an integrated and functional unit, an indivisible whole, controlled by the individual" [4] and leveraged through computation.

Our current societal system does not yet fully acknowledge the value of the infoindividual. The information of self is trampled chiefly because we are locked in our current set of behavioral predispositions. Individual information has value to self and society. Entities buy and resell it without individual knowledge, approval, or profit. Individuals lack the opportunity to leverage their own info because they cannot effectively aggregate it. Furthermore, individuals have no opportunity or legal precedent to defend their right to control their infoindividual. The individual lacks the basic right to assert intellectual property rights over the information of self.

The new viewpoint, a gestalt, of the infoindividual must first be identified and asserted in our own awareness, then demanded by each and every individual each and every day. As a defining characteristic of our information society in relation with computation, the gestalt of the infoindividual shall assert and protect our assets when information is the currency of exchange and computation is its lever. "Over three centuries after John Locke's essays set the foundation, later expressed in Jefferson's Declaration of Independence, for protection of life, liberty, and private property, the individual still does not deal on equal footing with the information powers. In general, institutions leverage their positions and change only when they are forced to change" [4].

20.4.1 Individual versus the Institution

Computation lies at the foundation of institutional control over the individual, chiefly because the aggregate was necessary to finance the powers of informational usage. "Until now, individuals have always been lumped into groups that form institutions with an agenda based on some subset of a full life lived; institutions have striven yet fall short of describing and fulfilling the life and legacy needs of the individual. You could be a Church member, but you might also have to become a union member, thereby asserting your moral structure as well as your work value through two separate institutions. By participating in these institutions, individuals have had to surrender some component of information autonomy in order to function in society" [5]. We may call upon John Locke to understand the role of the individual versus society: "to understand political power right, and derive it from its original, we must consider what state all men are naturally in, and that is, a state of perfect freedom to order their actions, and dispose of their possessions and persons as they think fit, within the bounds of the law of nature, without asking leave, or depending upon the will of any other man" [6].

As computation progresses under the mechanisms of the institution, the individual is sacrificed. "The institution exists as a subset of a controlling organization, such as management in a corporation, or as the dialectical opposite, such as members of a labor union. There is no means available to protect and assert the value of the information of self. In order to be an effective member of society, you have to conform to a variety of institutions. Your interface to the world compromises your individual choice" [7]. Technological change sets the stage for such a societal structure, and further technological change can therefore release the individual from such structure. "Men being, as has been said, by nature all free, equal, and independent, no one can be put out of this estate and subjected to the political power of another without his own consent, which is done by agreeing with other men, to join and unite into a community for their comfortable, safe, and peaceable living, one amongst another, in a secure enjoyment of their properties, and a greater security against any that are not of it" [8].

Computational relationships can best be understood through reflection upon humanity's recent past. "History may be viewed from the perspective of struggle between the information elite and those excluded from its glow. With this revelation, we gain clear perspective of key information drivers shaping the Millennium just past. The distribution of technology and capital changed rapidly with the onset of externalized and mass-distributed information. The Gutenberg Press liberated information, once an elitist asset controlled by wealth and proprietary language, so that intellectual nourishment could be spread across the new merchant class. Local versions of the Bible, vernacular printings, reinforced nationalist, mass community. As the Bourgeois outpaced the Nobles in societal influence, the institution of the Church experienced destabilizing erosion of its matrixed power structure. Martin Luther's 95 Theses were the lightening rod for an impending decentralization of Church power. The 95 Theses asserted the power of informational self-determination by challenging the Church's control of the informational path to theoretical or figurative salvation. Luther essentially called for a direct line to salvation, disintermediating the priests who controlled the personal information and approval necessary to speak to God. The Church's ability to control and tax salvation was abrogated" [8].

Thus, we see how computation as a mechanism of information organization was mechanistically harnessed and spread. "The Gutenberg Press liberated and expanded the base of the information 'haves,' allowing science and alternative viewpoints to flourish outside of proprietary language and financial sources. The individual was reaching for general pools of information as mass information dispersal began. The Gutenberg Press Millennium chartered fundamental human change as information in printed form became a ubiquitous externalized asset and method of advancement. That mass pooling information behavior now reverses itself with the onset of the Internet Millennium. Now, we make individual information an internalized asset as part of a system that recognizes information as the currency of exchange" [9].

20.5 Enterprise Architecture: Computation Harnessed in Context

Organizations such as corporations and governments create strategies and harness information tools as a result of their ability to federate resources. Organizations do so to extend reach and influence, and the overarching plan that describes the extent of information resources is called an enterprise architecture. Understanding the enterprise architecture of an organization offers the key to knowing the capabilities and weaknesses of said organization. Yet, who provides an enterprise architecture for the individual?

The broad concept of organizing the computational resources of an individual may be likened to the effort that a corporation or other large institutions must make. An enterprise architecture made for the individual human may thus be called a humanterprise. The humanterprise provides personal tools to help the individual form personal strategies. As the individual procures and produces, the humanterprise identifies the broad spectrum of provisioning, learning, value-exchanging information relationships with outside entities that each individual must establish, nourishing, and utilizing to survive and thrive within any environment. The human enterprise draws from all sources that generate information for the individual.

20.5.1 Humanterprise in Living Context

How does the humanterprise work in the context of our daily lives? Consider a visit to a physician; the individual pays for a professional to apply skills and training to diagnose and test. The results of those tests as well as the opinions of the physician are all information for which the individual has paid through a relationship with an outside provider. The individual does not choose to learn those skills personally, so they are provided by a specialist. The same could be said for procuring the approval to drive; information and capabilities must be revealed before governing authorities allow an individual to operate a motorized vehicle. Consider the information associated with any provider of goods. Consider all the information an individual generates personally through thoughts and actions. All constitute information that contributes to the complete picture of self. All the pieces of information that represent that picture of self may be cohesively organized through the humanterprise.

Pragmatically speaking, a human enterprise system, or humanterprise, can be described as the relationship of devices such as cell phones, laptops, workstations, and even handwritten notes. All comprise the wide field of tools that assist individuals in navigating their daily tasks and requirements, yet the list is by no means exclusive. All generate data that has specific, independent relevance as well as correlational value when compared and contrasted with other information from separate devices. The humanterprise, by its very purpose and design, allows for the constant expansion of computing capabilities that generate relevant data. If the humanterprise describes the broad field of computational capabilities, then every individual must embrace the resultant computational construct. In order to embrace, one must first understand.

20.6 Infofeudalism and Its Computation Remedy

The term *feudalism* suggests an economic and behavioral system in which individuals are bound to the land they serve as it sustains them. When we surrender our management of the information of self to third-party computational providers, we suffer from infofeudalism. "Trapped within a system that predisposes infofeudalism today, we suffer the same lack of capability to imagine that we are entitled to informational self-determination. The onset of computing power in the 1950s instigated the collection of personal information in a variety of corporate and governance entities, yet the information was relatively 'trapped' within these independent systems. With the ascendancy of a ubiquitous networking protocol, the Internet protocol, and simultaneously widespread distribution of inexpensive processing power, many entities are gathering a tremendous amount of your personal information without acknowledging your rights or discretion over it" [10].

Now, we consider it perfectly normal that corporations and governments should collect, hoard, analyze, investigate, and benefit from our information. "We exist now in a feudalistic information realm, where the individual's rights and self-interest are trampled upon with impunity. We now are confronted with a digital reality that threatens our informational self-determination: current digital information representation combined with ubiquitous networks debilitates individual control of self-information in favor of collecting institutions" [10].

We live in a time when aggregation of our thinking and behavior can most easily be accomplished by our online search engine and credit card providers. Third-party entities know more about us than we know about ourselves, chiefly because of this ability to aggregate. "Individuals must reassert that which is theirs, the information of self, or risk losing control of it forever. We find the feudal peasant's lack of personal freedom unconscionable. Future generations will find our infofeudalism equally unconscionable" [11].

By identifying the infoindividual, we proclaim to society and self the value of the information and knowledge of self. We live in a world that now constricts value chiefly as a measure of wealth and status, so the infoindividual must be harnessed as a means of production. At the intersection of our information and the ability to harness, there lies an information construct that has never before been possible. An information architecture—a set of guidelines and methods for information storage, retrieval, security, third-party access, development, and analysis—can provide structure for the raw data that is the infoindividual. This information architecture combines the capabilities of both computation and communication, but its value lies in its structure and leverage. What shall we call this information architecture? A Personal Dataspace.

20.7 New Computational Construct: The Personal Dataspace

What if our most important informational tool was, in fact, a system? Physical systems such as municipal water treatment and electrical provision support us in our daily roles and tasks. Our most important tools require systemic support. Systems such as electricity and prevalent communication networks based on the Internet protocol allow us to function better as individuals. These systems require computation, yet their sustenance is not an individual responsibility. All we have to do is connect to them. We use them each day, multiple times. The most important system we harbor is one that we nurture ourselves, and that is the system of information that describes ourselves. This special system, an information architecture, is known as a Personal Dataspace. "A Personal Dataspace (PD) is a safe place for individuals to electronically collect, keep, and use their own aggregated data" [12]. The Personal Dataspace provides the primary computational construct for life and liberty, present and future. Without the leverage and protection of a Personal Dataspace, we shall fail as individuals and surrender to overbearing and overmatched institutions.

20.8 Why Do Individuals Need a Personal Dataspace?

Aggregated, organized data has value. Individuals need a Personal Dataspace to solve the problem of personal data management. Personal data management means more than just collecting and organizing data; each and every individual can master the freedom of full informational self-determination. We assert our informational self-determination by exercising the daily right and action of controlling the existence, disposition, aggregation, presentation, and leverage of our information of self.

Maintaining data throughout its life span—so that it is not necessary to undergo a costly and time-consuming migration effort when a particular infodevice becomes obsolete—means individuals need a data construct. Within that construct, individuals are given a context within which they can leverage the value of their data to meet personal goals.

20.9 How Does the Personal Dataspace
Relate to Web Services?

Web services can rely on Personal Dataspaces for all customer-related information management. In this sense, the Personal Dataspace is a pillar upon which the web service economy rests. With the prevalence of a Personal

Dataspace information architecture, the individual shifts in power and control to the center of the provisioning web. Individuals may announce when they wish to be approached by vendors. Individuals may elect to share select views of their medical and behavioral patterns to health-care providers to enhance diagnosis accuracy and modality efficacy.

Social applications attempt to provide a platform for advertising and transaction but are flawed because of the coercive and dominant position provider companies retain over individual data. Credit bureaus utilize your data without permission or compensation. A Personal Dataspace provides the platform from which this wrong may be corrected.

20.9.1 Personal Dataspace and Personal Medicine

The Personal Dataspace provides a working, evolving foundation of computational raw material. An electronic medical record is a start, but so much information regarding ongoing nutritional and behavior indicators is simply not captured. An electronic medical record is a subset of a Personal Dataspace, sharing the beneficial characteristic of aggregation. The Personal Dataspace must be governed only by the individual, who shares information as is necessary for diagnosis and treatment. Baseline measurement on this raw material provides a context from which one can measure adherence to norms and deviation from same. Deviation from the norm can be studied versus population norms as well as demographic subsets as well as personal norms. Adjust for psychographic and genotypic specificity, and you begin to create a model tuned to environment and behavior.

20.9.2 Personal Dataspace and Peak Performance: Big Data

With a storehouse of personal information (raw material) cultivated in a Personal Dataspace, the challenge remains to deliver meaningful results. Insight can be derived from personal observation when you get the computation right. Treating all information as potentially relevant, then trying to find correlation and indication, can prove to be a daunting task. The analysis of large data sets, known currently as big data, requires the use of algorithms (specific computation). Rule sets known as heuristics provide guidelines to analysis for the human in the same way that large data sets for signal to noise ratios provide clarity for transmitted conversation in the telecommunications realm. The principle is the same: to find the relevant correlation in a sea of data.

Companies such as HumanLabs use algorithms from telecommunications to decipher the metrics of peak performance for specific individuals. Using a data construct that allows for any and all variables, from genetic code to daily behavior and nutritional input, one can create a personalized medicine model tuned for optimum performance. Want to know what to eat for your specific metabolism right before a big race or presentation? Consult your HumanLabs computed indicators.

20.10 Computation as a Sense

Sensing ourselves and our environment was traditionally simple: you saw, you heard, you smelled, you felt, you listened. If you did not, you suffered situational trauma, such as being eaten or poisoned. Our spoken lexicon even contains tacit understanding of this phenomenon: we often speak of *sensing danger*. Today, sensing has followed quickly the path led by computation. An MRI machine can sense what is happening within your body and show you. A tiny microelectromechanical systems (MEMS) device can sense chemical concentrations, electrical and magnetic anomalies, aberrant sounds, and unexpected forces. Our cell phones sense our position and signals from others trying to reach us from far away. All of these *senses* are backed by computation. Computation allows for interpretation of signals and reporting of the resultant information.

The development of new information devices and the constant onslaught from our infoenvironment assure us that we will daily consume more information than we can perceive and interpret comprehensively. Our brains are exceptionally well suited for integrating simultaneous input from a variety of senses. This is how we construct our perception of reality and maintain our capability for instantaneous reaction. Unfortunately, it is this physical survival programming that allows constant features of our environment to fade to the background as our nervous systems emphasize new stimuli. While sudden change is immediately grasped as a threat and processed quickly, gradual change perception is neglected. Ambient monitoring provides a passive mechanism to employ widespread sensors with data-gathering capabilities.

Everyone has experienced the onslaught of attention demand in modern society. "Because of both increased demands and our increased communicative and externalized processing capabilities, it is no longer sufficient to accept the imbalance between short and long-term sensing. We need to be able to synthesize all our information for better decision-making, creating a working, cumulative cognition. We see, we smell, we taste, we hear, we touch; all instances of environmental feedback" [13]. This environmental feedback is compounded by informational buzz, or noise, as well as the inherent transmission noise of systems delivering signals.

Thus, the human, still very physical in an informational epoch, must balance capabilities between physical and informational computation. "Proprioception is our awareness of our own movement. This physical awareness is essential to our survival, yet we lack a corresponding informational awareness. We might identify our new sensing capabilities as superior to our on-board, issued-at-birth sense. These new perceptive capabilities are above, or super to, normal perception and information gathering. This new informational awareness, or superception, is centered in our enterprise of self-information, our Humanterprise" [14], and leveraged through the mechanism of a Personal Dataspace.

Now, with our newly developed sense, we can perceive gradual, iterative change as an over-laying capability of standard perception. This new layer of capability, an overview that ties aggregate behavior to immediate cognition, is our superception, perception that is superior to, and supercedes, normal capabilities. By leveraging the voluminous data we will receive from new sensing devices, we can better incorporate change into our daily behavior based on this new channel of behavioral feedback.

Superception allows the individual to determine consequence from independent action, thus capturing value previously lost in iterative behavior. The self may be discovered, probed, and revealed in a way previously unknown—comprehensively and with full awareness of cumulative consequence. "Superception, a resultant capability of developing, nurturing, and leveraging an externalized portion of our Supercortex, (an externalized Computational Construct), comes entirely from the effort of the individual striving for the sublime state of human independence and liberating act of self-determination" [15].

20.11 Externalized Computation as a New Brain Functionality: The Supercortex

Can our brains handle what we face each day? "The individual conducts daily activities amidst a constant deluge of information. We sense so much in real time that our brains typically dump the complex, tedious, overflow data not related to immediate physical survival in favor of maintaining constant awareness, reasonable sanity, and continued conscious perception capabilities. This is fine for surviving minute to minute, yet the information we shed most easily seems to be that which is probably most vital to making long-term decisions about our own behaviors" [15]. We think as if we are hunting, therefore missing our own long-term trends.

In simplistic terms, organisms evolve higher layers of centralized processing control as more dynamism is required from the environment. As human cognitive capabilities have evolved, higher layers of our brains have grown over the more basic layers. Considering our human journey of the past 15,000 years, we arrive today with complex brains capable of superior signal processing and reasoned, symbolic representation and abstraction. "Our brains work because they specialize. We can identify layers of functionality that accompany our progress as thinking, rational mammals. Our involuntary vital functions, those basic to our existence, originate in the medulla oblongata at the top of our spine. The next component layer, the cerebellum, coordinates muscle and bodily functions. Our cerebrum is the expanded, upper layer portion of the brain responsible for consciousness and mental processes. The cerebral cortex is the surface layer of gray matter of the cerebrum that functions chiefly in coordination of sensory and motor information. The dorsal region of the cerebral cortex is the neocortex, the most recently developed area of our brains. These components functioning together have allowed us to successfully navigate a chiefly physical existence, yet our increasingly burdensome informational challenges beg for more. Figuratively, we require a new layer. We need the capabilities that a new layer of cortex might provide yet we need them now. We can't afford to wait for hundreds of years for evolution to develop the next layer. We have the ability to add it with technology and information architecture. This brain layer will help us to perceive and act on the iterative and cumulative, gaining benefit over the long term" [16]. We must start thinking not only as hunters but as gatherers of our own information.

What kind of computational model can meet this need? "The need to process the huge amounts of info that we generate calls for the development of an externalized Supercortex. This supercortex grants us enhanced ability to conceptualize. This new layer of externalized information processing, the supercortex, is a ubiquitously available and standardized information utility that supplements the processing now accomplished by our neocortex. We need to be able to store, aggregate, correlate, and analyze multivariate input in real time. Humans lead multivariate lives: complexities and contingencies to decision-making come from a variety of inputs, yet we lack the framework and modeling tools necessary to represent clearly in our minds the nature and consequence of our daily human decisions over the long term" [16]. We may use computation to provide this long-term capability.

Shall we wait for our onboard computation device, our brain, to meet our needs? "In essence, we choose to create an externalized thinking capability that far surpasses our biological provision. Consider the historical importance of the alphabet and writing. The ability to create and keep written records externalized working memory, thus more complex ideas and information could be stored, accessed and communicated. Aggressive theories could be modeled and tested while more risk-distributed transactions could be undertaken. As we consider our centralized role within our Humanterprise, it is only natural that we engage the partnership of a provider that delivers the functionality of survival capability, a supercortex" [16]. Our brains give us the ability to supplement their capabilities through the use of a supercortex.

Can this computational model help our everyday lives? "Humans can daily benefit from charting the course of their decisions. Infoindividuals must have the infofreedom to choose a suite of tools designed to support their own basic interests, not those of their provisioning sources. Humans need capabilities that far outstrip their current, biologically provided, on-board, always-available, processing and decision-making brains" [16]. Where does the supercortex exist? The power of an externalized supercortex is found in an individualized Personal Dataspace.

20.12 Intention Span: Provisioning as a Computational Mechanism

Intention span is the corollary of attention span; it is a "specific period of time in which an individual wishes to engage the surrounding provisioning environment" [17]. It is the span of time in which the individual intends to act. Computation and coordination are at the core of intention span. Establishing the appropriate relationship between individual and infoenvironment means relegating approaching providers to the structure of the individual's Personal Dataspace. The individual sets the parameters and specific access to pertinent information within the broad context of the Personal Dataspace, then issues a request for proposal at the desired instance. This respect for privacy and the value of attention is time efficient and resource efficient, for the individual can negotiate the best possible deal within the stated intention span, while provisioning entities can make their best proposal based on their capabilities at the right instance.

20.13 Computational Construct Established: Now Where Do We Go?

We understand the evolution and etymology of computation and human capability; we combine our onboard, brain-based systems with externalized capabilities into one dynamic human enterprise system or humanterprise. An individual humanterprise is manifested for daily use through a Personal Dataspace, a logical location upon which analysis can be performed. What next?

At this stage, the true design of self may take place. We start with genetic code, move to the physical manifestation of self and thinking capacity, and then on to externalized capability so that our daily choices become part of our very own laboratory experiment. We may constantly choose, assess, and improve upon the system of self. We become our own human laboratories. HumanLabs is an example of a contemporary company designed to use computation to support the individual.

How does a service such as HumanLabs work with the individual? Start with the information, apply the computation, and then draw the conclusion. The large pool of data that describes each individual, when organized within a Personal Dataspace, can be computationally analyzed with algorithms and heuristics in the same way that telecommunications systems must discern noise from true signals. Because the individual can be described as a complex, multivariate system, the most efficacious approach to personalizing risk mitigation and performance optimization is through the use of massive computation capability applied against big data sets. The aim is to tease out correlation.

Within the capabilities enabled by HumanLabs, we find a way to map genetics and epigenetics to behavior. Computation aligns the individual with genetics and behaviors. This Human Alignment provides a path for individuals to do the things their bodies are specifically designed to do at an optimum level with minimum risk. Considering as many inputs as possible, one may draw a correlation through repeated algorithms and heuristics until one finds the specific indicator that makes one better, faster, stronger, smarter; in summary, more capable of doing exactly what it is one wants to do. With Human Alignment, you are getting your behaviors in line with your best possible outcome, giving you control over your present day activities and thoughts so that you are optimized for best performance.

20.14 Computational Assistance: Externalized Recall

Thinking requires support data, so we call upon our powers of recall in order to form new thoughts. However, our powers of recall are also limited. Assistance is found in the form of computational access to storage devices comprised of microcircuitry or even magnetically stored signals on physical media such as a hard drive. Thoughts, words, images, videos, and sounds that would be extremely challenging to recall independently become part of our everyday lexicon, and it is computation that allows for easy and rapid access. Our best estimations of that which constitutes native intelligence are comprised of the capabilities we have been born with and those we have cultivated through learning. At the highest level, this problem-solving capability under time duress (a test situation) is celebrated through admittance to Mensa for high intellectual capacity. This type of mind power is suited for synthesizing and problem solving, yet recall is a task better relegated to externalized constructs.

Externalized recall can be localized, as in physically within reach, or situated in ideal locations far from the functioning individual. My computational constructs, such as the hard drive or microchip sets in my grasp, serve me and remain under my care. Microchips and hard drives far away from me but connected via networks such as the Internet serve me as well, yet are maintained by someone else. These recall resources are reached through networks such as the Internet. These networks are nothing more than computational agreements or protocols. The Internet is specifically the Transmissions control protocol/Internet protocol (TCP/IP), the fundamental agreement of computational exchange. Therefore, externalized recall is pervasive and not physically bound when computational constructs are shared and deployed. It is these shared computational constructs that enable the best implementation of Personal Dataspaces.

20.14.1 Computation in Defense of Storage

What kind of thinking can computation do? Current laptops have protective measures for the occasional bump or drop. What do we do to protect our onboard computational device, our brain? We wear a helmet. Computation upon which we depend can also be protected by computation, a proverbial helmet for the device. Therefore, the algorithm that senses and then locks the hard drive (using an accelerometer to tell when the machine rapidly changes positional vectors) represents computation protecting itself. The sensor has to process, interpret, and induce an act. The computational construct makes a helmet for a human's externalized brain, the precious, localized hard drive with so many thoughts and images upon it that a human's brain would be hard-pressed to recall them all independently.

20.14.2 Computation When Control Is Surrendered: The Automated Automobile

Consider the complexity of machine and human. A significant mixing of human brain-powered computation with externalized computation in the automobile creates a dynamic system that works on a large scale but with significant expenditure of attention as well as an appreciable accident rate. What if the human brain computation were removed from the system? What if roadway and surrounding automobiles might be consistently and immediately sensed via computational devices within the auto and the human were free to pay attention to completely separate tasks? This style of automation for the automobile, researched currently by Google, may completely change our perception of individualized transport for the foreseeable future. Individuals may completely trust their locomotion to computational constructs outside their own brains.

20.14.3 Computation When Signals Are Crossed: Synesthesia

Our onboard computational system, the brain, particularizes functions to various regions and specializes for optimized result. Functional MRIs can map the areas of the brain that are utilized for specific sensing as well as the computation in neural cells necessary to draw conclusion from the observed phenomenon. What if those functional regions were crossed? What if neural pathways confused or overlapped regions so that one might hear colors or see music? This phenomenon happens in the brains of those who experience synesthesia. In our struggle to further understand our computational construct, synesthesia may provide a valuable pathway to discovery.

20.15 Precursor to the Computational Construct

Bell Labs' Claude Shannon connected logic to communication by using computation to make up for noisy channel signal loss. In his seminal work, "A Mathematical Theory of Communication," 1948, Shannon engaged the central problem of transmission of data over a noisy channel, thus founding information theory. A broad range of computational disciplines, including artificial intelligence, systems biology, and informatics, and coding theory are all applications upon information theory. The simple method of data compression, not unlike the way our brains notice change, comes from Shannon's work to make a message at one point readable after transmission to another point. He paved the way for the second part of the computational construct, the Personal Dataspace enabled by the Internet P(IP)rotocol.

20.16 Life's Computational Construct for Sequencing and Transmission Error

Error correction in data transmission is also at the very heart of all organic life. Mistakes happen randomly in the transmission of information. Mistakes may also happen in the transmission of information from DNA code to protein synthesis; we can see the effects of this in the form of cancer, and we strive to find the basis for controlling this transmission error. The genetics of DNA interpretation that allows the human organism to be constructed and them to move within its environment and respond to stimuli also is subject to natural laws of channel noise and error rate. Internal mechanisms allow for cellular repair and apoptosis.

Why not apply Claude Shannon's theories of information transmission over a defined channel to genome sequencing. Essentially, researchers strive for better and quicker ways to interpret the information encapsulated in DNA into electronic format. In shotgun sequencing, the message is encoded for transmission by the process of reading, and then it is decoded by a reassembly algorithm to produce an electronic version. Shannon's theories give us a way to calculate channel capacity for the maximum rate of information flow during the process of sequencing.

20.17 Computational Direction: The Future of Information and Life

When the optimizations of physiological therapies are exhausted, when the algorithms can predict no more, when physical limitations have been met and yet the body, mind, and spirit still suffer, there is but one direction to turn: synthetic rendering of life. Only in its infancy, synthetic biology takes the scale and precision of engineering discipline and applies it to the computation of DNA. Already, a synthetic and computationally formed sequence of DNA has been manually engineered into a bacterial cell, with that cell able to *boot up*, or come to life, just as a software program makes a function come to life with input and output. This very fundamental control over the design of life affords us the final lever of computation. We can fully describe what we wish to be and do, then make it so.

20.18 In Summary: The Human Computational Construct

What do we measure when we consider computational effects on human beings? Can we look at physical manifestations when judging computational import? Is our way of computing, of thinking algorithmically, what makes us human, or do we compute merely because we are human?

Think back to the opening example of a primitive human hunting with his *onboard capabilities*—his senses and brain-based computation. Success was highly measureable, as was failure. Today, the effects of computation are buried in the process. We can measure the speed of the animal we hunt or control the animal's movement, yet both have computational dependencies. We can sense the soil composition of the greens that we raise, yet we still sell a product under the illusion of natural processes. We have hyper-arranged our environment to fit our need, celebrating our form, our drives and desires, our ability to construct and partition. We make cities that allow us to live in close proximity, yet maintain a modicum of separateness and individuality. We create connections to vast landscapes that host our food and provide our retreat. We, as a species, investigate all others and invest in our scientists a higher degree of credibility as we assess our role on this planet. All of these actions are informed by computation. The physical, mental, and spiritual construct of our lives is now highly dependent upon, highly formed by, and highly enhanced because of our computational construct. Read on, and see why, in this compilation of computational insights. See how. See more. See now. By the time these thoughts are compiled, the landscape will have changed once again, and our computational construct, our contextual view of our world, will have grown along with that change.

References

1. Shannon, C.E. A mathematical theory of communication. (July/October 1948). *Bell System Technical Journal* **27** (3): 379–423, Holmdel, New Jersey.
2. Boettcher, T. *The Infolocus Manifesto.* (Gutenberg Press, 2000, Wilmington, DE). p. 14.
3. Rand, A. *Fountainhead.* (Bobbs-Merrill, 1943, Indianapolis, Indiana), Section 1.
4. Boettcher, T. *The Infolocus Manifesto.* (Gutenberg Press, 2000, Wilmington, DE). p. 15.
5. Boettcher, T. *The Infolocus Manifesto.* (Gutenberg Press, 2000, Wilmington, DE). p. 16.
6. Locke, J. *Two Treatise of Government.* (Churchhill, 1690, London, England). State of nature, Chapter 2, Section 4.
7. Boettcher, T. *The Infolocus Manifesto.* (Gutenberg Press, 2000, Wilmington, DE). p. 17.
8. Locke, J. *Two Treatise of Government.* (Churchhill, 1690, London, England). State of nature, Chapter 2, Section 95.
9. Boettcher, T. *The Infolocus Manifesto.* (Gutenberg Press, 2000, Wilmington, DE). p. 18.
10. Boettcher, T. *The Infolocus Manifesto.* (Gutenberg Press, 2000, Wilmington, DE). p. 23.
11. Boettcher, T. *The Infolocus Manifesto.* (Gutenberg Press, 2000, Wilmington, DE). p. 24.
12. Boettcher, T. *Personal Dataspace: Information Architecture for the Individual.* (Gutenberg Press, 1997, Wilmington, DE). P. 13.
13. Boettcher, T. *The Infolocus Manifesto.* (Gutenberg Press, 2000, Wilmington, DE). p. 37.
14. Boettcher, T. *The Infolocus Manifesto.* (Gutenberg Press, 2000, Wilmington, DE). p. 45.
15. Boettcher, T. *The Infolocus Manifesto.* (Gutenberg Press, 2000, Wilmington, DE). p. 67.
16. Boettcher, T. *The Infolocus Manifesto.* (Gutenberg Press, 2000, Wilmington, DE). p. 68.
17. Boettcher, T. *The Infolocus Manifesto.* (Gutenberg Press, 2000, Wilmington, DE). p. 70.

Index